Soil Microbiology:
An Exploratory Approach

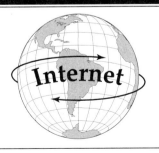

Soil Microbiology:
An Exploratory Approach

Mark S. Coyne

Delmar Publishers

an International Thomson Publishing company

Albany • Bonn • Boston • Cincinnati • Detroit • London • Madrid
Melbourne • Mexico City • New York • Pacific Grove • Paris • San Francisco
Singapore • Tokyo • Toronto • Washington

NOTICE TO THE READER

Cover Design: Elaine Scull

Delmar Staff
Publisher: Susan Simpfenderfer
Acquisitions Editor: Jeff Bernham
Developmental Editor: Andrea Edwards Myers
Production Manager: Wendy Troeger
Production Editor: Elaine Scull
Marketing Manager: Katherine M. Hans

COPYRIGHT © 1999
By Delmar Publishers
an International Thomson Publishing company **I** ⟨**T**⟩ **P** ®

The ITP logo is a trademark under license.
Printed in the United States of America

For more information, contact:

Delmar Publishers
3 Columbia Circle, Box 15015
Albany, New York 12212-5015

International Thomson Publishing Europe
Berkshire House
168-173 High Holborn
London, WC1V7AA
United Kingdom

Nelson ITP, Australia
102 Dodds Street
South Melbourne,
Victoria, 3205 Australia

Nelson Canada
1120 Birchmont Road
Scarborough, Ontario
M1K 5G4, Canada

International Thomson Publishing France
Tour Maine-Montparnasse
33 Avenue du Maine
75755 Paris Cedex 15, France

International Thomson Editores
Seneca 53
Colonia Polanco
11560 Mexico D. F. Mexico

International Thomson Publishing GmbH
Königswinterer Strasse 418
53227 Bonn
Germany

International Thomson Publishing Asia
60 Albert Street
#15-01 Albert Complex
Singapore 189969

International Thomson Publishing Japan
Hirakawa-cho Kyowa Building, 3F
2-2-1 Hirakawa-cho, Chiyoda-ku,
Tokyo 102, Japan

ITE Spain/Paraninfo
Calle Magallanes, 25
28015-Madrid, Spain

2 3 4 5 6 7 8 9 10 XXX 04 03 02 01 00 99

Library of Congress Cataloging-in-Publication Data
Coyne, Mark S., 1960–
  Soil microbiology : an exploratory approach / Mark S. Coyne.
    p. cm.
  Includes bibliographical references and index.
  ISBN 0-8273-8434-3
  1. Soil microbiology. I. Title.
QR111.C68 1999
579'.1757--dc21   98-39580
                                                        CIP

Dedication

To Sharon, Brian, and Lauren

Contents

Preface

There are many advanced soil microbiology texts, but the topics they cover are often too narrow or specialized for students new to the topic. The soil microbiology course I teach, for example, attracts students from agricultural engineering, animal science, biological science, crop science, forestry, geology, natural resource management, and soil science. Each major's preparation for a course in soil microbiology is completely different.

If you're new to soil microbiology, this textbook is for you. I intended it to be truly introductory and cover basic soil microbiology concepts that any undergraduate or first-year graduate student should know for their introduction to the subject. To follow most of the concepts you should have, at most, some beginning chemistry and biology, and be able to work a calculator.

I've arranged the textbook by sections. Section 1 provides an historical perspective of microbiology and soil microbiology and introduces or reviews basic concepts in microbial metabolism and enzymology that will be critical for you to understand later. Section 2 outlines the major functional groups of soil organisms, from the largest animals to the smallest viruses. In Section 3, I introduce the basics of soil as an environment and discuss how that environment influences the growth and activity of microorganisms. In Section 4, I discuss important mineral transformations and soil nutrient cycles in terms of microorganisms and their activities. You'll examine the critical aspects of the carbon and nitrogen cycles in soil in this section. Microbial interactions in terms of plant/microorganism interactions (symbioses) and microbial ecology are discussed in Section 5. Finally, in Section 6, I address microorganisms and environmental quality, and demonstrate how soil microorganisms influence our world.

While agricultural soils are the model for the soil ecosystem, this textbook includes examples from other environments (including people) illustrating microbial activities and demonstrating that soils and their microbiology are not isolated from the rest of the world. The graphs, tables, diagrams, and figures I use are drawn from original data and illustrate specific points that should help you examine and manipulate information the way soil microbiologists do.

I've used ideas from other soil microbiologists (teaching uses recycling just like the environment) and incorporated them into a format that highlights the topics I think a semester-long introductory course in Soil Microbiology should have. There are numerous chapters, but they're short, and they usually approximate the coverage of a single day's lecture. I begin each chapter with an overview of the major learning objectives and end with a summary of the key points and several review questions. A glossary of terms every soil microbiologist should know closes the textbook.

Instructors aren't and can't be encyclopedic. This textbook won't answer all your questions. Instead, it should give you and your classmates a common foundation and basic understanding of soil microbiology. Hopefully, this will excite your interest and help you to figure out additional questions to ask. This textbook will enable instructors to use class time to focus on specific topics at a more detailed and challenging level for you.

Acknowledgments

I am grateful to many people who helped bring this textbook to light. First, my thanks to the Soil Microbiology instructors who brought the subject alive for me: Tom Loynachan, Dennis Focht, and James Tiedje. Second, I wish to thank my colleagues in the Agronomy Department at the University of Kentucky who continually supported my interest in undergraduate education. In particular, John Grove spent many hours discussing the merits of various teaching approaches, Victoria Mundy read every page of the initial manuscript, and Jim Thompson told me what a Gelisol was. Third, my gratitude to the students who took my Soil Microbiology and related courses. Their response to the material was instrumental in its conversion to a text. Fourth, my editors at Delmar Publishing, for guiding me through the process of publishing a new work. Finally, special thanks goes to my family, immediate and extended, for their support in putting everything together.

1

INTRODUCTION TO SOIL MICROORGANISMS AND MICROBIOLOGY

*S*oil microbiology didn't spring out of the void. It developed from countless scientific discoveries in microbiology and soil science, which laid the groundwork for the discipline we refer to as Soil Microbiology. In this section, we look at pioneering microbiologists who laid that groundwork and approaches that soil microbiologists use to investigate their science. Next, we review fundamentals of microbiology that are basic to characterizing soil microorganisms and investigating their activities. Finally, we look at enzymes, which are the biochemical molecules that underlie most of the transformations in soil that soil microorganisms bring about.

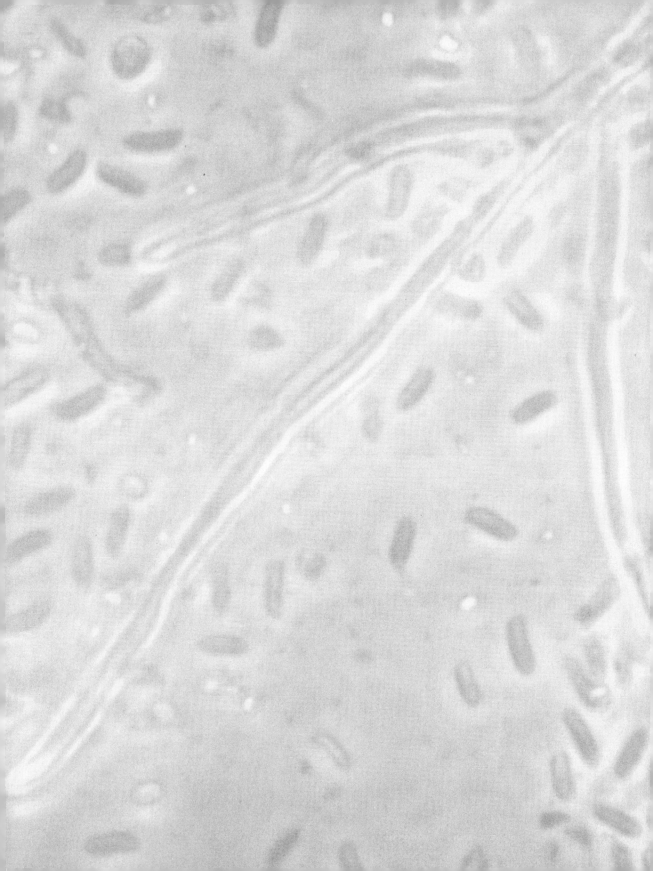

Chapter 1

A Historical Perspective of Soil Microbiology

Overview

After you have studied this chapter, you should be able to:

- Define soil microbiology.
- Describe soil as a complex biological system.
- Explain two basic approaches to the study of soil microbiology.
- Identify pioneering soil microbiologists in history and list their principal accomplishments.
- Discuss research areas in soil microbiology.

WHY STUDY SOIL MICROBIOLOGY?

A knowledge of soil microbiology is essential to understanding agricultural and environmental science. Without soil microorganisms, life as we know it couldn't exist on Earth. For instance, without soil microorganisms, we'd be overwhelmed by undecomposed organic matter. Virtually everything we do is influenced by microorganisms and their activity in soil. Why study soil microbiology? If we understand what's happening in soil, we get a better idea of how other biological systems work on Earth. From a soil microbiologist's viewpoint, if it doesn't happen in soil, or isn't influenced by soil microorganisms, it's probably not worth studying in the first place.

A PHILOSOPHICAL LOOK AT SOIL

Take a handful of soil. What is it? What does it represent to you? Soil? Dirt? Minerals? On first sight, it seems rather ordinary. But take a closer look. What's in soil? There's some air, some water, some minerals such as quartz and clay, and some organic matter, which is a general term for compounds that contain carbon. How is soil different from us? Would you believe that there really isn't that much difference—that soils are just as alive as we are?

How do you tell if something is alive? There are several criteria you could use. Living things move. Soils also move in wind and water, and part of the organic matter in soil is mobile. Living things, especially mammals like us, get hot because they generate heat. Soils also generate heat because they respire, which means they convert energy that is available in chemical bonds or light, into energy that can be used for work, and that work produces heat. Just take a look at a steaming compost pile on a cold day. Better yet, try sticking your hand in it.

Living things breathe, and so does the soil. If you measure soil with the appropriate instrument, you can observe soil absorbing oxygen (O_2) and releasing carbon dioxide (CO_2) and other gases. Living things change with time. So does soil. Change with time, what we call development or evolution, is one factor in soil formation. In addition, parts of the living organic matter in soil are growing, reproducing, and dying all the time. Soil microorganisms are dynamic, and different microbial groups dominate the soil at different times and seasons.

Living things eat, and so does soil. Proteins, nucleic acids, and simple and complex carbohydrates (molecules made out of carbon and water) are its food. Place any one of these compounds in soil and they will disappear just as assuredly as if you had eaten them yourself, which you do every day.

For the soil microbiologist, soil is a living organism—a mixture of living cells in an organic/mineral matrix. Not every cell is identical, or carries out the same function, or is active at the same time. Soil is an immensely complex and interesting organism. Our task in this textbook is to dissect the soil—to dissect this complex organism into its constituent parts and activities. We'll do this so that we can uncover basic principles about life and organisms in soil that we can apply to other soils and to other environments.

APPROACHES TO STUDYING SOIL MICROBIOLOGY

Let's begin with one of the first principles of microbiology—microorganisms are everywhere, almost without exception. We study them in soil because they are responsible for most of the activity that goes on there. Without soil microorganisms, the soil would be a sterile, largely inert rooting medium or perhaps a building or road foundation and little else.

A simple definition of soil microbiology is "the study of organisms that live in the soil; their metabolic activity; their roles in energy flow; their roles in nutrient cycling" (Atlas and Bartha 1993). The Soil Science Society of America (SSSA 1998) defines soil microbiology as "the branch of soil science concerned with soil-inhabiting microorganisms, their functions, and activities." You can study soil microbiology by taking two basic approaches. The first approach is to study the organisms in soil by examining their:

1. physiology—how they grow and metabolize
2. taxonomy—what they look like and how they are related to one another
3. pathology—how they cause diseases of plants, animals, and humans
4. symbioses—how they interact with more complex organisms

Another approach is to focus on what microorganisms do in soil—their microbial processes:

1. Biogeochemistry—how they affect our environment chemically
2. Nutrient cycling—how they recycle compounds in soil
3. Global change—how they affect global properties such as temperature and atmospheric chemistry
4. Ecology—how they interact with their environment and with other microorganisms

These categories overlap; you clearly can't study soil microbiology without taking both approaches. But if you hear about someone who is described as "a process-oriented soil microbiologist," you now know what that means.

A HISTORICAL OVERVIEW OF SOIL MICROBIOLOGY

Microorganisms are more than 3 billion years old, which is when the first fossilized evidence for bacteria appears. Soil microbiology as a discipline first began with agriculture and the manipulation of soils to improve crop production. The first soil microbiology observations can be conveniently attributed to ancient Roman agricultural writers such as Virgil, who noted that nodules (nitrogen-fixing structures) were present on legume roots.

During its infancy as a science, the main problem in soil microbiology (and microbiology in general) was that people couldn't actually see the organisms responsible for the changes they observed. Millennia ago, brewers in ancient Egypt, for example, fermented grain to make beer without knowing that microscopic yeast (fungi) were involved. This all changed in Delft, Holland, during the 17th century when Antonie van

Leeuwenhoek (1632–1723) made the first microscopes that clearly revealed microorganisms as small as bacteria. To put this in perspective, a typical bacterium is one micrometer (μm) in diameter, or 1/1000 of a millimeter (mm), which is approximately the width of the gap between your thumb and index finger if you put them just barely apart.

Leeuwenhoek had little scientific training. He was a draper by trade and had a part-time job as a city janitor—lens grinding was a hobby. The "animacules" he described swimming about in water were a mystery to him. However, Leeuwenhoek corresponded with the leading scientific organization of his day, the Royal Society of England. Robert Hooke (1635–1703), a member of that society, corroborated Leeuwenhoek's observations with his own description of microscopic protozoa, molds, spores, and plant cells. Hooke was the first person to use the term *cells* in this context. Hooke's book, *Micrographia,* was published in 1665 and described his observations. It can be considered the first microbiology textbook.

Surprisingly little progress was made in soil microbiology for the next 100 years. Then, in the mid-19th and early 20th centuries, a great era in microbiological research began with scientists such as Louis Pasteur, Robert Koch, and Serge Winogradsky.

Pasteur (1822–1895) a chemist by training, initially distinguished himself by separating crystals of tartaric acid into their mirror image isomers. Mirror image isomers are compounds that have the exact chemical formulation but a slightly different configuration—for example, your hands are like mirror image isomers. Pasteur soon became interested in microbiology. Pasteurization, the process of heating liquids to partially sterilize them, was one of his efforts to use microbiology to help France's wine and beer industry. Pasteur also studied yeast fermentation and developed a technique to demonstrate that microorganisms (or germs), rather than spontaneous generation, were responsible for biochemical changes such as food spoilage. In an elegant experiment, he made swan-necked flasks that allowed air, but not microorganisms, into sterile broth. Some of these flasks remain at the Pasteur Institute in Paris, still open to air, and still sterile. In addition, Pasteur was instrumental in developing vaccines for chicken cholera, rabies (a viral disease), and anthrax, a disease of animals caused by the bacterium *Bacillus anthracis.*

Some persons implied that Pasteur's many discoveries were due to luck. In response, Pasteur replied, "Chance favors the well-prepared mind."

One of Pasteur's contemporaries and scientific rivals in the mid-19th century was a German named Robert Koch (1843–1910). If Pasteur can be said to have pioneered microbial physiology, then Koch can be said to have pioneered microbial culture technique. Koch was a rural doctor who earned his reputation in microbiology by developing procedures for isolating and growing pure cultures of microorganisms.

The procedures for isolating and growing pure cultures in which every individual in the culture represents one original unique organism, were instrumental in demonstrating that microorganisms were disease-causing agents. The process for establishing this cause-and-effect relationship is called Koch's Postulates, named after Robert Koch. The postulates pertain to demonstrating without doubt that a particular microorganism causes a particular disease or symptoms. The postulates were developed from Koch's

A Historical Note

Koch's pure culture techniques were originally developed on inconvenient substances such as potatoes or gelatin. They would never have been so successful if it weren't for the help of Frau Angelina Hesse (1850–1934), the wife of one of Koch's assistants. Potatoes and gelatin tended to rot and liquefy when microorganisms grew on them. They also provided conditions in which all kinds of microorganisms developed, which made it difficult to keep individual cultures growing. Angelina Hesse knew how to make exquisite semisolid desserts from red algae extracts. The important component in the red algae is a carbohydrate called agarose that is relatively resistant to decomposition by microorganisms. Apparently, Walther Hesse, Frau Hesse's husband, decided to grow some microbial cultures on these desserts, and it worked spectacularly well. Agar media was born and Koch's reputation was made. Frau Hesse's opinion about this use of her talent is not recorded.

considerable work with anthrax. The fact that the two best microbiologists in the 19th century worked on anthrax indicates how serious a problem it was at that time. Even today anthrax is one of the most feared agents in biological warfare, so it still represents a serious potential pathogen.

Koch used his scientific approach to demonstrate that *B. anthracis* caused anthrax and that infection by *Mycobacterium tuberculosis* caused the disease we call tuberculosis.

Koch's postulates are quite simple:

1. A specific microorganism can always be found with a given disease.
2. The microorganism can be isolated and grown in a pure culture in the laboratory.
3. The pure culture will produce the disease when inoculated into a susceptible host.
4. The microorganism can be recovered from the infected host and grown again in a pure culture.

Koch's postulates are not always easy to follow. Some soil microorganisms cannot be isolated and grown in pure culture. These viable but not culturable microorganisms still can play important roles in soil processes. Some microorganisms, including fungal species found in plant roots and some bacterial species that permanently reside inside insects, can be observed but not cultured apart from their host.

While Pasteur and Koch were primarily concerned with disease-causing microorganisms, another microbiologist, Serge Winogradsky, was investigating microbial activities directly related to soil. Winogradsky (1856–1953) was a Russian microbiologist who is often called the "Father of Soil Microbiology" because of the outstanding diversity of soil processes he investigated (Figure 1-1). Winogradsky developed the Winogradsky column, a self-contained ecosystem for studying the sulfur cycle. He investigated microbial growth on CO_2 and inorganic ions, a process called chemoautotrophy. He studied nitrification, a microbial process in which ammonium (NH_4^+) is ultimately

Figure 1-1   Serge Winogradsky.
(Photograph courtesy of the American Society
for Microbiology Archives Collection)

Figure 1-2   Martinus Beijerinck.
(Photograph courtesy of the American
Society for Microbiology Archives
Collection)

converted into nitrate (NO_3^-). *Nitrobacter winogradsky,* one of the nitrifying bacteria, is named after Serge Winogradsky.

Winogradsky investigated microbial oxidation of ferrous iron, Fe^{2+}, the reduced form of iron to ferric iron, Fe^{3+}, the oxidized form, an essential component of rust. Winogradsky also isolated an anaerobic (growing without oxygen), spore-forming, nitrogen-fixing (converting nitrogen gas, N_2, into ammonia, NH_3) bacilli (rod-shaped bacterium). Because Winogradsky did much of his work at the Pasteur Institute, this bacterium was named *Clostridium pasteurianum.*

While Pasteur and Koch were working in Paris and Berlin, respectively, another great school of microbiology developed in Delft, Holland, Leeuwenhoek's old home. This effort was led by Martinus Beijerinck (1851–1931), who had nearly as wide-ranging an influence on soil and environmental microbiology as did Winogradsky (Figure 1-2). Beijerinck cultured the first nitrogen-fixing bacteria that grew symbiotically in association with legumes and the first aerobic nitrogen-fixing bacteria that grew asymbiotically as a free-living soil organism. These were *Rhizobium* and *Azotobacter,* respectively. An asymbiotic, nitrogen-fixing bacterium called *Beijerinckia* is named

Figure 1-3   Selman Waksman. (Photograph courtesy of the American Society for Microbiology Archives Collection)

after Beijerinck. He was the first person to recognize that there was a virulent biological agent smaller than a bacterium that could cause the plant disease Tobacco Mosaic Virus. Beijerinck coined the name "virus." He is often credited with the saying "Everything is everywhere, the environment selects."

If you've ever used an antimicrobial compound, you have the pioneering work of Sir Alexander Fleming (1881–1955) to thank. In 1928, Fleming announced his observation that a fungus, *Penicillium notatum,* contaminating an old plate of *Staphylococcus* in his lab, was surrounded by a zone of dead and dissolved cells. He had discovered evidence for the first antibiotic—penicillin. He received both knighthood, and the Nobel prize in 1945 for this critical observation.

Fleming's observations about the bactericidal products of microorganisms would not have seemed strange to Jacob Lipman (1874–1939), who founded the department for Soil Chemistry and Bacteriology at the New Jersey Agricultural Experiment Station at Rutgers in 1901. Lipman's textbook, *Bacteria in Relation to Country Life* (Lipman 1908), was the first attempt to popularize the science of soil microbiology for a wider audience, and it contains several photographs clearly demonstrating microbial inhibition.

One of Lipman's students, Selman Waksman (1888–1973), was born in Russia but emigrated to the United States and worked at Rutgers University (Figure 1-3).

Waksman, along with Lipman, might be considered one of the "Fathers of American Soil Microbiology," but we rarely hear about Waksman's early work on the microbial ecology of environments such as compost. Waksman was familiar with Lipman's and Fleming's research, and around 1944, Waksman and his research associate René Dubos isolated a soil actinomycete (a bacteria with filamentous growth) called *Streptomyces,* which has antibiotic properties much like those Fleming had found with *Penicillium.* Waksman had actually seen this many times before, but hadn't attributed much importance to it until two English scientists named Florey and Chain demonstrated during World War II that penicillin could be mass-produced industrially (Baldry 1965). It was Waksman who actually coined the term "antibiotic." He won a Nobel prize in 1952 for the discovery of this antibiotic compound, which we now call streptomycin. Waksman's laboratory ultimately became devoted to finding antimicrobial properties in soil microorganisms.

You should see a pattern here. Periods of slow, steady progress in soil microbiology are separated by great leaps in technology and discovery. A great leap occurred when Leeuwenhoek manufactured the first good microscopes. A great leap occurred when Koch developed pure culture techniques. Another great leap occurred in our era when scientists developed the first practical gas chromatographs, instruments that allow scientists to separate and measure the gases evolving during biological activity. The next great leap forward is undoubtedly the use of molecular biology techniques to create new organisms with unique properties and to search for relatedness among living things based on their genetic, not just their physical, characteristics.

CURRENT TOPICS IN SOIL MICROBIOLOGY

Soil microbiology is important enough that many large colleges and universities have microbiologists on their faculty who work in this field. Following are just a few examples of important research interests that these soil microbiologists pursue:

Symbiotic nitrogen fixation
Organic matter decomposition (waste removal and composting)
Mineral nitrogen transformations (nitrification, denitrification, and ammonification)
Rhizosphere studies (root/soil/microorganism interactions)
Soil enzymes (ureases, cellulases, ligninases, phosphatases)
Biodegradation and bioremediation
Metal transformation
Carbon cycling
Greenhouse gases and atmospheric pollution (production of methane, carbon dioxide, nitric oxide, nitrous oxide)
Release and monitoring of GEMS (genetically engineered microorganisms)
Microbial ecology
Subsurface microbial activity

As you can see, there are many different areas to study. Soil microbiology is not a subject limited for want of interesting topics. Let's begin our study by reviewing some of the basic physiology of living organisms in soil.

Summary

Soil microbiology is the branch of soil science concerned with soil-inhabiting microorganisms, their functions, and activities. The basic components of soil are gas, water, minerals, and organic matter. There are five characteristics that show that soil is a living system—movement, respiration, heat generation, digestion, and evolution. Soil microbiologists take two basic approaches to studying microorganisms—taxonomic and process-oriented. Soil microbiology is quite an old science, but its best-known practitioners, starting with Antonie van Leeuwenhoek, appeared after the 18th century. The most famous pioneering soil microbiologists were Louis Pasteur, Robert Koch, Serge Winogradsky, Martinus Beijerinck, Alexander Fleming, and Selman Waksman. Contemporary soil microbiologists work in many different areas, such as biodegradation, microbial ecology, and environmental quality.

Sample Questions

1. What are some of the criteria you could use to tell if something is alive?
2. What is soil?
3. Why can you draw an analogy between dissection and studying soil?
4. What is one of the first principles of soil microbiology?
5. Write a short, working definition of soil microbiology.
6. What are two approaches to studying soil microbiology? Which one do you prefer?
7. Who was Robert Koch and what are his postulates?
8. Why is Serge Winogradsky called the Father of Soil Microbiology?
9. What do Sir Robert Fleming and Selman Waksman have in common?
10. What are some topics that soil microbiologists study?

Thought Question

If you were designing a space probe to look for life on another planet, what sorts of things would you have it look for?

Additional Reading

There are excellent biographical materials that address the early history of soil microbiology. The most famous is *Microbe Hunters* by Paul De Kruif. It has influenced generations of microbiologists, although parts of the book may offend some people because De Kruif's commentary reflects the racism of his time. René Vallery-Radot, Pasteur's son-in-law, wrote a glowing biography of him called *The Life of Pasteur*. In contrast, Gerald Geison, in *The Private Science of Louis Pasteur,* wrote a much more critical assessment of Pasteur's work. P. E. Baldry gives a nice summary of Fleming and Waksman's work in *The Battle Against Bacteria*. Francis Clark wrote a historical perspective of soil microbiologists in the United States in *Soil Microbiology—It's a Small World.* You can get a real sense of some of these epic discoveries by reading *Milestones in Microbiology* by Thomas Brock. He has also written a new biography of Robert Koch, *Robert Koch: A Life in Medicine and Bacteriology.*

For those of you who want more challenging reading, three important journals that cover the current state of microbiological research are *Applied and Environmental Microbiology* (American Society for Microbiology), *Soil Biology and Biochemistry* (Elsevier Publications), and *The Journal of Environmental Quality* (American Society of Agronomy).

References

Atlas, R. M. and R. Bartha. 1993. *Microbial ecology: Fundamentals and applications.* 3d ed. Redwood City, CA: Benjamin /Cummings.

Baldry, P. E. 1965. *The battle against bacteria.* Cambridge, England: Cambridge University Press.

Brock, T. D. 1998. *Milestones in microbiology.* Washington, DC: American Society for Microbiology.

Brock, T. D. 1998. *Robert Koch: A life in medicine and bacteriology.* Washington, DC: American Society for Microbiology.

Clark, F. E. 1977. Soil microbiology—It's a small world. *Soil Science Society of America* 41:238–41.

De Kruif, P. 1953. *Microbe hunters.* New York: Harcourt Brace, & Co.

Geison, G. L. 1995. *The private science of Louis Pasteur.* Princeton, NJ: Princeton University Press.

Lipman, J. G. 1908. *Bacteria in relation to country life.* New York: Macmillan.

Soil Science Society of America. 1998. *Glossary of soil science terms.* Madison, WI: Soil Science Society of America.

Vallery-Radot, R. 1937. *The life of Pasteur.* Garden City, NY: Garden City Publishing Company.

Chapter 2

Microbial Growth and Metabolism

Overview

After you have studied this chapter, you should be able to:

- List seven microbial growth requirements.
- Classify microorganisms based on their carbon, energy, and oxygen requirements.
- Explain the role of oxidation and reduction (redox) reactions in microbial metabolism and list three basic metabolic types in soil.
- Define the concept of chemiosmosis and explain its relationship to cell respiration.
- Describe how the use of different substrates by microorganisms affects their growth and tell how it can be used to characterize them.

INTRODUCTION

In this chapter we review some basic microbial physiology. This is our only introduction to the basic modes of microbial life. It is difficult to understand a concept without understanding its terminology, so by necessity, we have to deal with many definitions. Microbial physiology is a course by itself, so you should look at more specific textbooks for additional details. For example, we do not go into depth about the synthesis of cell membranes and walls, proteins, and metabolites. For now, we will focus on how microorganisms grow, what they need to grow, what their methods of metabolism are, and how they generate energy.

WHAT DO MICROORGANISMS NEED TO GROW?

Microorganisms are about 70% to 85% water (Stolp 1988). The remaining dry matter consists of 50% protein, 10% to 20% cell wall material, 10% lipids (cell membrane material), 10% to 20% RNA, and 3% to 4% DNA. Like plants and animals, microorganisms require a few basic necessities of life: 1) a favorable environment including a favorable pH, an appropriate temperature, and appropriate redox conditions; 2) water; 3) mineral nutrients; 4) energy sources, which are often, but not always, organic carbon; 5) carbon sources, which are often, but not always, organic carbon; 6) electron donors and acceptors; and 7) growth factors. We discuss microbial environments including water availability in later chapters. For now, we begin by discussing mineral nutrition.

Mineral Nutrients and Growth Factors

The elemental composition of microbial dry matter would look something like this: 50% carbon (C), 20% oxygen (O), 14% nitrogen (N), 8% hydrogen (H), 3% phosphorus (P), 1% sulfur (S), 1% potassium (K), 0.5% calcium (Ca), 0.5% magnesium (Mg), and 0.2% iron (Fe) (Stolp 1988). To remember the most important mineral nutrients of microorganisms, use this memory aid:

C. HOPKNS CaFe—Mighty Good

The first seven elements in this list comprise 97% of microbial dry matter. However, we usually think of the last seven elements as the mineral nutrients that microorganisms require for growth. Nitrogen, for example, is found in both organic and inorganic forms:

Inorganic: NO_3^-, NH_4^+, N_2, N_3^-
Organic ($R-NH_2$): Proteins, amino acids, nucleotides, HCN

Ammonium (NH_4^+) is the preferred N source for microorganisms. Likewise, there are many forms of P. Organic P includes such compounds as inositol phosphate and nucleic acids. Phosphorus links the nucleosides together to make the polymers DNA and RNA. Inorganic P is largely P-containing mineral compounds; apatite is an example. However, microorganisms can't use the P in these mineral compounds until it is converted into a soluble inorganic ion such as HPO_4^{2-}.

Sulfur also has organic and inorganic forms that microorganisms use. For example, organic S consists of compounds such as proteins and amino acids (cysteine and methionine). Inorganic S includes compounds such as sulfate (SO_4^{2-}) and thiosulfate ($S_2O_3^{2-}$). In the memory aid just shown, Ca stands for calcium, Fe represents iron, Mighty Good stands for magnesium (Mg), and K stands for potassium (the original name for this element is Kalium). When microorganisms absorb these cations, they take them in as Ca^{2+}, Mg^{2+}, and K^+.

Microorganisms also require numerous micronutrients. They are called micronutrients because minuscule quantities are required by the microbial cell. Micronutrients are often metal cofactors required by enzymes. A cofactor is a nonprotein compound that is required to make enzymes, coenzymes, and other biochemical compounds function properly. Iron (Fe), for example, is used by cytochromes in electron transport. Manganese (Mn) is important in dismutases and photosynthesis. Zinc (Zn) is necessary for DNA polymerase. Copper (Cu) is found in a variety of reductases. Cobalt (Co) is essential to nitrogenase, the enzyme complex that prokaryotes use to fix nitrogen. Molybdenum (Mo) is a vital element that microorganisms use in nitrogenase and NO_3^- reductase. Nickel (Ni) is a cofactor of urease, the enzyme that decomposes urea (an important animal waste product and source of fertilizer N).

Other micronutrients that some microorganisms need for their metabolism include vanadium (Va), chlorine (Cl), sodium (Na), boron (B), selenium (Se), silicon (Si), and tungsten (W). Vanadium, for example, can be used to replace Mo during nitrogen fixation (Robson et al. 1986). Sodium is required by some marine microorganisms (Stolp 1988). Selenium and tungsten stimulate the growth of some methane-(CH_4) producing microorganisms (Stolp 1988). Silicon is an essential component in the cell wall of algae such as diatoms.

Some, but not all, microorganisms also need growth factors. A growth factor is a compound that is essential for microbial growth. Examples of growth factors include vitamins (thiamin, biotin, riboflavin, nicotinic acid, pantothenic acid, folic acid), amino acids (alanine, aspartic acid, etc.), or nucleotides (purines such as adenine and guanine or pyrimidines such as thymine, cytosine, and uracil). An auxotroph is a microorganism that requires one or more growth factors (something like choline or inositol, for example). A good definition of an auxotroph is "an organism that has additional nutritional requirements over and above the nutritional requirements of a completely self-sufficient organism." People are auxotrophs because we are unable to synthesize all of the vitamins and growth factors that we need to exist. Plants are not auxotrophs; they can synthesize all the growth factors they need; that's one reason why we eat plants.

Carbon and Energy Sources

We can characterize microorganisms, and all living things for that matter, by their carbon and energy sources. Microorganisms that use light to generate energy are called phototrophs. Microorganisms that break chemical bonds to generate energy are called chemotrophs. Chemotrophs can be either lithotrophs, which break inorganic bonds, or organotrophs, which break organic bonds. Autotrophic organisms obtain all their carbon for biosynthesis from CO_2 or HCO_3^- (bicarbonate). Heterotrophic organisms obtain their carbon for biosynthesis from organic carbon. Saprophyte is another name for a heterotrophic microorganism that obtains its carbon and energy from dead and decaying organic material.

It may be easier if these relationships are illustrated in a flowchart:

Energy source
    Chemical = Chemotroph
        Carbon source
            Organic carbon = Chemoheterotroph
            Inorganic carbon = Chemoautotroph
Energy source
    Light = Phototroph
        Carbon source
            Organic carbon = Photoheterotroph
            Inorganic carbon = Photoautotroph

People, for example, are chemoheterotrophs because we obtain our energy from metabolizing organic carbon bonds and we also obtain our carbon from ingesting organic carbon materials. Neither sunlight nor fresh air alone will sustain us because of our energy and carbon requirements.

Oxygen Requirements

You can make a further distinction among microorganisms based on whether they require O_2 for growth. This is important in determining the basis for microbial metabolism and where and when a microorganism can be active.

1. Obligate aerobes—O_2 absolutely required
2. Obligate anaerobes—O_2 not required and toxic
3. Microaerophiles—O_2 required but toxic at some low concentrations
4. Facultative anaerobes—O_2 the preferred electron acceptor but alternatives can be used

Obligate means the requirement is absolute, while facultative means the requirement is not absolute. A facultative microorganism is flexible; it has the faculty to carry out a process or utilize a compound.

Why is O_2 toxic to some organisms? Oxygen is a powerful oxidizing agent. When oxygen gains one electron during microbial metabolism it becomes unstable, and in an attempt to become a more stable form, it begins oxidizing everything around it including the nucleotides in genetic material. Oxidized nucleotides are one source of mutations in nucleic acid and are usually fatal.

Aerobes can't prevent toxic intermediates such as O_2^- (superoxide) and H_2O_2 (peroxide) from forming during oxygen metabolism. But they can limit harmful effects by disposing of these compounds as quickly as they are formed by the action of enzymes such as superoxide dismutase and catalase (Figure 2-1). Many anaerobes lack these enzymes, which protect aerobes from chemically reactive oxygen.

If O_2 is toxic to anaerobes, why can anaerobes always be isolated from aerobic soils? Two explanations have been given. The first is that even aerobic soils have some spots

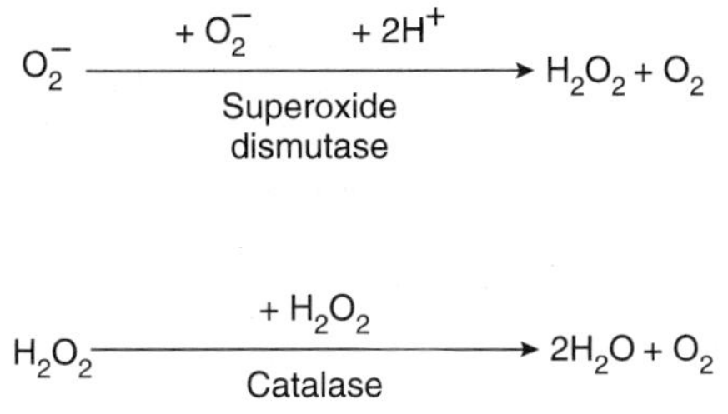

$$O_2^- \xrightarrow[\substack{\text{Superoxide} \\ \text{dismutase}}]{\substack{+ O_2^- \quad + 2H^+}} H_2O_2 + O_2$$

$$H_2O_2 \xrightarrow[\text{Catalase}]{+ H_2O_2} 2H_2O + O_2$$

Figure 2-1   Microbial removal of toxic oxygen compounds. (Adapted from Paul and Clark 1996)

where O_2 doesn't reach or where microbial activity rapidly consumes it. This is the concept of the anaerobic microsite. A second explanation is that the toxic compounds produced from O_2 are made only during electron transport. Electron transport occurs only in the presence of an electron donor (food). Consequently, obligate anaerobes persist for long periods in aerobic environments if they are not actively growing.

OXIDATION/REDUCTION (REDOX) REACTIONS

Oxygen plays a critical role in the redox reactions associated with microbial growth. The redox potential (E_h) is a measure of the tendency for electrons to move between oxidized and reduced compounds. You can also think of it as the potential energy available from moving electrons around. Oxidized compounds are electron poor, while reduced compounds are electron rich. Oxidations yield biological energy. So, for an organic or inorganic compound to be used as a microbial energy source, it must be capable of being oxidized.

What is an oxidation? In an oxidation, an electron is given up. If X and RH_2 are electron donors or reducing agents in this example, then:

$$X \Rightarrow X^{2+} + 2e^-$$
$$RH_2 \Rightarrow R + 2H$$

What is a reduction? In a reduction, an electron is accepted. R, Fe^{3+}, and Mn^{4+} are electron acceptors or oxidizing agents in these examples:

$$2H + R \Rightarrow RH_2$$
$$Fe^{3+} + e^- \Rightarrow Fe^{2+}$$
$$Mn^{4+} + 2e^- \Rightarrow Mn^{2+}$$

In a coupled redox reaction, one compound, the oxidant, oxidizes another compound and gains electrons (it becomes reduced). A second compound, the reductant, reduces another compound and loses electrons (it becomes oxidized). In the following coupled redox reaction, Fe^{3+} is the oxidant and H_2 is the reductant.

$$H_2 \Rightarrow 2H^+ + 2 \text{ electrons}$$
$$\underline{2Fe^{3+} + 2 \text{ electrons} \Rightarrow 2\ Fe^{2+}}$$
$$2Fe^{3+} + H_2 \Rightarrow 2Fe^{2+} + 2H^+$$

An electron donor or reductant is used to reduce carbon compounds for biomass production during microbial growth. For example, photosynthesis is a reduction in which CO_2 is reduced to carbohydrate $(CH_2O)_n$. An electron acceptor is required during oxidations. You have to deposit the electrons from reduced compounds somewhere, otherwise everything in the cell will become reduced and there will be no tendency for electrons to flow. Oxygen is the major electron acceptor in biological systems, which is one of the reasons why it is so important.

METHODS OF METABOLISM

Microorganisms can be grouped according to the type of electron acceptor they use and their mode of metabolic activity (Table 2-1).

The energy that can be generated from a redox couple, which is a chemical or biochemical reaction that has an oxidation step and a reduction step, depends on the magnitude of difference between an oxidizing agent's capacity to accept electrons and a reducing agent's capacity to donate them (Table 2-2).

For now, remember that the greater the magnitude of ΔG, the more energy is generated. Let's use the analogy of a flowing stream rather than a flow of electrons. The greater the change of elevation in the stream, the faster a waterwheel in the stream would turn and the more work the waterwheel could do.

There are three basic types of microbial metabolism: respiration, fermentation, and phototrophic growth (Figure 2-2).

Table 2-1 Characterization of organisms based on metabolic type and electron acceptor.

Metabolic Type	Electron Acceptor	Organisms
Fermentation	Organic	Prokaryotes; a few fungi
Aerobic respiration	O_2	Most prokaryotes; fungi; all animals
Anaerobic respiration	NO_3^-, NO_2^-, SO_4^{2-}, CO_2	Prokaryotes only

Table 2-2 Energy available in redox couples.

Electron Acceptor	Electron Donor	ΔG (KCal/mole electrons)
O_2	Organic	−30
NO_3^-	Organic	−20
O_2	Fe^{2+}	−20
O_2	NH_4^+	−10
O_2	Mn^{2+}	−7
SO_4^{2-}	Organic	−6
NO_3^-	NH_4^+	−1

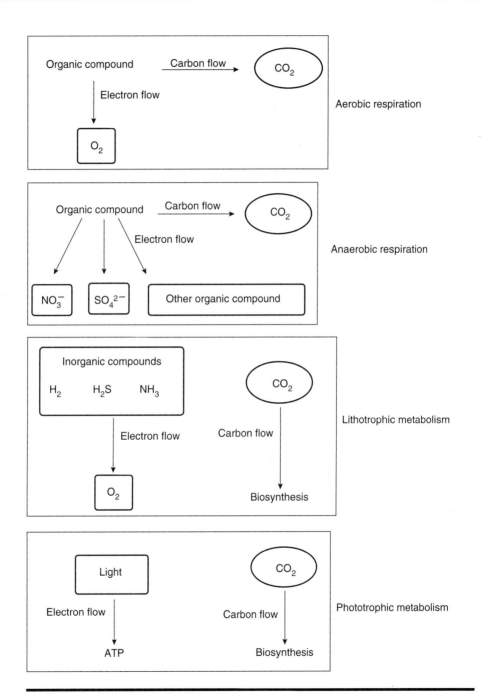

Figure 2-2   Types of microbial metabolism. (Adapted from Brock and Madigan 1991)

Aerobic respiration uses the oxidation of an organic or inorganic electron source coupled to the reduction of oxygen as the ultimate electron acceptor to generate energy; this is how it works. The chemiosmotic theory of microbial respiration postulates that the cytoplasmic membrane is impermeable to OH^- and H^+. A respiratory chain is located in the cell membrane that moves H^+ from inside the cell to outside the cell as electrons that come from oxidizing an electron donor are passed along to oxygen. ATP synthase, an enzyme, takes advantage of the pH gradient that develops. Because H^+ is moved outside the cell, the inside of the cell develops a higher pH relative to the outside of the cell. This gradient can be used to make ATP, which is the currency of cell energy—lots of ATP means lots of growth and activity. Proton movement from inside the cell to outside the cell also makes the membrane comparable to a miniature battery, with positive and negative poles corresponding to the outside and inside of the cell.

The net effect, which is all we care about, is that an electron donor, a reduced compound, is oxidized and an electron acceptor (usually O_2) is reduced. Protons are transported from inside to outside the cell membrane. The H^+ gradient can be used to do cell work, such as move the cell or take up nutrients, and to generate ATP (Figure 2-3).

The more steps in which H^+ can be moved outside the cell, the more potential there is for ATP to be generated. So we can modify our waterwheel analogy a little. The elevation of the stream from top to bottom (from reduced to oxidized compounds) is important, not because it can make the waterwheel spin faster, but because there is more opportunity to put several waterwheels in tandem to generate energy. It's the same

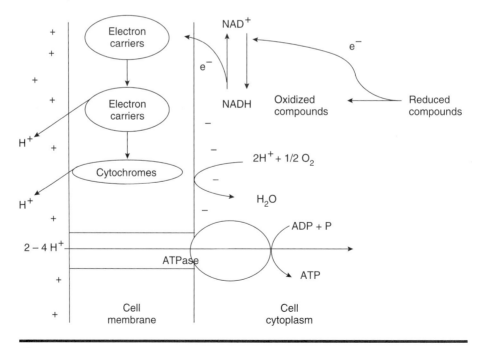

Figure 2-3   Schematic diagram of the chemiosmotic theory of energy generation. (Adapted from Gottschalk 1986)

principle underlying why large rivers, such as the Columbia, don't have a single dam; they have several.

Anaerobic respiration uses the same principle, except that the terminal electron acceptor is not O_2, but some other inorganic ion such as NO_3^-, Fe^{3+}, or Mn^{4+}.

Fermentation refers to internally balanced oxidation reduction reactions of organic compounds in cells that yield energy (Brock and Madigan 1991). Fermentation does not require an external electron acceptor. The substrates that are metabolized can be both electron donor and acceptor. Fermentation could, but usually doesn't, occur in the presence of O_2 and it does not require O_2. However, respiratory metabolism generates much more energy than does fermentation—so much more, that organisms that can both ferment and respire, such as yeast, will invariably respire if there is any available O_2. One consequence of a shift from fermentative to respiratory metabolism is that CO_2 evolution can actually decrease. This is called the "Pasteur Effect." Since more energy can be obtained through respiration than by fermentation, glucose metabolism declines three- to fourfold, and the amount of CO_2 correspondingly decreases. It also stops alcohol from being produced.

What's the net result of fermentation? ATP is formed. One part of an organic molecule is oxidized and usually released as CO_2. One part of the organic molecule is reduced and is usually released as:

1. An alcohol CH_3CH_2OH (ethanol)
2. An organic acid CH_3COOH (acetic acid)
                           $CH_3CH_2CH_2COOH$ (butyric acid)
3. A ketone CH_3COCH_3 (acetone)

How can you tell if something was fermented? Do a fermentation balance. Write the reaction and give all H a value of (1) and all O a value of (–2). A positive value means you have a reduced compound, and a negative value means you have an oxidized compound. For example, in the fermentation of glucose to make alcohol:

$$C_6H_{12}O_6 \Rightarrow 2CH_3CH_2OH + 2CO_2$$
$$C_6H_{12}O_6: (12H \times 1) + (6O \times -2) = 0$$
$$2CH_3CH_2OH: (12H \times 1) + (2O \times -2) = 8$$
$$2CO_2: (4O \times -2) = -8$$

Although part of the glucose was oxidized in the reaction to make CO_2 and part was reduced to make ethanol, there was no net change in the total number of electrons; they were just switched around and no external electron acceptor was used in the process.

PHOTOSYNTHESIS

Phototrophs are not as important to soil microbiology as are the chemotrophs because phototrophs can make their own carbohydrate and generate their own energy from sunlight in a process called photosynthesis. Thus, they do not need to carry out the sort of biological transformations that contribute to nutrient cycling. The problem

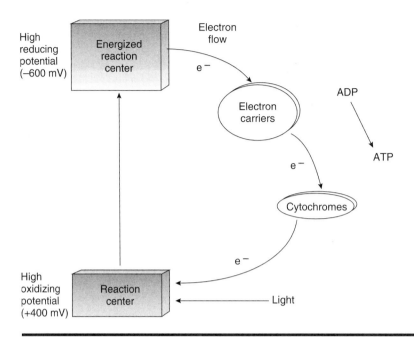

Figure 2-4   Schematic diagram of ATP production during cyclic photophosphorylation. (Adapted from Brock and Madigan 1991; Gottschalk 1986)

phototrophs face is how to generate energy and how to create enough reducing power (electrons to donate) to convert CO_2 into carbohydrate.

Let's look at two examples of photosynthesis and note their similarity:

Anaerobic (prokaryotes) $2H_2S + 2NADP^+ \Rightarrow 2S + 2NADPH + 2H^+$
Aerobic (plants/prokaryotes) $2H_2O + 2NADP^+ \Rightarrow O_2 + 2NADPH + 2H^+$

Prokaryotes include the organisms that don't have a cell nucleus. Plants are examples of eukaryotes, organisms that have a cell nucleus. In the first example, anaerobic photosynthesis, hydrogen sulfide is oxidized to elemental S. In the second example, aerobic photosynthesis, water is oxidized to oxygen. Cyclic photosynthesis produces energy in the form of ATP (Figure 2-4). Noncyclic photosynthesis produces electron donors such as NADPH that can be used to fix carbon (Figure 2-5).

Photoheterotrophs like *Rhodospirillum* get energy from cyclic photosynthesis and C from organic carbon. The ATP is generated by moving protons across a chemical gradient (i.e., chemiosmosis). During noncyclic photosynthesis in anaerobes, the electrons are diverted from the reaction centers and used to form NADPH. The electrons are replaced by the oxidation of hydrogen sulfide to elemental S.

Cyanobacteria and other aerobic photosynthetic organisms also use cyclic photosynthesis to generate ATP by a chemiosmotic mechanism. In contrast to anaerobes, they use two separate light-driven excitations of electrons during noncyclic photosynthesis

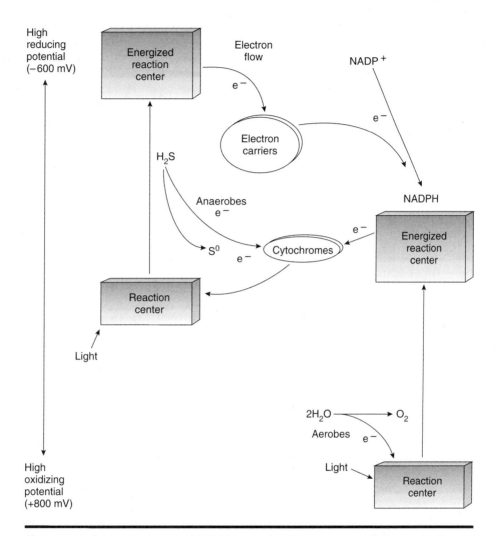

Figure 2-5 Schematic diagram of NADPH production during noncyclic photosynthesis. Only aerobes employ two light-driven reactions. (Adapted from Brock and Madigan 1991; Gottschalk 1986)

(Figure 2-5). The electrons from the first light-driven event are diverted to produce NADPH. These electrons are replaced by electrons from a different reaction center that has been energized by a second light-driven event. The electrons used to replace these diverted electrons come from the oxidation of water to O_2.

Cyanobacteria, like plants, use the NADPH that is produced to convert CO_2 into carbohydrate via the Calvin cycle. Many lithotrophs (autotrophs), which use CO_2 as their carbon source, also utilize the Calvin cycle to fix CO_2.

$$6CO_2 + 12NADPH + 18ATP \Rightarrow C_6H_{12}O_6 + 12NADP + 18ADP$$

MICROBIAL GROWTH—NUTRIENT AND SUBSTRATE UPTAKE

The cell membrane is differentially permeable, which means that it selectively excludes things. There are several uptake methods—passive diffusion if the molecule is small and uncharged; facilitated diffusion through proteins in the cell membrane; and active transport, an energy-dependent transport across a concentration gradient. Microorganisms do not typically absorb macromolecules, which means that large compounds must first be decomposed outside the cell before they can be utilized.

Microorganisms can be selective about the types of substrates on which they grow. This is an intrinsic factor that reflects both the biochemical capacities of the microorganism and the genetic regulation of its metabolism. Microbial growth on different substrates remains one of the basic methods for characterizing and classifying microorganisms, although it is rapidly being replaced by genetic methods. An example of how it might be used is shown in Figure 2-6. In this particular case, Streptococci (now called Enterococci) isolated from a soil sample were inoculated into broth containing one of several sugars and a pH indicator. If the sugar was utilized, acid was produced, the pH dropped, and the broth turned yellow. So, even though the streptococci isolates were indistinguishable on agar plates, they could be distinguished biochemically by the carbon sources they used.

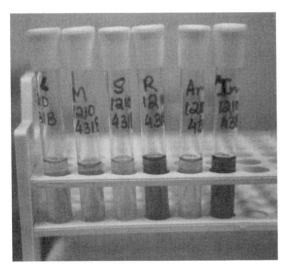
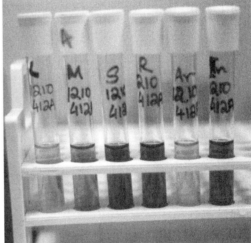

a   b

Figure 2-6 Differential sugar use by two streptococci isolates from soil. The sugar substrates in each case (from left to right) are: Lactose (L), Mannitol (M), Sorbitol (S), Raffinose (R), Arabinose (Ar), and Inositol (In). Note the differential color in (a) and (b), which represent two unique isolates. The tubes with lighter shading indicate that the sugar is metabolized. (Photograph courtesy of M. S. Coyne)

Summary

There are seven requirements for microbial growth in soil. A simple memory device, C. HOPKNs CaFe—Mighty Good, will help you remember the major inorganic nutrients that microorganisms require. Microorganisms can be classified on the basis of their energy sources and carbon requirements as phototrophs and chemotrophs, and heterotrophs and autotrophs. Microorganisms can further be described on the basis of their oxygen requirements as aerobes, facultative aerobes, and anaerobes.

Redox reactions are critical in microbial metabolism. Reduced compounds are oxidized to gain energy while oxidized compounds are used to accept electrons during metabolism and become reduced. Three basic types of microbial metabolism are respiration, fermentation, and phototrophic growth. Respiration occurs because microorganisms create proton gradients across their cell membranes that are used to do work. This process, called chemiosmosis, requires external electron acceptors. Fermentation requires no external electron acceptors; electrons are shuffled between different compounds in a cell. Phototrophic growth uses light to generate energy and is either cyclic or noncyclic. Microbial substrate use is one of the basic criteria for characterizing otherwise indistinguishable cultures.

Sample Questions

1. What are five requirements for microbial growth in soil other than inorganic nutrients?
2. Complete the following table:

Type of Metabolism	Energy Source	Carbon Source	Electron Donor
Chemoautotroph	?	?	?
?	Organic	?	?
Photoautotroph	?	CO_2	?
?	Light	Organic	Organic

3. Draw a diagram that illustrates the basis of energy generation during aerobic respiration.
4. What is the difference between a heterotroph and an autotroph?
5. In terms of energy and carbon source, describe your next-door neighbor.
6. Are humans aerobes, facultative anaerobes, or anaerobes? What does that mean?
7. What is a saprophyte?
8. How would you describe or characterize an obligately aerobic, facultatively photoheterotrophic chemoautotroph?

Thought Question

During his fermentation studies, Louis Pasteur noticed a very curious thing. When he reintroduced air into a fermenting culture, the production of CO_2 decreased. Since we know that microbial growth is better in aerobic conditions than in anaerobic conditions, it would seem logical to assume that CO_2 production should increase as a result of greater microbial activity. How do you explain this result? How would you demonstrate your conclusion?

Additional Reading

Here are some advanced textbooks that will do a more thorough job of explaining microbial physiology. *Bacterial Metabolism* by Gerhard Gottschalk (1986, Springer-Verlag, New York) is a classic text. *Biochemistry of Bacterial Growth, 3d edition,* edited by J. Mandelstam et al. (1982, Blackwell Scientific Publications, London), is, as the title suggests, a good look at biochemical aspects of bacterial growth. *Microbial Physiology, 2d edition,* by I. W. Dawes and I. W. Sutherland (1992, Blackwell Scientific Publications, London), is also a very readable text on the biochemistry of bacterial growth.

References

Brock, T. D., and M. T. Madigan. 1991. *Biology of microorganisms.* 6th ed. Englewood Cliffs, NJ: Prentice-Hall.

Gottschalk, G. 1986. *Bacterial metabolism.* New York: Springer-Verlag.

Paul, E. A. and F. E. Clark. 1996. *Soil microbiology and biochemistry.* San Diego, CA: Academic Press.

Robson, R. L., R. R. Eady, T. H. Richardson, R. W. Miller, M. Hawkins, and J. R. Postgate. 1986. The alternative nitrogenase of *Azotobacter chroococcum* is a vanadium enzyme. *Nature* 322:388–90.

Stolp, H. 1988. *Microbial ecology: Organisms, habitats, activities.* Cambridge, England: Cambridge University Press.

Chapter 3

Soil Enzymes

Overview

After you have studied this chapter, you should be able to:

- Discuss enzymes as the catalysts driving biological reactions in soil.
- Explain how enzyme activity depends on their sequence and structure.
- List several basic types of enzymes.
- Describe enzyme activity in terms of enzyme kinetics.
- Discuss the environmental conditions that affect enzymes in soil.
- Explain the role that extracellular enzymes play in microbial growth and biodegradation.

INTRODUCTION

The physiology and metabolism of soil microorganisms are driven by enzymes and the microbial habitat in soil is affected by these enzymes. Although enzymes come from living organisms, once those organisms die, some enzymes can persist and retain their activity in soil for long periods. The role of environmental enzyme research is to understand the state and behavior of cellular and extracellular enzymes in the environment and apply that knowledge to current environmental and agricultural studies. In chapter 3, we first review some basic aspects and terminology of enzymology. Next, we look at how enzymatic reactions at the cellular level are reflected in soil microbial activity. Finally, we discuss the topic of extracellular enzymes and examine how they occur in the soil environment.

WHAT IS AN ENZYME?

Enzymes are organic protein catalysts that transform inorganic and organic substances without themselves being changed. Catalysts lower the activation energy of chemical reactions and allow the reactions to proceed at temperatures and pressures at which they would not normally occur (Figure 3-1). The activation energy is the energy required to sufficiently destabilize a compound's chemical bonds to make product formation possible, or to bring chemical constituents in suitable proximity for a reaction to occur. Enzymes allow these chemical reactions to occur at temperatures and pressures suitable for life.

The measure of whether a reaction will or will not occur is ΔG, the free energy change. If ΔG is negative, the reaction will occur spontaneously, but that doesn't mean it will occur rapidly. For example, glucose oxidation has a ΔG of -686 kcal mole^{-1}. However, you could leave glucose on a bench for months without any appreciable

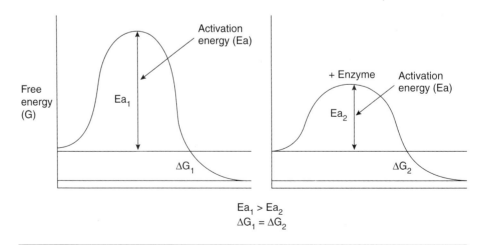

$$Ea_1 > Ea_2$$
$$\Delta G_1 = \Delta G_2$$

Figure 3-1 Schematic diagram of an enzyme-catalyzed reaction.

change. Glucose is stable in a kinetic or enzymatic sense. The oxidation of N_2 gas to NO_3^- in the atmosphere is also thermodynamically favorable. But the atmosphere only produces NO_3^- when enough energy is added to break apart the triple bond of the N_2 molecule. This is the sort of energy you find in lightning and it contributes a small amount of atmospheric NO_3^- to soil every year. The activation energy is high enough in glucose and N_2 that under normal conditions no transformation occurs.

These oxidations could be made to occur by raising the temperature. Paper, for example, is primarily composed of glucose polymers called cellulose. Paper sitting on a desk would remain there indefinitely if microorganisms were removed. However, raising the temperature to 233°C (451°F) by setting a match to the paper causes rapid oxidation, which is visible when the paper burns. No microorganisms could survive those temperatures, so microorganisms must utilize enzymes to make such reactions occur.

All enzymes are proteins, which are amino acid polymers, although not all proteins are enzymes. The easiest scheme to describe an enzyme-catalyzed reaction is:

$$S + E \Rightarrow ES \Rightarrow E + P$$

The substrate (S) and enzyme (E) combine to form an enzyme-substrate complex (ES). The substrate is transformed to release the product (P) and the enzyme, which can then be used to catalyze additional reactions. The interaction between enzyme and substrate is usually quite specific, but not always. Specificity, if it occurs, is caused by the precise interaction of amino acids at the enzyme's reaction site with the enzyme's substrate(s). The role for most of the enzyme is to make sure that this reaction site-substrate interaction is optimized. Mutations in even one amino acid close to the reaction site can have disastrous consequences for enzyme activity.

WHAT ARE ENZYMES MADE OF?

Enzymes are made of amino acids linked together by peptide bonds. Some enzymes also require coenzymes or cofactors to function. The basic structure of an amino acid consists of an amine group, a carboxyl group, and a distinctive side chain (Figure 3-2). The amino acids fall into six groups:

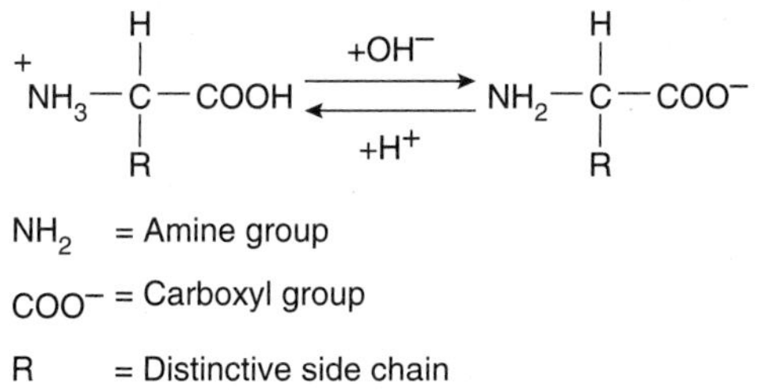

$$\overset{+}{N}H_3 - \overset{\overset{\displaystyle H}{|}}{\underset{\underset{\displaystyle R}{|}}{C}} - COOH \underset{+H^+}{\overset{+OH^-}{\rightleftharpoons}} NH_2 - \overset{\overset{\displaystyle H}{|}}{\underset{\underset{\displaystyle R}{|}}{C}} - COO^-$$

NH_2 = Amine group

COO^- = Carboxyl group

R   = Distinctive side chain

Figure 3-2   The basic structure of an amino acid.

Neutral Amino Acids
Glycine (5.97)
Alanine (6.02)
Leucine (5.98)
Isoleucine (6.02)
Valine (5.97)
Serine (5.68)
Threonine (6.53)
Glutamine (5.65)
Asparagine (5.41)

Acidic Amino Acids
Aspartic Acid (2.98)
Glutamic Acid (3.22)

Basic Amino Acids
Arginine (10.76)
Lysine (9.74)
Histidine (7.58)

Aromatic Amino Acids
Phenylalanine (5.48)
Tyrosine (5.65)
Tryptophane (5.88)

Secondary Amino Acids
Proline (6.30)
Hydroxyproline (5.83)

S-containing Amino Acids
Cysteine (5.02)
Methionine (5.75)

Amino acids are dipolar molecules, which means that they can be charged depending on the pH of the environment. The pI, or isoelectric point, is the pH at which the amino acid has no charge, or is neutral. This is characteristic of each amino acid. The pI of the amino acids listed above is in parentheses. If the pH of the environment is less than the pI, then the overall charge of the amino acid is positive. If the pH of the environment is greater than the pI, then the overall charge of the amino acid is negative. The implication of this is that the proteins in the cell walls of microorganisms have an overall net negative charge at the pH of most soil environments.

ENZYME CONFORMATION AND STRUCTURE

Enzyme function depends on primary, secondary, tertiary, and quaternary structure, which are pH and temperature dependent. Primary structure is the sequence of amino acids. Secondary structure is the arrangement of the amino acid chains into the form of aligned sheets or helixes. Tertiary structure is the intermolecular or interchain linkage of amino acids by hydrogen and sulfur-sulfur bonds—the protein twists, bends, and folds. Adding reducing compounds (compounds that donate electrons) to enzymes denatures or unfolds the enzymes because they reduce and break open the interchain bonds that keep the enzyme folded. Quaternary structure is the specific orientation of multiple enzyme subunits into more complex enzyme aggregates.

WHAT ARE THE TYPES OF ENZYMES?

Biological reactions consist of six basic types of enzymatic activity (Mandelstam et al. 1982).

1. Oxidoreductases $XH_2 + Y \Leftrightarrow X + YH_2$
2. Transferases $X\text{-}R + Y \Leftrightarrow Y\text{-}R + X$
3. Hydrolases $X\text{-}Y + H_2O \Leftrightarrow X\text{-}H + Y\text{-}OH$

4. Lyases $\qquad$ X-Y $\Rightarrow$ X + Y

5. Isomerases $\qquad$ X-R-Y-S $\Rightarrow$ X-S-Y-R

6. Ligases $\qquad$ X + Y + ATP $\Rightarrow$ X-Y + ADP + P$_i$ (or AMP + PP$_i$)

Oxidoreductases are enzymes that oxidize a substrate by removing a pair of electrons and the accompanying hydrogen atoms. The electrons have to go somewhere, otherwise the enzyme would be permanently reduced, so they are transferred to other compounds that become reduced in turn. Catalase is a good example of an oxidoreductase:

$$2H_2O_2 \xrightarrow{\text{Catalase}} O_2 + 2H_2O$$

In aerobic respiration, electrons are transferred from carrier to carrier by oxidoreductases through a series of oxidation and reduction steps that eventually reduce O_2 to H_2O in aerobic organisms.

There are many types of transferases: transacetylases transfer acetyl groups (CH_3COOH); transaminases transfer amine groups (NH_2); transmethylases transfer methyl groups (CH_3); transglycosidases transfer sugar groups; kinases transfer phosphate groups (PO_4^{3-}). The transferred group is frequently carried by a coenzyme; for instance, coenzyme A, CoA-SH, in this example:

$$CH_3CO-SCoA + P \xleftrightarrow{\text{Phosphotransacetylase}} CH_3CO-P + CoA-SH$$
$$\text{Acetyl CoA} \qquad\qquad\qquad\qquad \text{Acetyl phosphate}$$

Alpha-oxo-aminotransferase (alpha-keto-glutarate aminotransferase) carries out one of the critical steps in nitrogen cycling—transferring an amine group from glutamate to alpha-keto-glutarate to make two glutamate molecules.

Hydrolases break molecules apart by adding water across bonds. Hydrolases include: esterases, glycosidases, lipases, peptidases, phosphatases, and urease.

$$\text{Protein} + H_2O \xrightarrow{\text{Peptidase}} \text{Peptides} + \text{Amino acids}$$

Lipid
$$
\begin{array}{lll}
CH_2 - OCOR & & \\
| & & \\
CH_2 - OCOR & \xrightarrow{\text{Lipase}} & \\
| & + 3H_2O & \\
CH_2 - OCOR & & \\
R = -(CH_2)_n CH_3 & &
\end{array}
$$

Glycerol $\qquad$ Fatty acids
$$
\begin{array}{lll}
CH_2 - OH & & R - COOH \\
| & & \\
CH_2 - OH & + & R - COOH \\
| & & \\
CH_2 - OH & & R - COOH
\end{array}
$$

$$NH_2CONH_2 + H_2O \xrightarrow[+ H_2O]{\text{Urease}} 2NH_3 + CO_2$$

Lyases also break molecules apart. They include decarboxylases, deaminases, and aldolases.

$$CH_3COCOOH + TPP \xrightarrow{\text{Carboxylase}} (CH_3CHO-TPP) + CO_2 \rightarrow CH_3CHO + TPP$$
$$\text{Pyruvate} \qquad\qquad\qquad\qquad\qquad\qquad\qquad \text{Acetaldehyde}$$

Isomerases move things around on molecules. For example, one of the key enzymes in the metabolism of glucose in microorganisms is catalyzed by triosephosphate isomerase:

$$
\begin{array}{ccc}
\begin{array}{l}
\text{CH}_2\text{OP} \\
| \\
\text{CHOH} \\
| \\
\text{CHO} \\
\text{Glyceraldehyde-3 ---} \\
\text{phosphate}
\end{array}
&
\underset{\text{isomerase}}{\overset{\text{Triosephosphate}}{\longleftrightarrow}}
&
\begin{array}{l}
\text{CH}_2\text{OP} \\
| \\
\text{C} = \text{O} \\
| \\
\text{CH}_2\text{OH} \\
\text{Dihydroxyacetone} \\
\text{phosphate}
\end{array}
\end{array}
$$

Ligases or synthases put molecules together. Glutamine synthetase also catalyzes a critical step in the assimilation of inorganic N into microbial cells.

$$
\text{Glutamate} + \text{NH}_3 + \text{ATP} \xrightarrow{\text{Glutamine synthetase}} \text{Glutamine} + \text{ADP} + \text{P}_i
$$

ENZYME KINETICS

Anyone interested in soil enzymes is concerned with catalysis, and anyone concerned with catalysis is concerned with the velocity, or kinetics, of enzyme-catalyzed reactions (Tabatabai 1994). Most enzyme reactions occur at rates that are constant and do not change with substrate concentration (zero-order reactions) or at rates proportional to the substrate concentration (first-order reactions). These reactions can be described mathematically by the Michaelis-Menten equation:

$$
v = \frac{(V_{max})(S)}{K_m + S}
$$

where v = the velocity of the enzyme reaction
V_{max} = the maximum velocity of the enzyme reaction
K_m = the Michaelis constant, equivalent to the substrate concentration when $v = V_{max}/2$
S = the substrate concentration

A plot of a typical hyperbolic curve that results from plotting an enzyme reaction that follows Michaelis-Menten kinetics is shown in Figure 3-3.

You should be able to deduce from the Michaelis-Menten equation why the graphed data have two distinct regions. If the substrate concentration is very large relative to K_m (S >>> K_m), then the equation simplifies to:

$$
v = V_{max}
$$

and you have zero-order kinetics. In zero-order kinetics, the rate of a reaction is constant and doesn't depend on how much substrate is present because no matter how much substrate is added, you can't make the reaction go faster. Every active site is saturated.

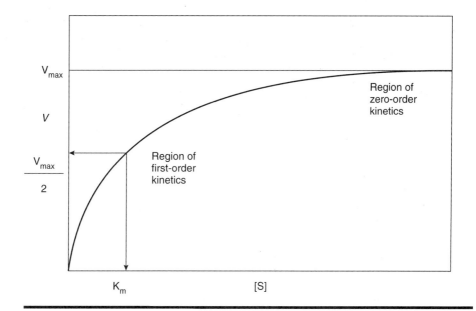

Figure 3-3 Idealized plot of a Michaelis-Menten reaction: the effect of substrate concentration on reaction velocity.

On the other hand, if the substrate concentration is low relative to the K_m ($S \ll K_m$), then the equation simplifies to:

$$v = (V_{max}/K_m)(S)$$

and the more substrate you add, the faster the reaction rate will be. This is first-order kinetics—the rate of a reaction depends on the amount of substrate present.

ENZYMES IN THE SOIL ENVIRONMENT

The processes that occur in the environment—mineralization, immobilization, nitrogen fixation, etc.—are all enzymatic reactions. There are many enzymes in soil. Oxidoreductases such as urate oxidase, peroxidase, and monophenol monooxygenase catalyze the oxidation and reduction of substrates such as uric acid, chloroaniline, and catechol, respectively. These reactions are important in decomposing organic wastes in soil. Hydrolases such as alkaline and acid phosphatase release inorganic P from organic P compounds, which is important in plant and microbial nutrition. Other hydrolases such as amylase and cellulase hydrolyze the bonds between the polymers starch and cellulose and make sugars available for microbial growth. Peptidases break peptide bonds and proteinases break down complex proteins such as casein and gelatin into amino acids that microorganisms can use as a source of carbon and nitrogen (Tabatabai and Fung 1992).

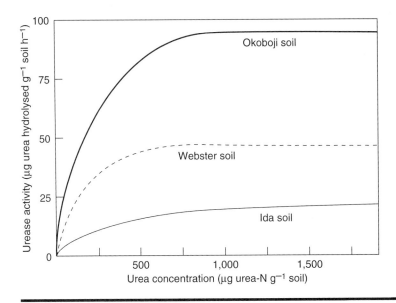

Figure 3-4   Urease activity in three different soils. (Adapted from Bremner and Mulvaney 1978)

Enzymes are associated with living cells, can be bound to the cell wall, and can be extracellular, or outside the cell. Enzymes can be associated with viable but dormant cells or attached to dead cells. The enzymes can be associated with substrates in the midst of an enzymatic reaction. Enzymes can be immobilized on minerals in soil and soil organic matter or enzymes may be polymerized into organic matter. The enzyme itself will have an effect on binding because of its mass, its isoelectric point, its potential binding sites, its solubility, and its concentration. These factors are influenced by the pH of the soil. Remember that enzymes have a variable charge depending on the pH of their environment. This is important in terms of the tertiary structure of enzymes because much of the enzyme structure is held together by hydrogen bonding between amino acids. If that structure is disturbed by suboptimal pH, the enzyme activity can be lost.

Enzyme activity varies with different soils (Figure 3-4). Enzyme activity is affected when the enzymes are adsorbed to soil. For example, the K_m of alkaline phosphatase increases and its V_{max} decreases when it is adsorbed to a variety of clay minerals (Table 3-1), which is to say that it takes more substrate to reach half of the maximum reaction rate and that maximum reaction rate is less than you would find with unadsorbed enzyme.

Table 3-1 Kinetic constants of alkaline phosphatase in the presence of clay minerals.

Clay Mineral	Clay Content (mg)	K_m(mM)	V_{max} (μg PNP mL^{-1} h^{-1})
Control	0	4.3	255
Kaolinite	50	4.6	244
	100	5.2	250
	150	6.0	247
Illite	50	4.0	175
	100	5.5	164
	150	3.9	98
Montmorillonite	50	5.7	275
	100	7.1	281
	150	7.9	279

PNP = paranitrophenol. When alkaline phosphatase hydrolyzes the substrate paranitrophenol-phosphate in this enzyme assay, it releases paranitrophenol, which has a yellow color at alkaline pH and can be quantified colorimetrically.

The final reaction volume was 5.0 mL.

(From Makboul and Ottow 1979)

EXTRACELLULAR ENZYMES

Here's a puzzler. How do microorganisms obtain nutrients when most organic C sources are polymers, and polymers are too large to diffuse through the cell membrane? How does an organism sense food and what are the mechanisms for sensing and obtaining food? To use an analogy, if you were blindfolded and in a room with a cookie 20 feet in diameter, how would you know it was there, let alone eat it? Extracellular enzymes like cellulases can be excreted into the environment by microorganisms to break down large polymers into more manageable subunits. For example, microorganisms excrete low levels of extracellular enzymes. Sometimes signals, (inducers) diffuse back, indicating that a food source is nearby. An inducer is a compound that stimulates, or induces, synthesis of more enymes. If you could smell the cookie in our analogy, your nose could guide you to it.

Unfortunately for this strategy, extracellular enzymes can be inactivated by adsorption to soil (Table 3-2), although not all soils have the same inactivating effect and not all extracellular enzymes are inactivated to the same extent. Extracellular enzymes can also be denatured by physical and chemical factors and serve as growth substrates for other microorganisms. After all, an enzyme is a protein.

Table 3-2 Influence of clay minerals on enzyme activity.

	Enzyme		
Clay Mineral	**Invertase**	**α-Amylase**	**β-Amylase**
	Percent of the Original Activity		
Allophane	20	57	12
Muscovite	33	96	1
Illite	9	27	1
Montmorillonite	0	4	0
Kaolinite	45	1	0

250 mg of clay in a final volume of 3 mL were amended with the respective enzymes.
Activities were measured within 1 hour.
(From Ross 1983)

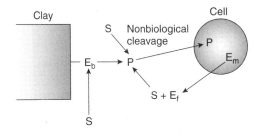

Figure 3-5 Contribution of immobilized extracellular enzymes to biochemical activity in soil. E_b = bound enzyme; E_f = free extracellular enzyme; E_m = microbial enzyme; P = product; S = substrate. (Adapted from Burns 1986)

So, how can microorganisms get the message that food is present without wasting time making extracellular enzymes? They can use extracellular enzymes that come from dead cells or they can start forming their own extracellular enzymes only after a sufficient number of inducers have been retrieved (Figures 3-5 and 3-6). Nonbiological cleavage of polymers can also occur in soil and can release inducers.

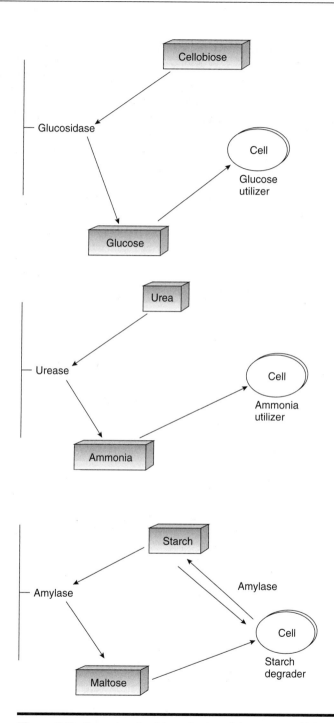

Figure 3-6   Examples of soil-bound extracellular enzymes. (Adapted from Burns 1986)

Summary

Enzymes are proteins (organic catalysts) that catalyze the transformation of inorganic and organic substances without themselves being transformed. They are made of amino acids linked together by peptide bonds. Enzyme function is dependent on primary, secondary, tertiary, or quaternary structure that is pH and temperature dependent. Biological reactions employ six basic types of enzymatic activity: oxidoreductases, transferases, hydrolases, lyases, isomerases, and ligases. Most enzyme reactions occur at rates that are constant and do not change with substrate concentration (zero-order reactions) or at rates proportional to the substrate concentration (first-order reactions). This can be described mathematically by the Michaelis-Menten equation. Enzymes are associated with living cells in the cytoplasm and periplasm; they are cell wall bound; they can be extracellular; they can be associated with viable but dormant cells or attached to dead cells, they can be complexed with substrates; and they can be immobilized on clays and humus or copolymerized into organic matter. Extracellular enzymes can be excreted into the environment to break down large polymers into more manageable subunits for absorption.

Sample Questions

1. What is the difference between a first-order reaction and a zero-order reaction?
2. Draw a diagram of a zero-order reaction.
3. What does it mean for enzyme activity if K_m increases and V_{max} decreases?
4. What are four states of enzymes in soil?
5. Based on the Michaelis-Menten equation, draw a theoretical plot that shows the following conditions:
   a. High V_{max}; low K_m
   b. Low V_{max}; high K_m
6. How do you measure extracellular enzyme activity in soil?
7. Why is it important to have high substrate concentrations when measuring maximum enzyme velocities in soil?
8. Why are there extracellular enzymes?
9. What is an example of an enzyme that might be found in soil?
10. In an enzyme reaction, there were 2 μg of product formed in 2 hours, 4 μg of product formed in 4 hours, and 8 μg of product formed in 8 hours. Based on this information, can you tell whether the reaction was first-order or zero-order?
11. Assume that the K_m for this reaction was 0.05 mM and that the substrate concentration used in the assay was 0.01 mM. Can you conclude what type of enzymatic reaction is now occurring?
12. The velocity of an enzyme reaction in soil was 0.5 μg of product formed per hour per gram of soil and the K_m was 0.025 mM substrate. What do you predict the V_{max} of the reaction will be?

13. If the velocity of an enzyme reaction is 1 μg product per hour when the substrate concentration is 0.001 mM, 2 μg product per hour when the substrate concentration is 0.004 mM, and 3 μg product per hour when the substrate concentration is 0.008 mM, what order of enzymatic reaction do you think the data represents?

14. The Lineweaver-Burk transformation is a way of making a linear form of the Michaelis-Menten equation. It has the form:

$$1/v = 1/V_{max} + (K_m/V_{max})\,(1/S)$$

Draw a theoretical plot of an enzyme assay using this transformation.

Thought Question

What are some structural characteristics of soil that make it possible for enzymes to persist without degrading?

Additional Reading

For anyone interested in an historical perspective of soil enzyme research, the relevant article to read is by J. J. Skujins, "History of Abiontic Soil Enzyme Research" (1978, *Soil Enzymes*, Academic Press, San Diego, CA). The editor of the book, R. G. Burns, has his own chapter in P. M. Huang and M. Schnitzer (eds.), *Interactions of Soil Minerals with Natural Organics and Microbes* (1986, Soil Science Society of America, Inc., Madison, WI), which will give you a more detailed look at soil enzymes.

References

Bremner, J. M., and R. L. Mulvaney. 1978. Urease activity in soils. In *Soil enzymes,* R. G. Burns (ed.), 149–96. London: Academic Press.

Burns, R. G. 1986. Interaction of enzymes with soil, mineral, and organic colloids. In *Interactions of soil minerals with natural organics and microorganisms,* P. M. Huang and M. Schnitzer (eds.), 453–96. Madison, WI: Soil Science Society of America.

Makboul, H. E., and J.C.G. Ottow. 1979. Alkaline phosphatase activity and the Michaelis constant in the presence of different clay minerals. *Soil Science* 128:129–35.

Mandelstam, J., K. McQuillen, and I. Dawes. 1982. *Biochemistry of bacterial growth.* Oxford, England: Blackwell Scientific Publications.

Ross, D. J. 1983. Invertase and amylase activities as influenced by clay minerals, soil-clay fractions, and topsoils under grassland. *Soil Biology and Biochemistry* 15:287–93.

Stotzky, G. 1986. Influence of soil mineral colloids on metabolic processes, growth, adhesion, and ecology of microorganisms and viruses. In *Interactions of soil minerals with natural organics and microorganisms,* P. M. Huang and M. Schnitzer (eds.), 305–428. Madison, WI: Soil Science Society of America.

Tabatabai, M. A. 1994. Soil enzymes. In *Methods of soil analysis, part 2: Microbiological and biochemical properties,* R. W. Weaver et al. (eds.), 775–833. Madison, WI: Soil Science Society of America.

Tabatabai, M. A., and M. Fung. 1992. Extraction of enzymes from soil. In *Soil biochemistry,* Vol. 7, G. Stotzky and J.-M. Bollag (eds.), 197–227. New York: Marcel Dekker.

2

THE SOIL MICROBIAL COMMUNITY

T here are many ways to begin studying Soil Microbiology. It's something of a paradox. You can't really know what goes on in soil unless you first understand what organisms are there, and you can't really describe the organisms in soil without bringing in a discussion of what they do. At some point, however, you have to examine the types of organisms that are in soil. In this text, it's going to be less important to know what the organisms are than to know what they do. Consequently, we start out with brief introductions of the major groups of organisms in soil and then, for most of the text, discuss the influence they have on soil processes.

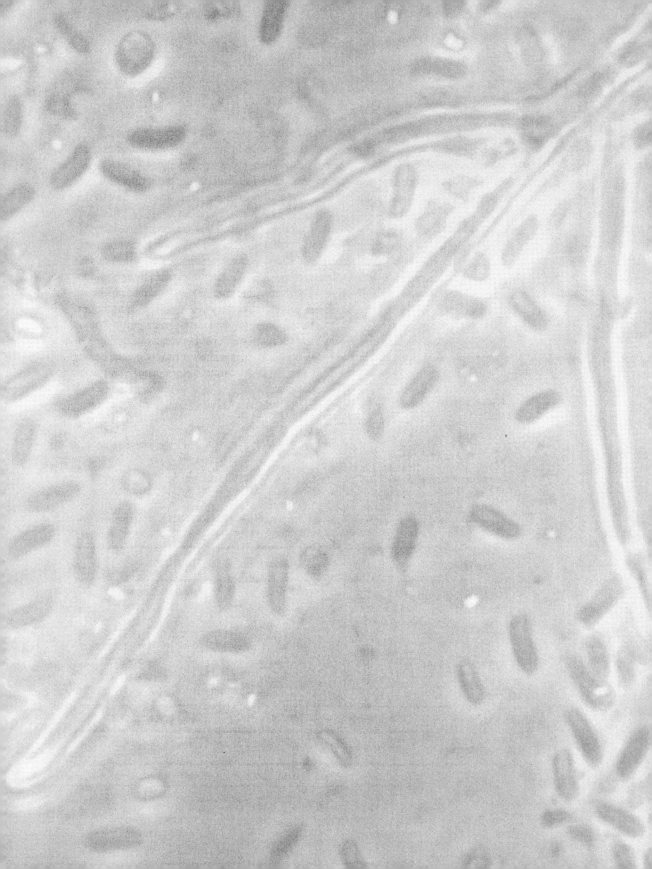

Chapter 4

The Macrofauna—Ants, Earthworms, and Other Creatures

Overview

After you have studied this chapter, you should be able to:

■ Explain four important roles of macrofauna.
■ Summarize the sequence of leaf litter decomposition in soil.
■ Identify the important macrofauna and explain how they assist in decomposition.
■ Discuss the different ways of classifying macrofauna.
■ Describe what earthworms are and tell why they are important in soil.

INTRODUCTION

Why bother studying macrofauna and large soil animals in a soil microbiology course? Macrofauna affect microbial predation, soil structure, and organic matter decomposition. For example, earthworms greatly affect soil physical structure—they promote root growth through channels, increase soil aeration, and move organic matter into the soil. Charles Darwin, the 19th-century naturalist who developed the theory of evolution, estimated that earthworms excavated over 44,000 kg of soil per hectare per year in agricultural soils. Here's what he had to say about them:

> "The plough is one of the most ancient and most valuable of man's inventions; but long before he existed the land was in fact regularly ploughed, and still continues to be thus ploughed by earthworms. It may be doubted whether there are many other animals that have played so important a part in the history of the world, as have these lowly organized creatures."

> Charles Darwin, 1881. *The Formation of Vegetable Mold Through the Action of Worms*

MACROFAUNA CONTRIBUTION TO BIOMASS

Macrofauna (earthworms, insects, etc.) are distinct from mesofauna (e.g., nematodes and rotifers) and microfauna (protozoa) because of their size and limited biochemical roles in soil. Table 4-1 shows a population pyramid of soil organisms; the larger the soil organisms get, the less numerous they become. Table 4-2 shows the relative soil biomass distribution among soil organisms. Biomass means living organic material. While macrofauna may be big, they don't necessarily have the greatest biomass in soil. If size and contribution to biomass are not necessarily related, neither are population density and importance to biological activity in soil. Two earthworms may have more effect on the soil than a billion protozoa because of the way earthworms alter the soil's physical characteristics by processes such as burrowing and mixing in organic matter.

THE ROLE OF MACROFAUNA

What do macrofauna do? Macrofauna accelerate organic matter decomposition. They mix organic matter and soil together. Macrofauna improve soil properties by increasing aggregation, partly as a result of mixing organic matter and soil together, and also by increasing aeration through channeling and burrowing in soil. Soils that have lots of aggregates (clumps of soil) are easier to manage and permit better plant root growth. Macrofauna act as microbial predators in soil.

If you had to pick the most important function of macrofauna in soil, it would probably be their role in accelerating organic matter decomposition. For example, during leaf litter decay in soil, the following stages occur:

1. Macrofauna (springtails, wood lice, and fly larvae) attack the litter. They puncture the leaf epidermis in a process called fenestration (Figure 4-1) and open the leaf to microorganisms.
2. Mollusks, wood lice (isopods), millipedes, ear wigs, fly larvae, and earthworms macerate the litter and pulverize it.

Table 4-1   Pyramid of soil organisms based on length and population (numbers in parentheses are in scientific notation).

Approximate Length (mm)[a]	Organism	Individuals per Thousand cm³ [b]
0.02–0.2	Protozoa	1,000,000,000 (1×10^9)
0.2–2.0	Nematodes	30,000 (3×10^4)
0.2–2.5	Mites (Acarina)	2,000 (2×10^3)
0.2–10	Springtails (Collembola)	1,000 (1×10^3)
0.2–1	Rotifers	500 (5×10^2)
1–30	Arthropods (insects, millipedes, and spiders)	100 (1×10^2)
1–60	Potworms	50
15–85	Earthworms	2

[a] Data from Wallwork (1970).
[b] A thousand cubic centimeters (cm³) has the same volume as a cube of soil measuring 4 inches on each side.

Table 4-2   Distribution of biomass in soil.

Group	Types	Percent of Soil Biomass
Microbes	Bacteria and fungi	80
Mesofauna and Microfauna	Nematodes, springtails and mites	2
Macrofauna	Potworms and earthworms	14
Others		4

(Adapted from Richards 1987)

Figure 4-1   Fenestration of leaves. (Photograph courtesy of M. S. Coyne)

Table 4-3 Mesh sizes that allow free entry of various soil organisms.

Mesh Diameter	Organisms with Free Entry
7.000 mm	All microorganisms and invertebrates
1.000 mm	All microorganisms and invertebrates except earthworms
0.500 mm	All microorganisms, mites, springtails, potworms, and small invertebrates
0.003 mm	All microorganisms

Note: A mesh that was 1–2 mm in diameter would have about the same spacing between individual strands as does a typical screen door.
(Adapted from Phillipson 1968)

3. Earthworms, insects, and other burrowing creatures transport the litter into soil.
4. Earthworms and potworms further macerate the litter and mix it with soil.
5. Microbial action turns the litter into an integrated component of the soil.

Macrofauna grind and macerate organic matter, increasing its surface area. Increasing a material's surface area increases its decomposition rate; the larger a material's surface area, the more available it is for microbial attack. If you exclude macrofauna from soil, organic matter decomposition slows. An interesting experiment that showed this effect involved incubating leaf tissue in mesh bags that excluded different organisms based on size (Table 4-3) and burying the bags in soil. About 95% decomposition occurred in the 7.0 mm mesh that didn't exclude any macrofauna, compared to 35% decomposition in the 0.5 mm mesh that excluded all but the smallest invertebrates and microbes (Figure 4-2).

Macrofauna mix organic matter and soil together. Dung beetles, for example, can bury up to 78% of the cattle feces in a pasture within weeks (Fincher et al., 1981). That's equivalent to adding about 175 kg of nitrogen per hectare; 168 kg N per hectare is a common fertilization rate for corn. Macrofauna improve soil physical properties. Some species of dung beetles bury dung 30 to 150 cm below the soil surface (Fincher 1972). Earthworms can burrow 3.6 m or deeper in suitable sites. When macrofauna make channels and tunnels, they improve aeration and water infiltration. Much of the soil in these channels is brought up to the soil surface. This recycles some of the nutrients, such as potassium (K), that leach from the soil surface. Anthills scattered on the soil surface are an example of this process in action (Figure 4-3). Soil excavation by ants can range from 940 to 25,000 kg of soil per hectare per year (Lockaby and Adams 1985) and soil excavation by scarab beetles (*Peltotrupes youngi*) can be as great as 7,800 kg per hectare per year (Kalisz and Stone 1984).

Macrofauna have considerable biological importance as microbial predators. Excluding the soil macrofauna from soil decreases overall soil respiration significantly more than can be accounted for by the macrofauna respiration alone. Why is this? Removing or eliminating the macrofauna indirectly, but significantly, affects the remaining microbial community. Predation by macrofauna forces the microbial popu-

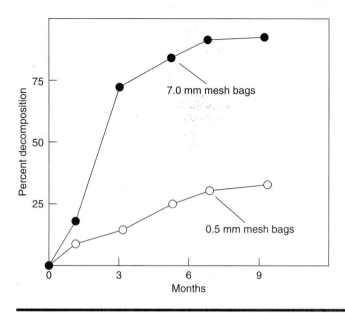

Figure 4-2   Decomposition of leaf disks by macrofauna. (Adapted from Phillipson 1968)

Figure 4-3   An anthill demonstrating how soil is circulated by macrofauna. (Photograph courtesy of M. S. Coyne)

lations to actively grow and reproduce. Macrofauna also increase the surface area of microbial food sources. So, with macrofauna present, overall biological activity and respiration are increased.

CLASSIFICATION OF MACROFAUNA

How do we classify macrofauna? Macrofauna can be classified by size. They typically vary in length from 10 to 100 mm or more. Macrofauna can be classified by whether they glide or crawl, burrow, or use existing channels. Macrofauna can be characterized by the amount of time they spend in soil, and by their habitat preference—air or water-filled pores or surfaces.

A better characterization, in terms of soil processes, is classification based on what macrofauna eat. Biophagous macrofauna consume living material. This group can be divided into macrofauna that are: 1) carnivores (animal eaters); 2) herbivores (plant eaters); 3) microbivores (microorganism eaters); and 4) omnivores (animal/plant eaters). It is estimated that microbivores consume approximately 50% of the annual production of microbial material.

Saprophagous macrofauna consume dead or decaying material. These macrofauna, likewise, can be broken down into more specific groups: 1) detritivores (detritus eaters); 2) cadavericoles (dead animal/carrion eaters); and 3) coprophages (dung eaters). The most important group of saprophages are the detritivores. Coprophages illustrate the principle that nothing in nature goes to waste—one organism's outhouse is another's castle.

TYPES OF MACROFAUNA

Some common soil macrofauna are listed in Table 4-4. Let's take a closer look at some of these groups.

Arthropods

There are more species of arthropods (insects, etc.) than any other animal phylum; they dominate the macrofauna in terms of diversity and numbers, but not biomass. Less than 10% of the soil biomass consists of arthropods (Moldenke 1994). The most common arthropods are wood lice (Figure 4-4), centipedes and millipedes, mites, insects, springtails (microarthropods), termites, and ants.

Arachnids (Mites)

Mites are members of the same group as spiders. The smallest mites are small enough to be classified as mesofauna. Predatory mites eat nematodes, potworms, insects, and other mites. Mites are also saprophagous (eat dead material), mycophagous (eat fungi), and coprophagous (eat waste). Mycophagous mite populations decline when fungicides are used, probably because their food source (fungi) decreases. Predatory mites are

Table 4-4   Common groups of soil and litter macrofauna.

Class	Order	Common Name
Turbellaria		Flatworms
Oligochaeta		Earthworms
		Potworms
Gastropoda	Pulmonata	Snails
		Slugs
Arachnida	Araneida	Spiders
	Acarina	Mites
Crustacea	Isopoda	Wood lice
Myriapoda		Millipedes
		Centipedes
Insecta	Collembola	Springtails
	Isoptera	Termites
	Lepidoptera	Moths
		Butterflies
	Diptera	Flies
	Coleoptera	Beetles
	Hymenoptera	Ants
		Bees
		Wasps

(Adapted from Paul and Clark 1989)

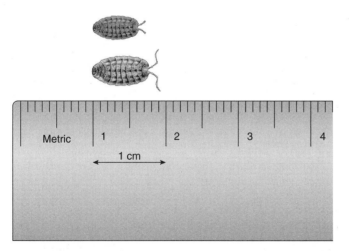

Figure 4-4   Isopods in soil (also called wood lice, sowbugs, and pillbugs). (Photograph courtesy of M. S. Coyne)

How Do You Study Soil Macrofauna?

Soil arthropods can be extracted from soil by changing the soil environment and inducing them to leave (a version of behavior modification) or by physical extraction. One of the most common methods used to cause arthropods to migrate from soil is the Berlese funnel extractor method (Moldenke 1994). This method uses a mesh-covered funnel on which the soil sample is placed. A light bulb above the funnel shines on the soil sample and makes light- and heat-sensitive arthropods migrate into the soil. The arthropods drop through the mesh into the funnel, and ultimately into a vial containing a preservative such as propylene glycol. This works well for soils that have a lot of organic matter.

Arthropods have a specific gravity slightly greater than that of water, so they can also be physically extracted from soil by flotation. In this method, the soil sample is mixed with a dense salt solution. The lighter material, including the arthropods, floats to the surface and can be skimmed off the top.

A wet funnel method can be used to collect enchytraeid worms from soil. Moist soil is placed on a mesh-covered, water-filled funnel. An electric light immediately above the funnel induces the enchytraeid worms to migrate through the soil where they ultimately drop through the mesh. Earthworms are big enough to see, and a simple way to find them is to dig up piles of rotting vegetation, or look beneath bags and boxes that are lying on the ground. Earthworms also respond to vibration, so pounding a ribbed post into moist soil will generally drive some earthworms above ground. Earthworm biologists are more systematic than this, however, since they want to quantify how many earthworms are in a known soil area. They typically either dig up a known volume of soil and handpick every earthworm they find, or they saturate the soil with a chemical that irritates earthworms, such as a dilute formalin solution (about 0.2% to 0.3% formaldehyde). When the soil is saturated in this way, the earthworms flee to the soil surface where they can be collected and sorted by hand.

abundant on buried residues. Mites are sensitive to drying. They fragment litter, but otherwise contribute little to plant decomposition. Mites transport fungal spores and move organic matter through soil. The feces (wastes) that mites produce are a habitat for microbes.

Myriapoda (Centipedes and Millipedes)

These arthropods have important roles as predators and carnivores (Figure 4-5). Their populations are higher in moist forest soils than in grassland soils. Some eat decaying plant material. They are important in fragmenting litter.

Figure 4-5   An example of a soil arthropod (Myripoda or millipede). (Photograph courtesy of M. S. Coyne)

Collembola (Springtails)

Springtails are small, primitive insects that are the most abundant and widely distributed of the arthropods. There may be greater than 10,000 individuals per square meter of soil. Springtails are usually less than 1 mm long and contribute little to biomass. There are surface and burrowing springtails. Springtails eat bacteria, fungi and spores, decomposing organic matter, feces, living plants, and animals. Springtails also eat detritus to obtain the microorganisms contained in it. Although springtails don't play a direct role in the turnover of soil nutrients, they are active in fragmenting litter.

Isoptera (Termites)

These insects are much more important to the biology of soil in the tropics and subtropics than they are in temperate climates. Termites eat wood, the main component of which is cellulose. Termites do not produce their own cellulase, the enzyme that breaks down the cellulose; instead some termite species rely on protozoa in their guts to do so. Other termite species live on humus and decaying litter. These termites lack protozoa, but eat a cultivated fungus growing on macerated cellulose that the termites prepare. Termites transport organic matter into their nests. In the tropics, termites play a role similar to that of earthworms in terms of moving organic matter into soil.

Mollusca (Snails and Slugs)

Mollusks are omnivorous—they have a varied diet that includes living, dead, or decaying materials; fungi, algae, or lichens; and earthworms, slugs, and snails (Figure 4-6). Mollusks are sensitive to changes in moisture and temperature. Because of this sensitivity, they are not uniformly distributed. Mollusk populations vary from 10 to 25 per square meter to sometimes greater than 50 per square meter. Some mollusks feed on the surface and move into the soil profile, thereby incorporating organic matter. Some produce cellulase or have cellulase producers in their gut. Some secrete mucoproteins that help form water-stable soil aggregates.

Enchytraeidae (Potworms)

Enchytraeid worms, or potworms, are smaller than earthworms, perhaps 50 mm in length and 0.2 to 0.8 mm in width. They are sensitive to drought and do not tolerate

Figure 4-6   A gastropod in soil—a slug. (Photograph courtesy of M. S. Coyne)

Figure 4-7   Earthworms burrowing into soil. (Photograph courtesy of M. S. Coyne)

drying, so their numbers are not uniformly distributed in soil. There may be up to 100,000 potworms per square meter in moist soil. Potworms eat microorganisms and some dissolved organic material. Since potworms have no enzymes for digesting complex polysaccharides, they do not digest any polymerized organic matter they ingest.

EARTHWORMS

There are approximately 8,000 species of earthworms in soil. Some of the most important are in the family Lumbricidae (Figure 4-7). Common genera include *Allobophora*, *Apporectodea*, *Dendrobaena*, *Diplocardia*, *Eisenia*, and *Lumbricus*. In grassland and woodland soils of the Northern Hemisphere, there may be greater than 1,000 earthworms per square meter in the upper 15 cm of soil. In agricultural soils, there may be 100 to 200 earthworms or fewer. Earthworms are not as numerous in the tropics as they are in the Northern Hemisphere, but they can adapt to acid soils and play an important role in decomposing organic matter (Fragoso and Lavelle 1992).

Earthworms are active only if there is water present. Water allows earthworms to secrete mucus that eases their movement through soil; earthworms also need a turgid body for locomotion and burrowing. Earthworms need moist skins to respire, or breathe. Consequently, earthworm populations are highest in spring and fall and lowest in summer (Hendrix et al. 1992). During summer, earthworms tend to burrow to where soil moisture remains or move to locations that remain cool and moist. Earthworms

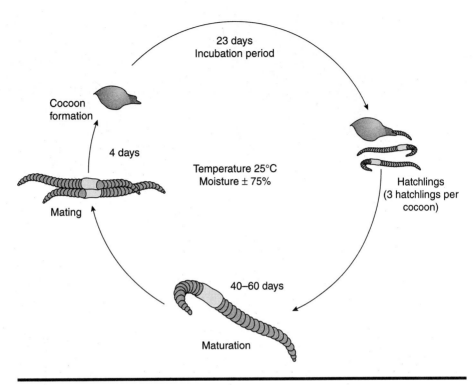

Figure 4-8 The life cycle of a cocoon-forming earthworm, *Eisenia fetida*. (Adapted from Venter and Reinecke 1988)

such as *Lumbricus rubellus* form cocoons that persist during dry or cold periods that are unfavorable to their activity. Figure 4-8 shows the life cycle of an earthworm, *Eisenia fetida,* that forms cocoons (Venter and Reinecke 1988). In winter, earthworms may also burrow below the frost line. *Apporectodea* species use this strategy (Marinissen 1992).

As a general rule, when the organic carbon in soil increases, earthworm populations also increase (Hendrix et al. 1992). Consequently, no-tillage soil, which leaves a lot of crop residue and minimally disturbs the soil, has much higher earthworm populations than does tilled soil. Finer textured soils, such as silt loams, which tend to have more organic carbon and better water-holding capacity than sandy soils, also tend to have more earthworms.

Earthworms dominate the macrofauna biomass. The dry weight of earthworms may exceed 100 g per square meter (Marinissen 1992). There are two general groups of earthworms—those that live in surface organic horizons and ingest little mineral matter (they primarily ingest decayed plant material, sometimes with great selectivity), and those that live on mineral soil and rarely come to the soil surface. Some earthworms do both. Some earthworms tunnel several meters into the soil, but most tunnel only a few

Figure 4-9   A worm cast. Notice the consistency of these soil deposits compared to those deposited by ants. (Photograph courtesy of M. S. Coyne)

centimeters below the soil surface. Most earthworms do not emerge on the surface. An exception is the European earthworm *Lumbricus terrestris,* which feeds on the soil surface at night.

Earthworms have a direct role in litter decomposition. They fragment plant litter and mix it with soil. They bring litter from the surface into the soil. They secrete mucus in tunnel channels, which may improve soil structure. The lining of the earthworm burrows, called the drilosphere, is rich in organic matter and inorganic nutrients (Edwards et al. 1992). Worm casts, the mineral material that passes through a worm's gut and is deposited on the soil surface during its burrowing activity, are somewhat water stable and provide a good environment for microorganisms (Figure 4-9). Although poorly aerated, worm casts are rich in NH_4^+ and partially digested organic matter. Worms have cellulase and chitinase in their guts, which helps degrade cellulose and chitin polymers, and continue to work even after wastes are excreted.

What's the evidence that earthworms actually transport surface material into the soil? Daane et al. (1996) did a nice experiment to show how burrowing by earthworms helps to transport surface bacteria into the soil (Table 4-5). They began the experiment by inoculating each gram of the soil surface with 10^8 cells of a prokaryote called *Pseudomonas fluorescens.* Each soil had a different treatment: no earthworms (the con-

Table 4-5   Transport of *Pseudomonas fluorescens* through soil by earthworms.

Soil Depth (cm)	Control	Lumbricus rubellus	Lumbricus terrestris	Apporectodea trapezoides
	Pseudomonas fluorescens per gram of soil ($\times 1,000$)			
0–5	5,248	5,888	3,236	6,760
5–10	35	616	813	724
10–15	0	19	93	178
15–20	0	0.2	64	96
20–25	0	0	5	63
25–30	0	0	5	10
30–35	0	0	0.9	1.6
35–40	0	0	0.9	1.4

(Adapted from Daane et al. 1996)

trol); *Lumbricus rubellus*—an epigeic, or nonburrowing surface feeder; *Lumbricus terrestris*—an anaeic, or burrowing surface feeder; and *Apporectodea trapezoides*—an endogeic earthworm that feeds, burrows, and casts within soil.

The control shows that without earthworms, the *Pseudomonas fluorescens* traveled only about 10 cm into the soil. *Pseudomonas fluorescens* traveled a little farther with the surface feeder, *Lumbricus rubellus* (down to 20 cm). However, in treatments with the earthworms that burrowed into soil, *Pseudomonas fluorescens* was found all the way down to 40 cm. The *Pseudomonas fluorescens* may have been digested and excreted at these depths, or they may have simply been attached to the body of the earthworm and went along for the ride. The point is that earthworms burrowing in soil helps transport microorganisms to greater depths, and the same is true for other types of organic matter. It's this organic matter that provides food for other organisms in soil.

Summary

The four major roles of macrofauna in soil are to accelerate organic matter breakdown, mix organic matter and soil, contribute to improved soil structure, and act as predators of soil microorganisms. Macrofauna can be characterized by size, time of residence in soil, feeding habit, or method of locomotion. Arthropods, mollusks, enchytraeid worms, and earthworms are the major macrofauna involved in soil processes. Some important genera of earthworms include *Diplocardia* and *Lumbricus*. The environment affects their distribution in soil. By burrowing and forming cocoons, earthworms can escape unfavorable environments. Earthworms, through circulating soil and mixing organic matter with soil, are essential members of the soil environment whose effect far outweighs their numbers.

Sample Questions

1. What are some of the ways you can classify macrofauna? Which classification system do you think is best, and why?
2. What is the relationship between the size and number of macrofauna in soil?
3. How do macrofauna accelerate the breakdown of organic matter in soil?
4. How do macrofauna influence soil structure?
5. Based on the data from a buried bag experiment in the following table, what is the relationship between litter decomposition and mesh size? What do you think causes this relationship?

Percent (%) decomposition of leaf litter buried in two bags of different mesh size.

	Mesh Size (mm)	
Day	0.005	1.00
	% Decomposition	
7	2	4
14	4	8
28	8	14
56	20	28
112	30	60
224	35	85

6. How many cubic decimeters are in 1 cubic meter (m^3) of soil?
7. Based on the data in Table 4-1, how many rotifers would be in this 1 m^3 of soil?
8. In Table 4-5, why do you think the population of the prokaryote *Pseudomonas fluorescens* decreases as soil depth increases?
9. Graph the data in Table 4-5 for the column under *Lumbricus terrestris*. How does the graph look compared to the same data plotted on semi-log paper?

Thought Question

One day Louis Pasteur was visiting the farm of St. Germain, near Chartres. Sheep on this farm appeared to be especially susceptible to anthrax, a disease caused by *Bacillus anthracis*. Pasteur noticed a spot in a recently harvested field where the color of the soil was visibly different from the neighboring earth. The owner of the farm indicated that sheep killed by anthrax had been buried there. On closer examination, Pasteur noticed a mass of small earthen mounds on this spot. After further study, Pasteur recommended that animals should never be buried in fields intended for pasture (no pun intended) or hay production. Furthermore, burying grounds for diseased animals should be chosen in sandy or chalky soils that were poor and dry. What do you think Pasteur suspected, and what was the rationale behind his recommendations? How would you design an experiment (including appropriate controls) to test Pasteur's suspicions?

Additional Reading

By the time you read this textbook, earthworm biologists will be publishing around seven new research papers on earthworms every week. If you want to catch up on all the published work about earthworms, you might have to read over 9,000 papers. Fortunately, there are a couple of publications that can provide you with a glimpse of earthworm biology in a nutshell. The first is *Earthworm Ecology,* edited by Clive Edwards (1997, St. Lucie Press, Boca Raton, FL). The second is a collection of the papers presented at the 4th International Symposium on Earthworm Ecology, which appears in *Soil Biology and Biochemistry,* Vol. 24, Number 12.

References

Daane, L. L., J. A. E. Molina, E. C. Berry, and M. J. Sadowsky. 1996. Influence of earthworm activity on gene transfer from *Pseudomonas fluorescens* to indigenous soil bacteria. *Applied and Environmental Microbiology* 65:515–21.

Edwards, W. M., M. J. Shipitalo, S. J. Traina, C. A. Edwards, and L. B. Owens. 1992. Role of *Lumbricus terrestris* (L.) burrows on quality of infiltrating water. *Soil Biology and Biochemistry* 24:1555–61.

Fincher, G. T. 1972. Notes on the biology of *Phanaeus vindex* MacLeay. *Journal of the Georgia Entomological Society* 7:128–33.

Fincher, G. T., W. G. Monson, and G. W. Burton. 1981. Effects of cattle feces rapidly buried by dung beetles on yield and quality of coastal bermuda grass. *Agronomy Journal* 73:775–79.

Fragoso, C., and P. Lavelle. 1992. Earthworm communities of tropical rain forests. *Soil Biology and Biochemistry* 24:1397–1408.

Hendrix, P. F., B. R. Mueller, R. R. Bruce, G. W. Langdale, and R. W. Parmelee. 1992. Abundance and distribution of earthworms in relation to landscape factors on the Georgia piedmont. *U.S.A. Soil Biology and Biochemistry* 24:1357–61.

Kalisz, P. J. and E. L. Stone. 1984. Soil mixing by scarab beetles and pocket gophers in north-central Florida. *Soil Science Society of America Journal* 48:169–172.

Lockaby, B. G., and J. C. Adams. 1985. Perturbation of a forest soil by fire ants. *Soil Science Society of America Journal* 49:220–23.

Marinissen, J. C. Y. 1992. Population dynamics of earthworms in a silt loam soil under conventional and "integrated" arable farming during two years with different weather patterns. *Soil Biology and Biochemistry* 24:1647–54.

Moldenke, A. R. 1994. Arthropods. In *Methods of soil analysis, part 2: Microbiological and biochemical properties,* R. W. Weaver et al. (eds.), 517–42. Madison, WI: Soil Science Society of America.

Paul, E. A., and F. E. Clark. 1989. *Soil microbiology and biochemistry.* San Diego, CA: Academic Press.

Phillipson, J. 1968. *Ecological energetics.* Great Britain: Edward Arnold Publ. Ltd.

Richards, B. N. 1987. *The microbiology of terrestrial ecosystems.* Essex, England: Longman Scientific & Technical.

Venter, J. M., and A. J. Reinecke. 1988. The life-cycle of the compost worm *Eisenia fetida* (Oligochaeta). *South African Journal of Zoology* 23:161–65.

Wallwork, J. A. 1970. *Ecology of soil animals.* London: McGraw-Hill.

Wallwork, J. S. 1976. *The distribution and diversity of soil fauna.* London: Academic Press.

Chapter 5

The Mesofauna— Nematodes

After studying this chapter, you should be able to:
- ■ Name the important nematode groups.
- ■ Discuss the role nematodes play in organic matter decomposition.
- ■ Describe what nematodes look like and tell how they live and grow.
- ■ Explain why nematodes are important to the soil microbial community.
- ■ Discuss how to extract nematodes from soil.

INTRODUCTION

The organisms in soil range in size from large to small, and individual sizes within a group may vary greatly. Consequently, the use of terms like mesofauna is somewhat arbitrary. Mesofauna, on average, are smaller than macrofauna, yet not so small as to be considered microscopic. Mesofauna could be defined as all organisms in soil 200 to 1,000 μm in length. They may or may not be visible to the naked eye. Mites and springtails, for example, might just as well be categorized as mesofauna. For the purpose of this discussion, we use nematodes as the most representative and most ecologically important mesofauna group.

NEMATODES—GENERAL INTRODUCTION

Alternate names for nematodes are roundworms, threadworms, and hairworms. Nematodes are present in virtually all environments, although you usually can't see them; most soil nematodes are microscopic and transparent—less than 50 μm wide by 2,000 μm long (marine nematodes can be up to 9 m long) (Figure 5-1). After the protozoans, which we discuss in Chapter 6, nematodes are the most numerous animal group in soil. There can be greater than a million (10^6) nematodes per square meter of soil. Numbers do not necessarily equate to mass and nematodes do not represent a significant portion of the soil biomass.

Figure 5-1   An electron micrograph of a soil nematode about 500 μm in length (0.5 mm). (Photograph courtesy of G. Koening)

Nematode characterization is part art and part science. Although there are some 10,000 total species of nematodes, only about 1,000 species are found in soils. In a single soil sample, there may be only 10 to 25 different species. Nematodes parasitize higher plants and animals. Nematodes are responsible for root knot of grapevines and soybean cyst formation. Most pathogenic infestations of plants by nematodes are caused by only a few species of plant pathogenic nematode.

Free-living nematode forms are mostly found in the upper 10 cm of the soil profile; 90% of nematodes are in the upper 15 cm of soil. Nematodes are active in water films, although some nematodes persist through drying cycles, either by forming cysts or by entering a dormant state. Nematode populations are highest in organic environments at neutral pH, but nematodes are not restricted to these environments. Nematode populations are higher in the vicinity of plant roots than elsewhere in the soil. Some nematode species parasitize living roots while others feed on the rich microbial population found near roots (the rhizosphere population) compared to the rest of the soil. The larger populations reflect the higher organic matter content near roots. Nematodes have high reproductive rates, producing five to six generations per year. Thousands of eggs may be deposited in egg packets.

Nematodes do not directly participate in organic matter decomposition; rather, the nematodes are saprophytes or predators. Free-living nematodes consume microorganisms, other nematodes, rotifers, and protozoa. Because nematodes can consume up to 5,000 cells per minute, they help regulate microbial populations in soil. Nematodes are themselves parasitized by other soil organisms and consumed as part of the soil food chain, which regulates their populations.

NEMATODE CLASSIFICATION

Nematodes are generally grouped into five categories based on their food sources and selective feeding habits: pathogenic, herbivorous, microbivorous, omnivorous, and carnivorous. Additional characteristics are morphological features such as the structure of the esophagus and whether they have a stylet, an organ used to puncture food (Yeates 1971) (Figure 5-2). There are nematodes that feed on microorganisms (microbivores) and nematodes that feed on larger organisms such as protozoa, rotifers, tartigrades, other nematodes, and enchytraeid worms. Omnivorous nematodes feed on living and dead plant and animal tissue. Many nematodes are human and animal parasites. These nematodes typically don't live in soil, but are present in soil for part of their life cycle and pose an environmental and human health concern.

Plant-feeding nematodes are frequently studied because of their role as parasites. The genus *Meloidogyne,* for example, contains many nematode species that are important plant parasites of global distribution and cause root knot in economically important plants such as tomato and tobacco. Fungal-feeding nematodes are often included with plant feeders for ecological studies. They also have stylets, with which to suck out the insides of their prey. The nematode *Aphelenchus avenae,* for example, feeds on the fungus *Rhizoctonia solani.* Interestingly, some fungal species such as *Arthrobotrys*

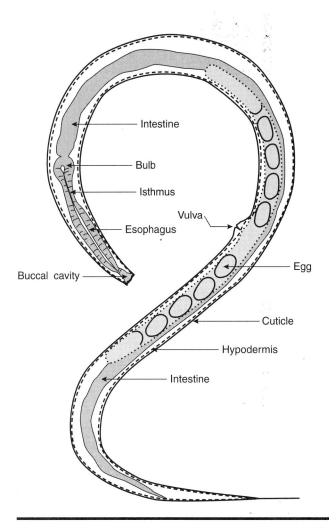

Figure 5-2   Basic morphological features of a soil nematode. (Diagram courtesy of G. Koening)

oligospora, are nematode predators. This relationship is detailed when we discuss antagonism between soil organisms.

NEMATODE ECOLOGY

Nematode populations constantly fluctuate in soil. Likewise, the nematode community composition existing in a soil changes with time depending on the site management and age. Plant parasites, for example, fluctuate in response to their host—increasing during plant cultivation and decreasing after plants are harvested. Nematode populations are

How Do You Extract Nematodes from Soil?

Several methods are used to extract nematodes from soil. Unfortunately, the extraction efficiency for nematodes is about 50%, which illustrates the difficulty nematologists have in quantitatively assessing soil populations. It's difficult to estimate the size of a population if you know you can only count half of its members. In wet sieving, a soil sample is resuspended in water and progressively sieved through finer and finer screens (down to 25 μm mesh) until the nematodes that have been extracted can be backwashed into a separate container. This method is often combined with a density centrifugation of the backwashed nematodes in sucrose (sugar) once most of the soil has been removed. Ideally, in a solution of 1.25 M sucrose, nematodes will float to the surface of a centrifuge tube while sediment will precipitate. Unfortunately, the method doesn't distinguish between live and dead nematodes.

The Baermann funnel method (Figure 5-3) involves placing a soil sample in a basket that rests inside a funnel. A pinch clamp at the base of the funnel allows it to be filled with water until the soil resting in it is saturated. Nematodes migrate from the saturated soil into the funnel and gradually accumulate at the bottom near the pinch clamp, which can be released to collect them. The disadvantage of this method is that only live nematodes are collected and the soil has to permit the ready movement of nematodes.

Nematode cysts can also be counted. The advantage of counting cysts is that the soil samples can be dried before analysis. The recovery of cysts relies on their buoyancy in soil suspensions. Since cysts float in water, they can be skimmed off the surface of the suspensions and concentrated by sieving through progressively smaller filters.

typically lowest in winter and early spring in temperate climates and lowest in summer and fall in warmer climates. Low populations may simply indicate that nematodes have migrated farther into the soil profile to escape inclement soil conditions (Ingham 1994).

From an ecological perspective, nematodes regulate microbial populations through predation in addition to their role as plant parasites. They serve as a link in the food web between the microbial world and more complicated organisms. Nematode populations reflect the availability of organic material on which they live and typically are lowest in deserts (with little organic matter) and highest in permanent pastures (with well-developed rhizospheres), as Table 5-1 illustrates.

As shown in Table 5-1, there isn't a very good relationship between nematode population and biomass. For comparison, the earthworm biomass measured on a dry weight basis is at least 10 times higher.

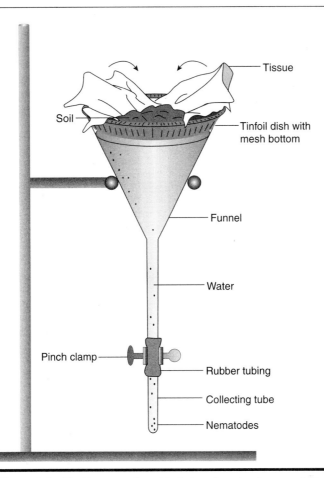

Figure 5-3   The Baermann funnel technique for extracting nematodes from soil. (Courtesy of G. Koening)

Table 5-1   The density and biomass of nematode populations in selected environments.

Environment	Population (millions/m²)	Biomass (g fresh weight/m²)
Desert, USA	0.4	0.1
Beech forest, England	1.4	0.4
Tropical forest, Africa	1.7	—
Pasture, Poland	3.5	2.2
Tundra, Sweden	4.1	1.1
Pine forest, Sweden	4.1	0.6
Potato field, Poland	5.5	0.7
Rye field, Poland	8.6	1.1
Mixed forest, Poland	7.0	0.7
Grassland, Denmark	10.0	14.0

(Data from Sohlenius 1980).

Summary

Mesofauna can be arbitrarily defined as those soil animals 200 to 1,000 μm in length. Small mites, springtails, rotifers, and tartigrades are all mesofauna by this characterization. The most important of the mesofauna are the nematodes, which are unsegmented roundworms with pathogenic, herbivorous, microbivorous, and carnivorous species. Nematodes are small, between 0.25 and 5.5 mm long, and numerous, but don't contribute much to the soil biomass. Nematodes are important plant pathogens. There are also many nonpathogenic species that inhabit the soil and the rhizosphere and regulate microbial populations. Nematodes in soil are found as living worms, dead worms, eggs, and cysts.

Sample Questions

1. What are some typical examples of mesofauna?
2. What is the rationale behind the way nematodes are classified?
3. What do you see as some of the problems with quantifying nematodes in soil?
4. Are nematode populations static? Explain your answer.
5. Why do you think nematodes in soil are smaller than nematodes in marine environments?
6. Based on the data in Table 5-1, how many nematodes would you expect to find in one hectare of soil?
7. If nematode extraction methods are only about 50% efficient, what do you predict would be the true nematode populations in Table 5-1?
8. Use the data in Table 5-1 to estimate about how much an individual nematode weighs.
9. How many nematodes do you predict a soil must have to equal the biomass of earthworms in soil?
10. If nematodes consumed an average of 5,000 bacteria per minute, how many microorganisms would a population of 100,000 microbivorous nematodes consume in one day?

Thought Question

You have been asked to be the scientific consultant for a remake of the movie "Squirm," which was about killer worms in Georgia. The director proposes to film a new horror movie based on giant, mutated nematodes 20 meters long. What would be some of the consequences for humans and the environment if nematodes this large were actually alive? How would you propose that they be killed in the movie's climax?

Additional Reading

A thorough discussion of nematodes is in *Soil and Freshwater Nematodes,* 2d edition, by T. Goodey (1963, John Wiley & Sons, New York). This text describes most of the soil and freshwater genera of importance.

References

Ingham, R. E. 1994. Nematodes. In *Methods of soil analysis, part 2: Microbiological and biochemical properties,* R. W. Weaver et al. (eds.), 459–90. Madison, WI: Soil Science Society of America.

Sohlenius, B. 1980. Abundance, biomass, and contribution to energy flow by nematodes in terrestrial ecosystems. *Oikos* 34:186–94.

Yeates, G. W. 1971. Feeding types and feeding groups in plant and soil nematodes. *Pedobiologia* 8:173–79.

Chapter

6

The Microfauna— Protozoa and Archezoa

Overview

After you have studied this chapter, you should be able to:

- Distinguish between protozoa and archezoa, and tell how they differ from higher organisms.
- Identify the four classifications of protozoa and give representative examples of each.
- Describe where protozoa are found in soil, how many there are, what they eat, how they grow, and what their ecological role is.
- Explain several methods for studying protozoa.
- Discuss some insights into the danger that pathogenic protozoa pose for drinking water supplies.

INTRODUCTION

"I Wonder," said Noah,
"Did we bring the Protozoa?"
"Don't worry," said his daughter,
"We've some in the drinking water."
William D. Barney

What are protozoa? They are unicellular animals with a cell nucleus and mitochondria that are phagotrophic, meaning they ultimately ingest their prey by surrounding and enveloping it in a cell membrane (Cavalier-Smith 1981). Archezoa are similar to protozoa, but they lack mitochondria, so they are more primitive. Protozoa and archezoa differ from single-celled plants in that they are generally not photosynthetic, and those that are (*Euglena,* for example) do not have starch-containing organs. They differ from fungi because they don't have chitin in their cell walls (chitin is a structural polymer) and don't have filamentous growth.

Protozoa and archezoa are the simplest animals and the most abundant invertebrates. There are more than 30,000 species ranging in size from 10 to 100 μm. We study protozoa and archezoa because every arable soil (soil on which crops can be grown) contains them and they are important predators of bacteria and algae in soil. Two species, *Cryptosporidium* (a protozoa) and *Giardia* (an archezoa), are widespread pathogens and important drinking water contaminants.

MORPHOLOGY AND CLASSIFICATION

There are four major classifications of protozoa and archezoa that are primarily based on the way they move (Figure 6-1):

1. Mastigophora (flagellates, e.g., trypanosomes)
2. Sarcodina (ameboids, e.g., *Amoeba* and foraminifera)
3. Ciliophora (ciliates, e.g., *Paramecium*)
4. Sporozoa

Mastigophora

Mastigophora dominate soil ecosystems. They are about 5 to 10 μm long and have one or more flagella (Figure 6-2). Flagella are whiplike organs that mastigophora use for propulsion. Mastigophora may also have pseudopodia (false feet), which are simply extensions of the cytoplasm that permit the organism to move from one location to another. Mastigophora weigh between 0.2 and 28 nanograms (a nanogram, ng, is 1 billionth of a gram; a raisin weighs about 1 gram). They reproduce asexually by splitting apart (fission) or sexually by fusion (two cells join together and exchange genetic information). An example of Mastigophora are the trypanosomes (*Trypanosoma*) that cause African sleeping sickness. *Euglena* is also one of the Mastigophora; it is a photosynthetic protozoa.

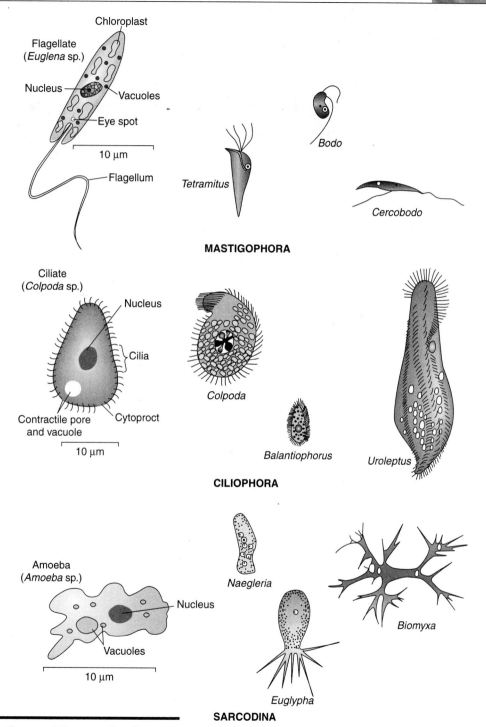

MASTIGOPHORA

CILIOPHORA

SARCODINA

Figure 6-1   Representative examples of protozoa: flagellates, ciliates, and amebas. (Adapted from Killham 1994; Sandon 1927)

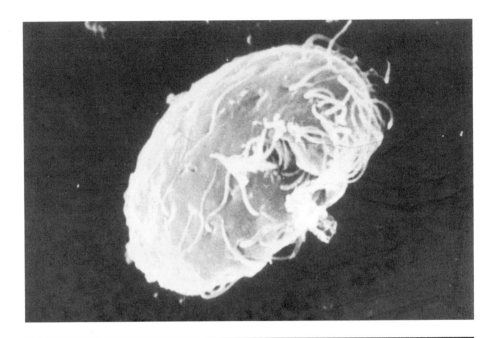

Figure 6-2   *Giardia,* a flagellated archezoa. (Photograph courtesy of the Soil Science Society of America)

Sarcodina

Sarcodina are protozoa that also move by means of pseudopodia. Some sarcodina have an exoskeleton and some do not. If the sarcodina have an exoskeleton, in which case they can also be called "testate" protozoa, the pseudopodia emerge through distinct holes in the exoskeleton. Sarcodina reproduce by fission of single cells and can also reproduce sexually by fusion. Asexual reproduction is the norm, as it is with most protozoa. A good example of the sarcodina is the amebas. Amebas may be up to 600 μm in diameter, but a more typical size in soil is 10 μm, with a mass of 0.8 to 6.0 ng. *Entamoeba histolytica* is an important water-borne sarcodina pathogen that causes amebic dysentery.

Foramanifera are testate sarcodina that typically have exoskeletons containing silica, chitin, or calcium carbonate ($CaCO_3$). They are primarily marine animals. Over geologic time, the insoluble shells accumulate in enormous sedimentary deposits. The white cliffs of Dover, England, for example, are largely composed of foraminifera shells.

Ciliophora

Ciliophora are protozoa that are ciliated. Cilia are numerous, short, hairlike appendages that pulsate synchronously and allow ciliates to move through water films or sweep food into their cell mouth or cytosome. A good example is *Paramecium,* although one you're more likely to see in soil is *Colpoda.* Asexual reproduction is by binary fission and budding. Sexual reproduction is by fusion or conjunction, which is a temporary fusion and swapping of genetic material. The largest ciliates are 100 µm long, but most, like *Colpoda,* are only about 20 to 30 µm long. The mass of ciliates ranges from 1.5 to 750 ng per individual.

Sporozoa

Sporozoa are completely parasitic organisms, so they have a limited role (if any) in soil. Sporozoa have no organelles for locomotion. Asexual reproduction is by multiple fission and sexual reproduction is by fusion. Sporozoa are important human pathogens that cause infection either through insect bites (malaria is caused by *Plasmodium,* which is carried by mosquitoes) or by ingestion of food and water contaminated by fecal wastes. *Cryptosporidium,* for example, which occurs in some cattle, has been found in surface water samples worldwide. A 1991 U.S. EPA report indicated that 90%

Protozoa in the News—Cryptosporidium and Giardia

If it's summertime, it must be *Giardia* season. That's what many backpackers and campers say when they head out for the trail. Experience tells them that any untreated water they get from a campground has to be filtered or boiled at least 1 minute before they drink it, otherwise, the archezoa *Giardia lamblia* is likely to infect them and cause giardiasis—cramping, bloating, and diarrhea.

Imagine what happens when an entire city comes down with similar symptoms. That's precisely what occurred in Milwaukee, Wisconsin in April 1993, when over 400,000 people started to experience severe diarrhea, cramps, nausea, and dehydration. The culprit? A microscopic protozoan called *Cryptosporidium,* which occurs in wild animals, birds, and domesticated cattle.

Unfortunately for the residents of Milwaukee, a series of incidents showed them just how fragile the safety of their drinking water was, and just how much they had come to take that water quality for granted. Heavy spring rains caused runoff of animal wastes containing *Cryptosporidium* oocysts from the watersheds surrounding Milwaukee into Lake Michigan, where the city obtains much of its water. Typically, Milwaukee's water treatment plants add a coagulant to the water to settle the suspended solids, filter the water, and then chlorinate it before use. In this instance, however, they were caught using a different coagulant, which was not as effective at removing suspended solids—solids that were higher than normal due to the excessive spring rains. The higher turbidity levels made the filtration of the water less effective, particularly since the *Cryptosporidium* oocysts are only 4 to 6 µm in diameter. In addition, both *Giardia* and *Cryptosporidium* are resistant to chlorination.

As few as 10 oocysts of *Cryptosporidium* can cause cryptosporidiosis, and its detection in water samples is a time-consuming and costly process, so analyses are not routinely done (Current and Haynes, 1984; Current 1986). As with other waterborne diseases, by the time the problem was recognized, many people were infected. To most healthy people, cryptosporidiosis is an inconvenience. However, for the very young and old, and for those with weakened immune systems, it is a very serious disease, and several people died as a result of this outbreak.

Table 6-1   Numbers of protozoa in soil.

Ecosystem	Flagellates	Amebas	Ciliates
	Average cells per gram of dry soil		
Agricultural			
Uncultivated soil	7,000	30,000	155
Unfertilized maize	833,000	1,295	2,230
Manured maize	740,000	144,690	11,265
Grassland			
Tallgrass (Minnesota)	400,000	80,000	150
Semiarid prairie	8,000	5,000	80
Mountain meadow	28,000	24,000	138
Forest			
Lodgepole pine	30,000	25,000	225
Douglas fir	25,532	3,018	74

(From Ingham 1994)

of untreated water and 30% of treated water nationwide was contaminated with *Cryptosporidium*. The *Cryptosporidium* oocyst, which is the cell type voided in wastes, is heat and cold resistant, tolerant of chlorine, and only 4 to 6 µm in size, which makes it difficult to filter.

POPULATIONS

How many protozoa of each type are in soil? Archezoa are typically lumped in with the protozoa with respect to population studies. Total protozoa may range from 100,000 to 300,000 per gram in the upper 15 cm of the soil surface, though a typical range would be closer to 10,000 to 100,000 (Alexander 1977) (Table 6-1). There may be daily population changes in the protozoa. Fewer than 1,000 of these protozoa are ciliates; the rest are flagellates and sarcodina. Flagellates are the most numerous of the protozoa in a typical soil and they dominate acid soils. Protozoa biomass can be 5 to 20 g m^{-2}.

GROWTH

The life cycle of protozoa has two distinct phases—an active phase and an inactive, cyst phase. Cysts are thick coatings generated by the protozoa to protect themselves from unfavorable soil conditions. Protozoa can persist for years as cysts. Lack of prey (food) or water typically causes protozoa to encyst. The return of ideal conditions causes the encysted protozoa to return to an active growth phase. This shift may occur in as little as 24 hours. During active growth, protozoa may divide once or twice per day, although protozoa that form shells (testate protozoa) may take up to a week to replicate.

Table 6-2   Effect of manure on the number of amoeba in barnfield soil.

| Month | Control Soil | | | Manured Soil | | |
	Active	Encysted	Percent Encysted	Active	Encysted	Percent Encysted
	Number per gram of soil			Number per gram of soil		
April	530	1,790	77%	16,040	2,060	11%
May	13,900	4,100	23%	49,550	3,550	7%
June	6,540	1,940	23%	34,130	5,870	15%
July	4,040	4,020	50%	52,500	11,000	17%
August	8,770	3,730	30%	22,300	10,500	32%

(From Singh 1949)

At any one time, many of the protozoa in soil can be encysted rather than active, as Table 6-2 shows. There are several important points you can draw from these data. First, the number of active and encysted amebas (and, we assume, other protozoa) varies widely during the year—from as few as 530 to as many as 52,500 per gram of soil depending on the season and the soil treatment. Second, active amebas usually out-number encysted amebas. Third, manure addition increases the total number of active amebas. Fourth, even though the manure treatment seems to increase the number of encysted amebas, they are a small proportion of the total population. This means that overall, amebas are more active in the manure-amended soil than in the control soil.

Environmental conditions favoring bacterial growth, such as a rich organic matter source like manure, favor protozoa growth. Protozoa also thrive when the soil is well aerated and the soil pH is between 6 and 8. Protozoa have been found in soils with pH that ranges from 3.5 to 9.0, but tolerance varies, and individual protozoa species have a more restrictive pH range. High temperatures are detrimental to protozoa because the protozoa are typically mesophiles (requiring moderate temperatures). Water films around soil are required by all protozoa, particularly by ciliated forms; flagellated forms are more drought resistant. The absence of water and lack of prey are two factors that definitely induce a survival mechanism like encystment to occur.

FEEDING

Protozoa are saprophytic, feeding on dissolved inorganic and organic substances, and phagotrophic. They feed directly on other organisms by grazing and predation. The ingested food is surrounded by a vacuole where digestion occurs. Ciliates and flagel-lates consume bacteria, yeasts, other protozoa, and rotifers. Amebas feed on bacteria, protozoa, yeasts, spores, and algae. The consumption rate can be 100 to greater than 1,000 bacteria per hour depending on the protozoa and the size of the bacteria. Proto-zoa can be selective grazers. Bacteria may be edible to some protozoa and not to oth-ers. Susceptible bacteria may be eaten in presence of bacteria that are not prey.

ECOLOGY

Protozoa probably have a limited role, if any, in affecting soil environmental conditions such as structure and pH. However, protozoa affect the structure and function of microbial communities. We know that eliminating protozoa decreases the overall decomposition rate of organic matter in soil. This suggests that protozoa, like macrofauna, help stimulate decomposition. Some soil microbiologists have proposed that protozoa regulate microbial populations in soil because they are predatory (Alexander 1977). Predators have beneficial effects on their prey collectively, but not individually. That is to say, even though individual prey may suffer, the overall population of the prey benefits from predation. So, decomposition is faster in the presence of protozoa because microbial populations are kept young and active, even if individual members of the population get eaten.

Predation may keep one microbial group from dominating an ecosystem. Figure 6-3 illustrates that in a simple soil system with the prokaryote *Xanthomonas campestris* and protozoa, an increase in the protozoa population corresponds to a dramatic decrease in the *X. campestris* population. In the absence of *X. campestris,* the protozoa population does not change.

Notice that the protozoa population increased by only 2 million whereas the *X. campestris* population decreased by over 4 million. A lot of prey have to be consumed to support protozoa growth. Also notice that the *X. campestris* population in Figure 6-3 didn't disappear. Why didn't the protozoa completely eliminate their prey? There are several reasons. As the proportion of edible cells in the prey population declines, protozoa may switch to other prey, or their own activity may slow. Prey may

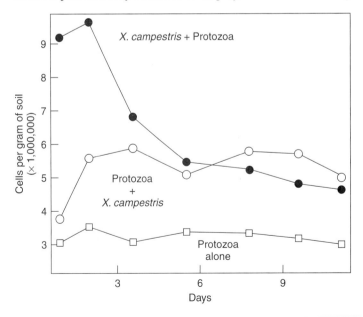

Figure 6-3 Control of a prokaryote population by protozoa. (Adapted from Habte and Alexander 1975)

also be protected in pores too small for protozoa to enter (micropores). Adding clay to a mixture of bacteria and protozoa in soil increases the survival of the bacteria because the clays aggregate and form micropores. Protozoa may be attacked by nematodes and fungi as their population grows. It's also possible that the prey population growth may keep ahead of predation. The most obvious reason is that most soils aren't closed systems. Prey are never eliminated because they constantly migrate into the area from other sites.

Protozoa also play important ecological roles outside the soil. Flagellated protozoa, for example, are important inhabitants of the rumen in cows and other livestock, where they help to digest cellulose. An even more interesting relationship is the three-member symbiosis of bacteria, protozoa, and termites that leads to cellulose digestion in nature. Higher termites such as *Zootermopsis angusticolis* harvest cellulose as a food source, but they can't digest it. The termites contain *Tricercomitis,* a protozoa about 15 to 20 µm in length that breaks down the cellulose into organic acids that the termites can absorb. In turn, the protozoa coexist with bacteria called methanogens that turn the protozoa waste products into methane.

Summary

Protozoa and archezoa are single-celled invertebrates and the smallest soil animals. Archezoa principally differ from protozoa in that they have no mitochondria. The protozoa and archezoa are studied much the same way—by serial dilution and plating onto media that contains a food source such as bacteria or yeast, then looking for evidence of protozoa growth. Counting protozoa in soil is difficult because at any one time much of the population is encysted (in a resting state) for lack of either sufficient water or food for them to grow on. When protozoa are actively growing there are usually between 10,000 and 100,000 cells per gram of the upper soil surface.

Protozoa are grouped into four broad categories based on movement: Mastigophora (flagellates), Ciliophora (ciliates), Sarcodina (ameboids and foraminifera or testate protozoa), and Sporozoa (immobile obligate parasites). Protozoa play an important ecological role in regulating bacterial populations in soil. Protozoa also play an important role as pathogens in drinking water. Two of the most important are *Giardia* (a flagellate) and *Cryptosporidium* (a sporozoa). *Cryptosporidium* caused the largest outbreak of waterborne disease in the history of the United States when it contaminated the Milwaukee water system in 1993.

Sample Questions

1. How would you define what a protozoan is?
2. Name the four major groups of protozoa in soil using either the technical or common designation.
3. What are some of the reasons why protozoa populations vary so much during the year?
4. Why are *Giardia,* an archaezoa, and *Cryptosporidium,* a protozoa, frequently mentioned in the news?

5. What do protozoa eat?

6. Suggest reasons why protozoa can't completely eliminate their prey.

7. Graph the data for the percent encysted protozoa in Table 6-2. Explain why the percentages change during the season.

8. Assuming that the top 15 cm of soil in a hectare weighs about 2 million kilograms (kg), how many total protozoa would you expect to find in a hectare of land?

9. If the biomass of protozoa may be as high as 20 g per square meter, how much would the total protozoa biomass be in 1 hectare (10,000 square meters)?

10. If you enumerated 100 protozoa per gram of a soil sample that weighed 8.5 g, but had to dilute it 100-fold initially, how many protozoa are actually present?

Thought Question

How could you design an experiment to show that protozoa can never completely eliminate their prey?

Additional Reading

An exhaustive review of the current status of classifying protozoa is in an article by T. Cavalier-Smith called "Kingdom Protozoa and its 18 Phyla" (*Microbiology Reviews,* 1993, Volume 57, pp. 953–994). The traditional classification system is outlined by R. H. Whittaker in "New concepts of kingdoms and organisms" (*Science,* 1969, Volume 163, pages 150–160). Lee and Kugrens (1992, Relationship between the flagellates and the ciliates. *Microbiological Reviews.* Vol. 56, pp. 529–542) have some nice diagrams that show the structure of these protozoa in great detail and that illustrate their feeding habits.

References

Alexander, M. 1977. *Introduction to soil microbiology.* 2d ed. New York: John Wiley & Sons.

Cavalier-Smith, T. 1981. Eukaryote kingdoms: seven or nine? *Biosystems* 14:461–81.

Current, W. 1986. *Cryptosporidium*: Its biology and potential for environmental transmission. *Critical Reviews in Environmental Control* 17:21–31.

Current, W. L., and T. B. Haynes. 1984. Complete development of *Cryptosporidium* in cell culture. *Science* 224:603–605.

Habte, M., and M. Alexander. 1975. Protozoa as agents responsible for the decline of Xanthomonas campestris in soil. *Applied Microbiology* 29:159–64.

Ingham, E. R. 1994. Protozoa. In *Methods of soil analysis, part 2: Microbiological and biochemical properties,* R. W. Weaver et al. (eds.), 491–516. Madison, WI: Soil Science Society of America.

Killham, K. 1994. *Soil ecology.* Cambridge, England: Cambridge University Press.

Sandon, H. 1927. *The composition and distribution of the protozoan fauna of the soil.* London: Oliver and Boyd.

Singh, B. N. 1949. The effect of artificial fertilizers and dung on the numbers of amoebae in Rothamsted Soils. *Journal of General Microbiology* 3:204–10.

The Chromista—Algae

Overview

After you have studied this chapter, you should be able to:

■ Describe a new classification kingdom called the Chromista.

■ Explain why the Chromista includes some soil organisms that traditionally have been called algae.

■ Identify the major categories of algae, tell how numerous they are, and state where they are found.

■ Explain why eutrophication represents algae growth run amok.

INTRODUCTION—WHAT ARE CHROMISTA?

The classification of organisms has been ever changing since Aristotle attempted to neatly organize living things into categories for study in ancient Greece. The Kingdom Chromista, first proposed by Cavalier-Smith (1981), represents an attempt to group together organisms with similar properties. Consequently, this kingdom includes organisms that were previously characterized as protozoa, fungi, and plants.

The common features of the Chromista are that they are eukaryotes (have a cell nucleus), they have chloroplasts associated with the endoplasmic reticulum, and they have one or more flagella. One or both of the last two features exists in every member of the Chromista. Oomycetes, for example, which were previously categorized as fungi, have biflagellate zoospores (a zoospore is a reproductive motile cell). Two examples of oomycetes are *Pythium*, which causes damping-off disease in plants and *Phytophthora infestans*, which causes potato blight and was responsible for the Irish potato famine in the early 19th century. The organisms that we call algae in this chapter are all photosynthetic and some have flagellated zoospores.

All classification schemes are adopted by consensus, since only the most vain believe they understand the order of nature. The adoption of the Kingdom Chromista as the standard method for classifying certain organisms is by no means universally accepted or applied. We treat all of the algae in this chapter as though they are microscopic plants, which is the traditional way of looking at them, and which fits their common ecological roles regardless to which classification kingdom they belong.

WHAT ARE ALGAE?

Algae are oxygen-evolving photosynthetic organisms that have "plantlike" chlorophylls and other photosynthetic pigments. They range in size from marine versions such as seaweed (kelp being an example) that form filaments 40 meters long, to microscopic species living in soil. *Chlorella,* for example, is 2 to 3 μm in diameter. Algae are the most common and widespread plants. Although they are mainly aquatic, they inhabit all terrestrial environments. Algae are important in soils where light and water are available even though water may be only briefly present in these environments.

In the course of geologic time, algae were important because they caused a major change in the geochemistry of Earth. The oxygen they evolved during photosynthesis converted Earth's atmosphere from one that was oxygen poor to one that was oxygen rich. Algae contributed to the removal of carbon dioxide (CO_2) from the atmosphere via the formation of $CaCO_3$ and the formation of organic matter during photosynthesis.

Algae have important ecological roles as pioneer colonizers in hot and cold desert ecosystems where higher plants struggle to survive. Algae contribute to primary productivity, the formation of organic carbon compounds, and to the formation of soil structure. *Chlamydomonas,* a green algae, have been added to sandy soils to aid erosion control and clay soils because the extracellular carbohydrates they produce improve aggregation and infiltration (Metting et al. 1988).

Algae associate with fungi to make complex organisms that we call lichens. Lichens contribute to silicate bioweathering by excreting organic acids. This is the first step in

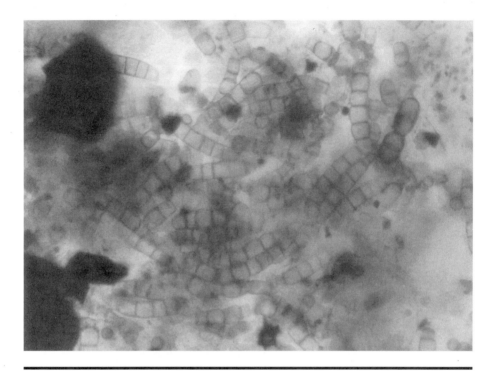

Figure 7-1 A filamentous algae. (Photograph courtesy of M. S. Coyne)

turning rocks into soil. Algae are near the bottom of the food chain. They are consumed by protozoa, nematodes, mites, and earthworms. Consequently, pesticides and insecticides can stimulate algal blooms by eliminating predatory insect and microfaunal grazers.

IMPORTANT ALGAE GROUPS IN SOIL

There are three major groups of algae:

1. Green algae (Chlorophyta)
2. Diatoms
3. Yellow-green algae (Xanthophyta)

Green Algae (Chlorophyta)

Green algae contain chlorophyll *a* and *b* and have a cellulose cell wall that also contains other polysaccharides such as alginic acid. They may be unicellular, form colonies of many cells, or form long filaments of cells like beads on a string (Figure 7-1). Sexual reproduction is by fusion and asexual reproduction is by division (fission), zoospore formation, or fragmentation. Green algae store their photosynthetic products as starch, much like higher plants. This last point is one of the major distinctions between green

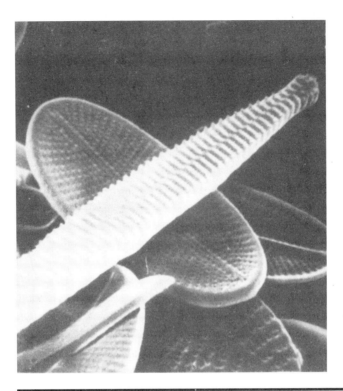

Figure 7-2   Electron micrograph of a diatom. (Photograph courtesy of the Soil Science Society of America)

algae, which may properly be called plants, and both diatoms and yellow-green algae, which can be considered members of the Chromista. The photosynthetic organisms in the group Chromista do not store starch.

Diatoms

Diatoms contain photosynthetic pigments such as chlorophylls, carotenes, and xanthophylls. They have an overall brown appearance because the chlorophyll is masked by the other pigments. They store their photosynthetic products as oil. Diatoms are unicellular and reproduce sexually by fusion and asexually by fission and zoospore formation. The diatoms are interesting because of their unique structures (Figure 7-2) and reproduction—not necessarily because they significantly affect soil properties or productivity. They have a silica and pectin exoskeleton that resists degradation. When the diatoms die, the cell walls remain relatively intact, and can accumulate with time. Diatomaceous earth is so called because it is rich in these empty cell walls. Most deposits of diatomaceous earth are the result of uplifted ancient seabeds on which the remains of diatoms accumulated.

Diatomaceous earth is commercially valuable as a filtering material, an insulating material, and a polishing agent. It is the principal component of China clay. Diatomaceous earth is also used as a nonchemical insecticide. The silica cell walls are quite hard—harder than the exoskeleton of insects. When diatomaceous earth is spread around plants, and insects travel through it, the silica cell walls cut into the exoskeleton and expose the internal fluids of the insects to bacteria or to leakage. In other words, the insect pests cut themselves to death on the diatoms.

The cell wall of the diatom is known as a frustule. It is composed of two halves, called valves, that fit together like a hatbox or the top and bottom of a Petri dish. When the diatom reproduces, it splits apart and synthesizes two new valves that fit inside the existing valves. One consequence of this reproductive strategy is that the size of individuals in the diatom population progressively decreases with each reproductive cycle. Naturally, this can't go on forever. At some point the diatom starts a process called auxospore formation. Essentially the diatom, a diploid organism (it has two sets of chromosomes—people are also diploid organisms), undergoes meiotic division to form haploid nuclei (with one set of chromosomes—eggs and sperm, our reproductive cells, are haploid cells). The cell wall is shed, and the haploid nuclei from the same or different diatoms fuse (becoming diploid again), enlarge, and then reconstitute the cell wall (Dodd 1977).

Yellow-Green Algae (Xanthophyta)

Yellow-green algae contain xanthophyll, carotenes, and chlorophyll. They have a pectin cell wall, although some also have a silica cell wall. Yellow-green algae are found as single cells and as cells that have formed filaments. They reproduce by fusion of cells and by fission of cells or by forming zoospores or simply fragmenting. The yellow-green algae store their photosynthetic products as oil.

WHERE ARE ALGAE FOUND?

In general, algae are more numerous than protozoa and macrofauna, but less numerous than prokaryotes. Algal colony-forming units vary between 10^9 and 10^{10} per square meter of soil or between 10^3 and 10^6 colony-forming units per gram of soil (Metting 1993). Algae typically contribute between 7 and 300 kg of biomass per hectare. Algae blooms may contain 1,500 kg per hectare. Algae populations vary, depending on soil type. In temperate soils, green algae are more numerous than diatoms, which in turn are more numerous than yellow-green algae. Green algae dominate in acid soils, while diatoms grow best in neutral soils.

Algae are found in the upper layers of soil where there is both water and light. Algae may be endolithic (within rock). That is, they're found just under the surface crust of sandstone and limestone where water and light can penetrate. Algae are also found at depths to 100 cm where light does not penetrate. How can this be? Do they proliferate, or are they carried by seepage? Are these algae active or dormant? In all likelihood, the algae obtained from deep in the soil were carried there by moving water or burrowing animals and stayed there, dormant, until they were brought to the surface again; rather than grow, they simply persisted.

ENVIRONMENTAL EFFECTS

What factors control algae populations? Temperature is one controlling factor—algae become less active in winter, although some cold-loving algae can grow at temperatures of 2°C. They turn the snow pink in alpine regions. Water is a controlling factor—algae blooms occur after rainfall. Salt is another controlling factor—algae can adapt to environments with high salt contents by pumping sodium ions (Na^+) out of the cell and pumping potassium ions (K^+) into the cell. The K^+ is much less damaging to the algae, and by providing osmotic balance, it helps the algae cells from losing water to their salty environment. The green algae can convert starch into glycerol (up to 30% of the algae dry weight). Glycerol is a water-soluble nonionic solute. One single-celled green algae, *Dunaliella salina,* grows in evaporation pans used to prepare sea salt. *Dunaliella* can grow in solutions containing up to 44% salt (Postgate 1994).

Soil fertility has a direct effect on the abundance of algae. Since algae can produce their own carbohydrates, it is the lack of soil nutrients such as N, P, and K that help to control algae growth. Herbicides also reduce algae populations in the soil environment. The response of algae in surface water to limiting nutrients, particularly P, is one of the factors leading to eutrophication of these environments. Figure 7-3 shows an example of eutrophication in an agricultural environment.

Figure 7-3   A eutrophic environment receiving runoff from a swine production facility. (Photograph courtesy of M. S. Coyne)

What Is Eutrophication?

Eutrophication is the overenrichment of water by nutrients. It causes excessive plant growth, stagnation, and death of other aquatic life such as fish. It is a major problem in watersheds and waterways surrounded by urban centers, such as the Great Lakes and Chesapeake Bay.

The word "eutrophic" comes from the Greek *eu,* which means "good" or "well" and *trophikos,* which means "food" or "nutrition." Eutrophic waters (ponds, lakes, and streams) are well-nourished and nutrient rich; so rich, that plants and algae grow uncontrollably. When the plants die and decompose, the water becomes depleted of oxygen by microbial action.

The stagnation that occurs during eutrophication happens because microorganisms decomposing the dead and dying plant material in water consume oxygen faster than it can be resupplied by the atmosphere. Fish, which need oxygen in the water to breathe, become starved for oxygen and suffocate. Noxious gases such as hydrogen sulfide (H_2S) are also released. So, the hallmark of a eutrophic environment is one that is plant-filled, littered with dead aquatic life, and smelly.

Eutrophication is a natural process that occurs as lakes age and fill with sediment, deltas form, and rivers seek new channels. Human activity can accelerate the process and cause it to occur in previously clean, but nutrient-poor water. This is sometimes referred to as cultural eutrophication. The nutrients that cause eutrophication usually come from surface runoff of soil and fertilizer in mismanaged agriculture, or from domestic and industrial wastes discharged into rivers and lakes.

Phosphorus (P) and nitrogen (N) are two of the nutrients that most limit plant growth in freshwater. When they are supplied, plant growth can explode and eutrophication can occur. Phosphorus was one of the major causes of eutrophication in Lake Erie during the 1960s. Before preventative action was taken, the lake was considered to be dying. These preventative actions included banning phosphates in laundry detergent and imposing stricter conservation practices on farmers to reduce soil erosion in the watersheds draining into Lake Erie. Many areas now restrict the total amount of P that can be applied to land in sensitive watersheds.

Summary

Although algae have been treated as photosynthetic plants, new classification schemes include some of them in the Kingdom Chromista. Algae are photosynthetic and aquatic, so they play a role in soil only where there is available water and light. There are several algae groups: green algae (chlorophyta), yellow-green algae (xanthophyta), and diatoms. Algae are pioneer colonizers and can contribute to primary productivity. They are at the bottom of the food chain. They contribute to soil aggregate stability because they add organic material to soil. Some algae form relationships with fungi to make complex organisms called lichens. Algae and lichens contribute to soil formation because they help in the bioweathering of rocks and minerals. Diatoms are interesting types of algae because they have a hard, silicate cell wall. Diatomaceous earth is made up of these algae cells and has many uses from pottery, to filters, to biocidal agents. Algae blooms develop in wet environments that have plentiful nutrients. These blooms can contribute to a process we call eutrophication.

Sample Questions

1. What is a lichen?
2. What are the major groups of algae in soil and where are they found in the soil profile?
3. What is "diatomaceous earth"?
4. In what ways are soil algae different from other soil organisms?
5. Why can you find soil algae at great depths where there is no light for them to grow?
6. Draw a graph showing the relative distribution of algae in soil as a function of soil depth.
7. There are between 1 billion (10^9) and 10 billion (10^{10}) algae per square meter of soil. Based on that number, how many algae are in a square decimeter of soil?
8. If you suspected that there were between 10^3 and 10^6 colony-forming algae per gram of a soil sample, how much would you dilute 10 g of the sample so that you would only have to count 100 colony-forming units?

Thought Questions

1. What steps could you take to prevent eutrophication of a pond in a farmer's field?
2. Could glowworms provide enough light for algal growth in soil?

Additional Reading

Everything you probably ever wanted to know about growing and studying algae can be found in a four-volume series called *The Handbook of Phycological Methods,* sponsored by the Phycological Society of America and published by Cambridge University Press: Vol. 1—Culture Methods and Growth Measurements; Vol. 2—Physiological and Biochemical Methods; Vol. 3—Developmental and Cytological Methods; Vol. 4—Ecological Field Methods: Macroalgae.

References

Cavalier-Smith, T. 1981. Eukaryotic kingdoms; seven or nine? *Biosystems* 14:461–81.

Dodd, J. D. 1977. *Course book in general botany.* Ames, IA: The Iowa State University Press.

Metting, F. B. 1993. Structure and physiological ecology of soil microbial communities. In *Soil microbial ecology. Applications in agricultural and environmental management.* F. B. Metting (ed.), 3–25. New York: Marcel Dekker.

Metting, F. B., W. R. Rayburn, and P. A. Reynaud. 1988. Algae and agriculture. In *Algae and human affairs,* C. A. Lembi, and J. R. Waaland (eds.), 335–70. Cambridge, England: Cambridge University Press

Postgate, J. 1994. *The outer reaches of life.* Cambridge, England: Cambridge University Press.

Chapter 8

Fungi

Overview

After you have studied this chapter, you should be able to:

- Discuss the historical importance of fungi.
- List the major differences between fungi and prokaryotes.
- Name the different fungal groups in soil and the most common fungal genera.
- Describe how the environment influences fungi.
- Summarize the ecological roles of fungi.
- Define "competitive saprophytic growth" and tell why it's important.
- Explain how fungi disperse and survive in soil.

INTRODUCTION

We get our names for the study of fungi from the original Latin and Greek words for mushroom, the most obvious manifestation of fungi: mycology (from the Greek, *mykes*) and fungus (from Latin). A soil microbiologist who studies fungi is therefore a mycologist. Common English names for fungi are molds, mildews, rusts, smuts, yeasts, mushrooms, and puffballs. This is not a text about mycology, so we won't spend much time identifying and characterizing fungi. Instead, we focus on generalizations about basic groups, habitats, and functions.

Why are fungi important? They may cause more joy and grief than any other microorganism. Yeasts leaven our bread. They ferment sugar to alcohol in wine and beer (examples being *Saccharomyces cerevisiae* and *Saccharomyces carlsbergensis*). They hydrolyze and flavor milk to make Brie, Camembert, Limburger, Roquefort, Gorgonzola, and blue cheeses (*Penicillium* species). They ferment plant products to make flavorings such as soy sauce (*Aspergillus oryzae*). Fungi also cause pestilence and crop loss. Most plant pathogens are fungi. Chestnut blight and Dutch elm disease robbed our forests of two majestic tree species. Rusts destroy thousands of hectares of wheat every year.

CHARACTERIZATION

What sets fungi apart from bacteria? The most important difference is that fungi are eukaryotes and bacteria are not. Bacteria are prokaryotes, meaning they do not have a cell nucleus. In addition, fungi are usually filamentous in contrast to most bacteria. Fungi are, in general, much larger than bacteria. Table 8-1 lists some other differences.

Typical fungi have the following characteristics: They are composed of slender filaments, called hyphae, that are 3 to 8 μm in diameter. Hyphae are either septate (divided into compartments by cross walls) or nonseptate (coenocytic; not divided by cross

Table 8-1 Differences between fungi and prokaryotes.

Fungi	Prokaryotes
Eukaryotic	Prokaryotic
Nuclear membrane	No nuclear membrane
Multiple chromosomes	Single chromosome
Mitochondria, organelles	Few internal structures
Polysaccharide-type wall (cellulose, chitin)	Peptidoglycan walls
Two types of ribosomes	70S ribosomes only
80S	
70S	
Multicellular and differentiated	Usually unicellular
Sexual reproduction	Mostly asexual reproduction
Cells >5 μm in diameter	Cells <5 μm in diameter
Structural diversity	Metabolic diversity

a   b

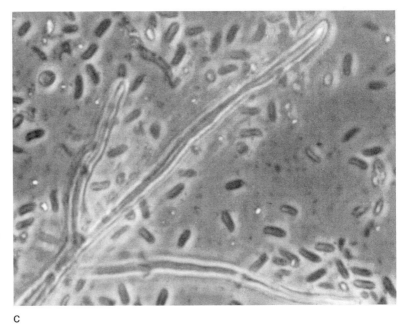

Figure 8-1 A fungal
fruiting body (a), mycelium
(b), and hypha (c).
(Photograph courtesy of
M. S. Coyne)

c

walls). Hyphae may be vegetative or fertile. Fungi reproduce by sexual spores, asexual spores, and fragmentation (Figure 8-1a,b,c). Collectively, hyphae compose the mycelium. The mycelium may be undifferentiated or organized into a fruiting body that can be greater than 0.1 m in diameter. A plate of spaghetti is like undifferentiated mycelia; spaghetti molded into the shape of the Statue of Liberty is like the organization of a fruiting body. It is still a mystery how single, supposedly nonthinking cells can coordinate in such complex ways.

TYPES OF FUNGI

Slime Molds—Myxomycetes

Slime molds have long been grouped with the fungi in most classification schemes. However, many classification schemes now include the slime molds, such as *Dictylostelium,* with the protozoa because of their ameboid-type growth and other genetic characteristics. Slime molds are found beneath decaying organic matter. They are multinucleated organisms that are ameboid-like, but they can group together to make more complex fruiting structures.

Flagellated Fungi—Oomycetes

Oomycetes have been traditionally classified as fungi. As noted before, the oomycetes are now classified in a new kingdom—the Chromista. Two species characterize this group: *Pythium,* which causes damping-off disease, and *Phythophthora,* which was responsible for potato blight in Ireland during the mid-19th century and ultimately led to millions of Irish dying from starvation. The immigration of the survivors to places like Australia, Canada, and the United States was a direct result of this fungus.

Sugar Fungi—Zygomycetes

The best example of this group is *Rhizopus nigricans,* the bread mold (Figure 8-2a,b).

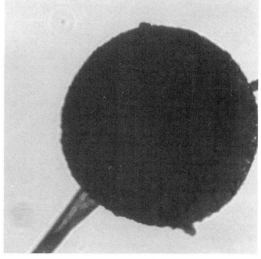

a   b

Figure 8-2 *Rhizopus nigricans,* the bread mold, on waste food and in a microscopic view. (Photograph courtesy of M. S. Coyne)

Figure 8-3 Fruiting bodies of higher fungi. (Photograph courtesy of M. S. Coyne)

Higher Fungi

Higher fungi (Figure 8-3) include the ascomycetes (30,000 identified species), which are fungi such as yeasts, morels, and truffels. It also includes basidiomycetes (25,000 species). These are characterized by *Agaricus,* rusts, mycorrhizae, and wood rot fungi.

Imperfect Fungi

Imperfect fungi includes groups like the deuteromycetes; *Penicillium* is an example.

Sterile Mycelia

Sterile mycelia are fungi that reproduce solely by hyphal fragmentation. This means that if the hyphae are broken into fragments, each fragment can begin forming a new mycelium. When spores are found, sterile mycelia are then usually reclassified as ascomycetes. This illustrates the difficulty of finding appropriate growth media for many fungi. Table 8-2 summarizes some of the significant differences and similarities between fungal groups.

Table 8-2   Major fungal groups in soil.

Group	Appearance	Soil Genera
Myxomycetes	Plasmodium	*Physarum*
Oomycetes	Unicellular	*Pythium*
	Aseptate hyphae	
Zygomycetes	Aseptate hyphae	*Mucor*
	Septate hyphae	*Rhizopus*
Ascomycetes	Septate hyphae	*Saccharomyces*
Basidiomycetes	Septate hyphae	*Boletus*
Deuteromycetes	Septate hyphae	*Aspergillus*
		Fusarium
		Penicillium
Sterile Mycelia	Septate hyphae	*Rhizoctonia*

Table 8-3   Relative contribution of soil organisms to the biomass of a temperate grassland soil.

Organism	Biomass (kg per hectare)
Plant roots	20,000–90,000
Fungi	2,500
Bacteria	1,000–2,000
Actinomycetes	0–2,000
Protozoa	0–500
Nematodes	0–200
Earthworms	0–2500

(From Killham 1994)

HABITATS AND ENVIRONMENTAL INFLUENCES

In most well-aerated soils, fungi are the largest fraction of the microbial biomass. Table 8-3 shows a comparison of biomass for various soil organisms in a temperate grassland. There are from 2×10^4 to 1×10^6 fungal propagules per gram of soil. A propagule is any part of the fungus, spore or hyphae, that can form a new fungal colony. There are probably fewer fungi than there are bacteria and actinomycetes. However, because of spore production and hyphal fragmentation, it is impossible to get an accurate estimate of fungal populations because the original organism has been multiplied through these various progeny.

Fungal distribution is determined by the availability of organic carbon because soil fungi are primarily saprophytic organisms growing on dead and decaying tissue (a few fungi parasitize living animals, and plant pathogenic fungi parasitize living plants). Usually fungi are found in the upper soil profile (0 to 15 cm). The composition of

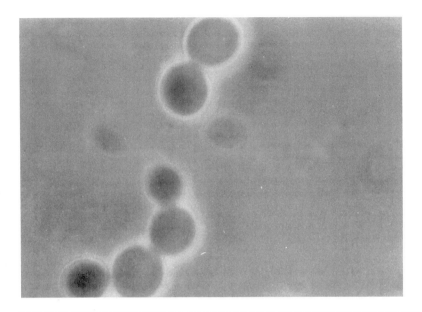

Figure 8-4   Photomicrographs of yeast (*Saccharomyces cerevisiae*), a single-celled fungi that reproduces by budding. (Photograph courtesy of M. S. Coyne)

vegetation also affects fungal species distribution. Fungi exist over a wide pH range but are more tolerant of acid soils than are other microorganisms. Consequently, to select fungi instead of other microorganisms, people use a media that has an acidic pH. Fungi are also mesophylic, meaning they grow best at temperatures between 6°C and 50°C.

Fungi require water for metabolic activity, but they persist in semiarid climates by either entering a dormant state or by leaving behind their spores. Their populations decline in wet soils because most fungi require O_2. Some fungi, primarily yeasts, can survive in anaerobic environments by fermenting sugars into alcohol (Figure 8-4). The position of the species in the soil profile can be affected by its sensitivity to CO_2: A species will live in the upper profile if it is inhibited by CO_2; in the middle profile (15 to 30 cm) if it is insensitive to CO_2; and in the lower profile if its growth is enhanced by high CO_2 (> 30 cm). Some fungi appear to grow in poorly aerated systems because the hyphae are exposed to air pockets.

Some fungi are endophytic parasites of grasses. An endophyte is an organism that lives within another organism. Fescue (*Festuca arundaceae*), for example, is infected by a fungal endophyte. The endophyte apparently gives the fescue additional resistance to drought and insect predation compared to other grasses. That's why you see fescue incorporated into so many lawns and pastures. Unfortunately, the fungus also releases alkaloids (poisons) that are detrimental to cattle grazing on the fescue. Other fungi, the mycorrhizal fungi, form beneficial symbiotic associations with the roots of plants and trees.

How Do You Study Fungi in Soil?

It is difficult to qualitatively and quantitatively study fungi in soil because they exist in so many morphological and physiological states (Parkinson 1994). To compound the problem, there is no single growth medium that allows the isolation of all fungal types from soil. The most common method you will likely see is the soil dilution plate method, in which soil is diluted in buffer until only a few fungi are left, and then representative samples from each dilution are spread (plated) onto solid media. The most commonly used medium is Czapek-Dox medium, which consists of sucrose (a carbon source), nitrate, phosphate, magnesium sulfate, a trace of iron sulfate (inorganic nutrients), agar (a solidifying agent), and some yeast extract to provide complex growth factors that the fungi may need. To stop bacteria from growing, the media is acidified to between pH 3.5 and 5 and a bacteriostatic compound such as crystal violet or rose bengal is added. Martin's Rose Bengal (Figure 8-5) is another common medium used to isolate fungi (Martin 1950).

Another approach is the direct observation and isolation of fungi from incubating organic debris (Seifert 1990). A Petri plate is lined with filter paper or moist paper towels. Organic debris (leaves, stems, wood, bark) is placed in the center of the plate, covered with a loose-fitting lid, and incubated at room temperature. After 2 days, it is examined for evidence of fungal sporulation using a dissecting microscope. Individual spores can be picked out with an agar-covered needle and incubated in growth media to get a pure culture.

There are also several staining methods that use fluorescein diacetate (FDA) or calcafluor to stain fungal cell components and allow them to be visualized with a fluorescence microscope.

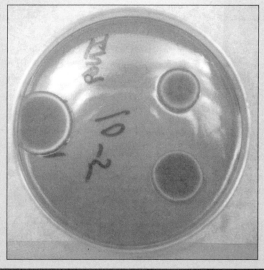

Figure 8-5 Fungal colonies on Martin's Rose Bengal agar. (Photograph courtesy of M. S. Coyne)

FUNCTIONAL ROLE

Fungi are the primary agents of organic matter decay. Fungi degrade complex molecules like cellulose, hemicellulose, pectins, starch, and lignin. They decompose resistant compounds. Cellulose, for example, is broken down by brown rot fungi while lignin is broken down by white rot fungi. An example of a white rot fungi is *Phanaerochaete chrysosporium.*

What other roles do fungi play? They are a nutrient reservoir since they form the bulk of microbial biomass. They help in the binding of soil aggregates. They make nutrients available by decomposing organic matter. They can be plant and animal pathogens. For example, *Histoplasma capsulatum* is an endemic fungal pathogen of people living in the Ohio River valley. Athlete's foot is caused by *Trichophyton* species. It's unfortunate, but many different unicellular fungi that reproduce by budding are commonly called yeasts, even though they can be wildly different organisms. *Candida* is an infectious pathogenic fungus that is often called a yeast. Baker's yeast and the yeast used to ferment alcohol are members of the genus *Saccharomyces.* Unlike *Candida, Saccharomyces* is nonpathogenic and beneficial.

People also eat fungi. *Agaricus bisporus,* the button mushroom, appears on many dinner tables. These mushrooms are grown primarily on composted horse manure. Shitake is a mushroom delicacy that is cultivated from fungi that decompose oak logs. Portabella is another edible mushroom. Truffles and morels are two types of fungi highly prized by gourmets.

COMPETITIVE SAPROPHYTIC GROWTH

Fungi are saprophytes, so a term that you should learn is "competitive saprophytic growth." Competitive saprophytic growth is essentially the sum of physiological characteristics that allow a saprophyte, such as a fungus, to successfully compete with other soil organisms in the colonization and metabolism of organic matter.

These physiological characteristics include, but are not limited to: 1) a fast growth rate; 2) a set of enzymes that allow an organism to grow on diverse substrates; and 3) the ability to produce inhibitory or toxic compounds and exclude competitors. The sequence of colonization of dead tissue by fungi is a good illustration of competitive saprophytic growth. Initially, pioneer fungi such as zygomycetes colonize the tissue because they germinate rapidly and use multiple compounds for growth (another term for this is "zymogenous growth"). Next, secondary saprophytes become established. They grow on leftover products from the metabolism of pioneering fungi. Production of inhibitory compounds and tolerance to the same are characteristic of this group. Finally, tertiary saprophytes become established. They are slow-growing, cellulose and lignin degraders. They have many different enzymes that allow them to attack resistant compounds and are often resistant to inhibitory agents produced by other microorganisms. Another term by which they are known is "autochthonous fungi."

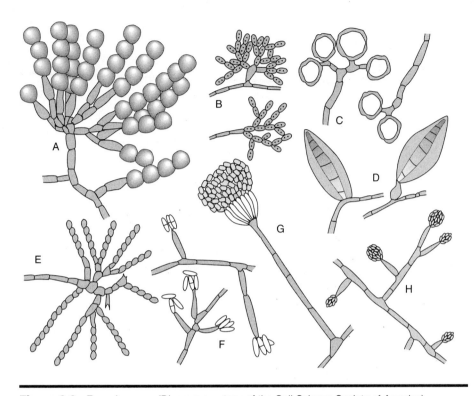

Figure 8-6   Fungal spores. (Diagram courtesy of the Soil Science Society of America)

SURVIVAL AND PROPAGATION

Fungi have several types of survival and dispersal mechanisms (Figure 8-6). Spores of various types, sclerotia, and rhizomorphs are cell structures used for both dispersal and survival. Conidia are asexual spores borne on the end of the hyphae. Depending on the fungal species, these spores survive from weeks to years in soil. Normally conidia are short-lived. Chlamydospores are thick-walled cells that develop from hyphae or conidia. They survive in soil for several months. Sclerotia are hard, mycelia-packed resting structures that can persist for years in soil. Oospores are thick-walled sexual spores. Sporangiospores are asexual spores borne in saclike sporangia that persist for 9 to 10 weeks in soil. Ascospores are formed by meiosis in special saclike structures that are the result of sexual fusion. Rhizomorphs are thick strands of undifferentiated hyphae.

Microorganisms must either compete for food, disperse, and die, or become inactive and persist until growing conditions become favorable again. Dispersal can be active or passive, effective or ineffective, and in time or in space. Dispersal significantly extends a microorganism's range, maintains its population in that range, and provides for genetic improvement by outbreeding.

Figure 8-7   A fairy ring in soil. (Photograph courtesy of the Soil Science Society of America)

What are some mechanisms of dispersal that fungi use? The fairy ring is a good example of active growth (Figure 8-7). Fungi radiate from some infection center, such as an old tree stump, in an ever-widening circle that can reach diameters of up to 3 kilometers. A fairy ring is marked by a ring of visible mushrooms and lush growth compared to surrounding grass. The mushrooms are the fruiting bodies of the underground fungi, while the lush growth indicates that previous fungal growth is dying, decomposing, and releasing nutrients that can be taken up by plants.

Fungi can be windblown without any active mechanism for their dispersion, as in the case of the soredia of lichen (soredia are sporelike structures). Wetting and drying cycles may release fungi. Ascomycetes can use expansion and collapse of turgid cells to disperse. Spores spread by this mechanism are called ballistospores. *Pilobolus* is an example of a fungus that produces ballistospores. A turgid cell, capped by a spore mass, eventually bursts and blows the spores to another location. The turgid cell is light sensitive and turns in the direction of light so that the spore mass is blown to the open air. Fleshy basidiomycetes such as *Nidularia* (bird's nest fungi) may form cups that spread spores by a water splash mechanism. *Lycoperdon* (puffballs) use a bellows mechanism (just like a bellows—if you press it, air or spores blow out). *Phallus* uses insect dispersal. The fruiting body smells like something dead and attracts flies that carry off the spores as they explore the fungi's surface.

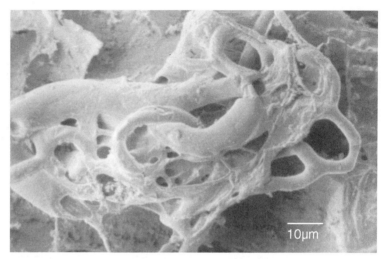

a

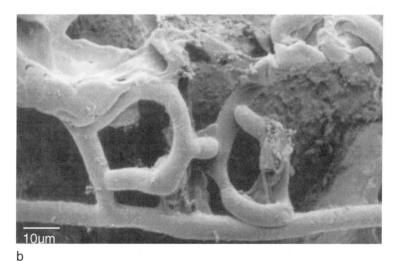

Figure 8-8 Examples of nematode-trapping fungi. (Photograph courtesy of G. Koening)

b

NEMATODE-TRAPPING FUNGI

If there is such a thing as justice in the microbial world, it comes from the observation that nematodes are trapped for food by some fungi (Figure 8-8a,b). Fungi do this in different ways. *Arthrobotrys* forms inducible constricting rings. The rings are induced, or caused to form, when nematodes are in the presence of the fungi. Fungi can recognize that nematodes are present because there are receptors on the fungus that recognize special proteins, called lectins, on the nematodes. These rings are pressure sensitive. When nematodes pass through the rings, the rings constrict and immobilize the nematode.

Fungal intrusions called haustoria then penetrate the nematodes. Nonconstricting rings are also formed. They act as a maze into which nematodes wander, but cannot escape. Adhesive knobs are formed that break off on nematodes and from which haustoria grow. Fungi also produce conidia that germinate inside the nematodes once they are consumed.

Summary

There are both beneficial and detrimental aspects of fungi. Fungi are different from bacteria in that they are larger, eukaryotic, and generally filamentous. Fungi are very diverse morphologically. They can be microscopic and unicellular like the yeasts, or form huge fruiting bodies like the puffball. The fungal phyla are myxomycetes, oomycetes, zygomycetes, ascomycetes, basidiomycetes, and deuteromycetes.

Well-aerated environments with lots of organic matter favor the growth of most soil fungi because they are primarily saprophytic. Fungi grow competitively by rapid growth, by producing antibiotics, or by digesting resistant plant compounds. Isolating them is a problem because there are so many different kinds. Growing fungi in pure culture often involves using bacteriostatic agents and acid environments.

A major role of fungi in nature is to decompose organic matter, allowing other microorganisms to consume it. Fungi are the greatest fraction of microbial biomass in soil, so they represent a great store of nutrients for potential microbial growth. Fungi persist in soil through a variety of mechanisms including spores and resting structures. The spores are of numerous types and are spread by numerous mechanisms. A few fungi are adapted to feeding on nematodes, while other fungi are harvested or grown by people to be eaten.

Sample Questions

1. What are ways by which soilborne plant pathogenic fungi survive in the absence of their host?
2. Why are fungi poor competitors in wet soils?
3. What is competitive saprophytic growth and why is it important?
4. What is the reasoning behind adding bacteriostatic compounds and acidifying media to isolate fungi?
5. What is the principal role of fungi in the environment?
6. How small would a mesh bag have to be to exclude soil fungi from decomposing leaves within it?
7. Assuming you have 1 cm^3 of soil that was completely packed with fungal mycelia with an average diameter of 8 μm, what do you calculate the total length of fungal mycelia to be?
8. How would you determine the total hyphal length in soil and how would you convert this to the total fungal biomass in soil?

9. Draw a graph that shows the number of fungal colony-forming units that may grow during a dilution plate count as a function of: a) the time of shaking a soil sample; and b) the vigor of shaking a soil sample. Explain the assumptions behind your graph.

10. Draw a graph that shows the relationship between fungal biomass or fungal population with soil depth. Explain your reasoning.

11. If the biomass of fungi is assumed to be 200 nanograms (ng) per μm cubed (μm^3), calculate the total fungal biomass in 1 cubic centimeter.

12. Since you can think of fungi as enormously long cylinders, calculate the volume of 300 μm of fungal hyphae that have an average diameter of 5 μm.

Thought Question

The largest organism in the world is said to be a fungus called *Armillaria*. It may be spread throughout a land surface area of up to 8 hectares. How do you think researchers determine the extent of this organism?

Additional Reading

Fundamentals of the Fungi by E. Moore-Landecker (1996, Prentice-Hall, Upper Saddle River, NJ) provides a thorough investigation of mycology. Consulting numerous field guides to mushrooms in the environment is an excellent way to start investigating fungi in nature. An example is *A Field Guide to Mushrooms, North America* by K. H. and V. B. McKnight (1987, Houghton Mifflin, Boston, MA). If you feel compelled to eat the mushrooms you find, apprentice yourself to an aged mycologist before embarking on this potentially deadly hobby. One taste of *Amanita*, the death cap mushroom, is enough to send you on an emergency visit to the hospital.

References

Killham, K. 1994. *Soil ecology*. Cambridge, England: Cambridge University Press.

Martin, J. P. 1950. Use of acid, rose bengal, and streptomycin in the plate method for estimating soil fungi. *Soil Science* 69:215–32.

Parkinson, D. 1994. Filamentous fungi. In *Methods of soil analysis, part 2: Microbiological and biochemical properties*. R. W. Weaver et al. (eds.), 329–350. Madison, WI: Soil Science Society of America.

Seifert, K. A. 1990. Isolation of filamentous fungi. In *Isolation of biotechnological organisms from nature,* D. P. Labeda (ed.), 21–51. New York: McGraw-Hill.

Chapter 9

Filamentous Prokaryotes— Actinomycetes

Overview

After you have studied this chapter, you should be able to:

■ List the characteristics of the actinomycetes and the names of the most common actinomycetes in soil.

■ Describe how environmental factors affect actinomycetes in soil.

■ Define what antibiotics are, and explain when they are produced, why they are produced, and the ecological consequences of their abuse.

■ Summarize how actinomycetes and other microorganisms resist antibiotics.

INTRODUCTION

Take a handful of garden or field soil, hold it close to your nose, and breathe deeply. What do you smell? It's not obvious that you should smell anything based simply on the composition of soil, because most soils are primarily made up of inert materials such as sand, silt, and clay. But you probably do smell something: an earthy, musty smell. Maybe it's a smell that brings back old memories of cutting grass in spring or burning leaves in fall. The smell is real even if the images it evokes are just memories. What your sense of smell detects are microbial products called geosmins (1,10 dimethyl-9-decalols). Geosmins produce the smell of freshly plowed soils and musty cellars—the smells that remind city folk of country life. Geosmins are produced by the next group of microorganisms we're going to study—the actinomycetes.

CHARACTERISTICS OF ACTINOMYCETES

Actinomycetes are prokaryotes that may look like fungi—prokaryotes with an identity crisis, if you will. Actinomycetes were originally called ray fungi when the methods of classification were mostly based on the appearance (morphology) of isolates growing in pure culture (one of Koch's legacies). Actinomycetes grow as filamentous mycelia and form spores (Figure 9-1). Unfortunately, what an organism looks like in pure culture may bear little resemblance or consequence to what it looks like in its natural

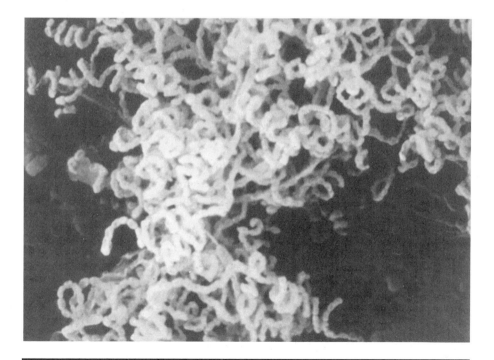

Figure 9-1   Electron micrograph of an actinomycete, *Streptomyces,* showing abundant spore formation. (Photograph courtesy of the Soil Science Society of America)

environment. There are two important characteristics that distinguish actinomycetes from fungi: 1) actinomycetes have no cell nucleus, so they are prokaryotic, and 2) actinomycetes form hyphae that are from 0.5 to 1.0 μm in diameter, which are much smaller than fungal hyphae (which are 3 to 8 μm in diameter).

Actinomycetes are not photosynthetic. Most actinomycetes are saprophytes, growing by decomposing organic matter. Some actinomycetes are human pathogens—*Mycobacterium leprae* causes leprosy and *Mycobacterium tuberculosis* causes tuberculosis. Other actinomycetes are animal and plant pathogens. Like most other microorganisms, however, actinomycetes are usually harmless soil organisms. Some actinomycetes are particularly beneficial. Actinomycetes in the genus *Frankia* form associations with woody shrubs and trees and fix nitrogen.

ENVIRONMENTS AND POPULATIONS

Actinomycetes compose 10% to 50% of the total microbial population in soil. They are found in soil (most commonly), composts, and sediment. They are second in abundance to bacteria in soil with populations that range from about 10^5 to 10^8 propagules per gram of soil. A propagule is any part of a microorganism that can grow and reproduce. Actinomycete populations are higher in pastures than in cultivated soils and higher in cultivated soils than in fallow soils. Since most actinomycetes are saprophytes, it's likely that their population depends on the available decomposable organic matter in each system. As you go deeper in soil, the actinomycete population declines less than does the population of other microbial groups, probably because actinomycete spores are being recovered at the lower depths.

Actinomycetes, with some exceptions, are aerobic—requiring O_2 for growth. As a result, they do not grow well in wet soils. Actinomycetes are not tolerant of desiccation, but the spores they produce can tolerate desiccation. Consequently, after a drought, actinomycetes may make up 30% to 90% of the microbial population that is recovered. This is because, unlike most other prokaryotes, they have formed spores that germinate after water becomes available.

Actinomycetes grow very little at 5°C. They are isolated more commonly from hotter soils than colder soils. In all likelihood, this too is due to spore recovery, since hotter soils are drier soils. It doesn't mean that actinomycetes love heat (although some do). Optimum growth is between 28° and 37°C, but some actinomycetes grow at 55° to 65°C in compost heaps. Composting manure piles have actinomycete numbers that can increase to 10^{10} propagules per gram of compost, but this is an exceptional population density.

Actinomycetes are tolerant of alkaline conditions. In alkaline soils, 95% of the microbial isolates may be actinomycetes. On the other hand, actinomycetes are acid intolerant for the most part, although acid-intolerant species exist. At a pH of less than 5, actinomycetes make up less than 1% of the microbial population. Actinomycete intolerance to acid is used to control some plant pathogenic actinomycetes. Potato scab, caused by *Streptomyces scabies,* can be controlled by acidifying soils (Figure 9-2). The graph in this figure shows that as pH declines, the *Streptomyces scabies* population and the incidence of scabbed potatoes also decline. When the soil has been acidified to a pH

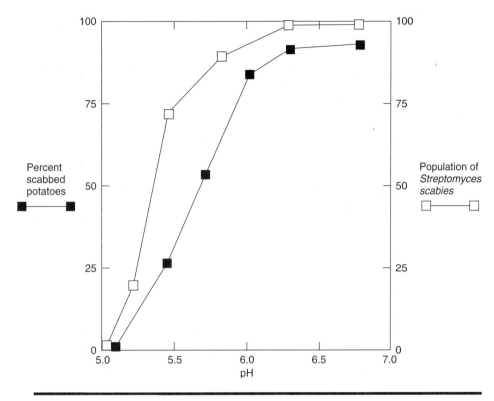

Figure 9-2   The relationship between scabbed potatoes and *Streptomyces scabies*. *Streptomyces* population is on a relative scale. (From Alexander 1977)

of < 5.8, fewer than 50% of the potatoes are scabbed. *Streptomyces scabies* growth appears to be affected as soon as the pH drops below 6.3.

Actinomycetes are saprophytes. As organic matter increases, the actinomycete population increases. However, actinomycetes do not proliferate until the later stages of organic matter decay. Actinomycetes are poor competitors for readily degradable substrates, but they use cellulose and chitin. Almost all actinomycetes degrade chitin, and chitin degradation can be used as a selective procedure to isolate actinomycetes.

CLASSIFICATION

Actinomycetes are classified based on such things as cell wall chemistry, whole-cell sugar composition (which is diagnostic for groups of actinomycetes), and DNA or RNA hybridization (which characterizes relationships among actinomycetes based on their genetic or genotypic similarity). Some important groups (genera) of actinomycetes are *Micromonospora, Nocardia, Streptomyces, Streptosporangium,* and *Thermoactinomyces.*

Streptomyces are more prevalent than *Nocardia,* which in turn are more prevalent than *Micromonospora.* Ninety percent of the actinomycetes from soil may be

Table 9-1   Actinomycete genera found in soil.

Genus	Environment
Actinomadura	Wide distribution in cultivated soils
Actinomyces	Human and animal tissue
Actinoplanes	Worldwide distribution
Agromyces	Wide range of soil types
Amycolata	Forest soils and the rhizoplane
Amycoatopsis	Forest and cultivated soils
Arthrobacter	Numerous and widely distributed in soils
Aureobacterium	Several different soil species isolated
Catellatospora	Woodland soils
Cellulomonas	Mostly soil isolates
Corynebacterium	Farmland soils and manure
Dactylosporangium	Sediment, water, and plant litter
Frankia	Root nodules of N-fixing shrubs and trees
Geodermatophilus	Desert soils
Glycomyces	Global distribution in soils
Gordona	Isolated from soil and sediment
Kibdellosporangium	Desert and tropical soils
Microbispora	Widely distributed in cultivated soils
Micromonospora	Common in most soil and sediment
Microtetraspora	Cultivated soils
Mycobacterium	Saprophytes in many soils and an animal pathogen
Nocardia	Widely distributed in soils
Nocardiodes	Clay soils, savanna grassland, and plant litter
Oerskovia	Organic-rich soils
Pilimelia	Wide range of soil types
Planobispora	Worldwide distribution in diverse soils
Pseudonocardia	Compost, manure, cultivated soils
Rhodococcus	Pastures, marine sediment, and farmland
Sacharomonospora	Manure, compost, and peat
Saccharothrix	Common in many soils
Streptomyces	Widespread and numerous in most soils
Streptosporangium	Widely distributed in cultivated soils and plant litter
Thermonospora	Composts, bagasse, manure, and soils

(From Wellington and Toth 1994)

Streptomyces and this genus alone may represent 5% to 20% of the total microbial isolates in nonspecific dilution plate counts. Table 9-1 gives some of the different genera of actinomycetes found in various environments.

ANTIBIOTICS

In 1940, before antibiotics were developed, five genera of actinomycetes were described; there are now over 80. Our increased knowledge has mainly occurred

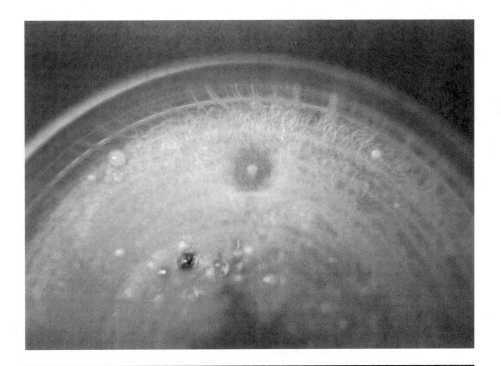

Figure 9-3   Antimicrobial activity of an actinomycete on a dilution spread plate. Note the inhibition of the surrounding *Bacillus* (the feathery-looking growth). (Photograph courtesy of M. S. Coyne)

because actinomycetes are the greatest known natural source of antibiotics, and antibiotics can be incredibly valuable compounds. Antibiotics are organic molecules that have antimicrobial properties (Figure 9-3). They are extremely diverse compounds with molecular weights that can vary from less than 100 to greater than 12,000. The building blocks of antibiotics come from intermediate metabolites. They form through multistep biosynthetic pathways. For example, up to 30 steps are involved in streptomycin biosynthesis.

More than 5,000 antibiotics have been identified and they continue to be identified at the rate of 300 per year. About 75% are isolated from actinomycetes. Table 9-2 shows the approximate number of antibiotics produced by different microbial groups. Fungi and prokaryotes such as actinomycetes make antibiotics, but far more antibiotics have been isolated from prokaryotic species.

About 75% of the antibiotics identified in actinomycetes have come from members of the genus *Streptomyces*. For example:

1. Streptomycin → *Streptomyces griseus*
2. Chloramphenicol → *Streptomyces venezuela*
3. Aureomycin, tetracycline → *Streptomyces aureofaciens*

Table 9-2 Relative number of antibiotics produced by different microbial groups.

Microbial Group	Number of Antibiotics Produced
Fungi	
Phycomycetes	14
Ascomycetes	299
Penicillium	123
Aspergillus	115
Basidiomycetes	140
Imperfect Fungi	315
Bacteria	
Pseudomonas species	171
Enterobacteria	36
Micrococci	16
Lactobacilli	28
Bacilli	338
Miscellaneous bacteria	274
Actinomycetes	
Mycobacterium species	4
Actinoplanes species	18
Streptomyces species	3,872
Micromonospora species	41
Thermoactinomyces species	17
Nocardia species	48
Other actinomycetes species	2,078

(From Crueger and Crueger 1982)

Most antibiotics are useless for human use because they are very toxic. They may cure the disease, but they also kill the patient.

Antibiotics are not constantly produced. Antibiotic synthesis begins late when growth is slowing or has stopped. So, antibiotic production can be repressed by a readily available C source, abundant N, or high levels of P, all of which contribute to keeping the actinomycetes actively growing. In other words, more nutrients means less antibiotic production. Antibiotic production may be activated by hormonelike compounds. There is also a potential link between differentiation (sporulation) and antibiotic production.

ANTIBIOTIC RESISTANCE

How do actinomycetes adapt to the antibiotics they produce? Basically, by two means—temporal separation of antibiotic production from growth, and development of resistance mechanisms (Piddock 1990). Temporal separation means that the actinomycete doesn't produce the antibiotic while it is actively growing. What are some

A Short Biography of Selman Waksman

Selman Waksman (1888–1973) was born in Novaia-Priluka, a small peasant village in the Ukraine. His family was Jewish, and in Czarist Russia, this meant Waksman had to fight a perpetual struggle against anti-Semitism to obtain a meaningful education. He was devoted to his mother, but when she died in 1909, Waksman foresaw dismal future prospects in Russia, and emigrated to Philadelphia in 1910.

His first employment was on a cousin's farm in New Jersey, close to Rutgers College in New Brunswick. The head of the Bacteriology Department, Jacob Lipman (another Russian immigrant), persuaded Waksman to study there, and in 1911 Waksman began college. His senior research project was to enumerate different microorganisms in soil. Shortly after Waksman graduated in 1915, Lipman appointed him research assistant at the New Jersey Agricultural Experiment Station, where he spent virtually his entire scientific career.

Much of Waksman's work was on the failure of foreign microorganisms, particularly pathogens, to survive in soil. Many scientists shared his feeling that indigenous microorganisms were destroying the pathogens. René Dubos, a Frenchman who worked closely with Waksman in the 1920s, 1930s, and 1940s, isolated several soil bacteria with antimicrobial activities.

By the 1940s, Waksman's testing of antimicrobial properties in microorganisms increasingly focused on the actinomycetes he had first observed as an undergraduate 25 years before. The actinomycete *Streptomyces griseus,* from which Waksman's group first extracted streptomycin in 1943, is an organism Waksman first isolated from soil in 1915.

By the time streptomycin was isolated, Waksman's laboratory had painstakingly examined approximately 100,000 different soil isolates for their antimicrobial properties. Commercial production of streptomycin to control tuberculosis began in 1946 and Waksman won the Nobel prize in 1952. The Nobel prize specifically noted Waksman's discovery of streptomycin, but in reality, it was recognition of a lifetime of work on the ecology of soil microorganisms and their influence on human activity and health.

resistance mechanisms? The target (whatever molecule the antibiotic affects) can be modified and rendered insensitive. Resistance can be made a part of biosynthesis—the antibiotic is not actually active until it is excreted. The antibiotic can be actively exported. The easiest mechanism of resistance is to become impermeable to the antibiotic and not take it up at all. Antibiotic-producing microorganisms are insensitive to the antibiotic they produce when they are growing slowly. Growing cultures are most sensitive to antibiotics—their own and others'. *Streptomyces* can produce actinomycin to concentrations of 120 mg mL^{-1} even though their own growth is inhibited by 4 mg mL^{-1} and stopped by 50 mg mL^{-1}.

One consequence of the ready and sometimes unwarranted use of antibiotics is that the incidence of antibiotic-resistant microorganisms is increasing worldwide.

Antibiotics against human pathogens are becoming ineffective in many cases. For example, cephalosporin antibiotics are the chemicals most widely used to treat the sexually transmitted disease gonorrhea, which is caused by the bacterium *Neisseria gonorrhoeae*. However, as many as 50% of the isolated *Neisseria gonorrhoeae* in some countries have developed resistance to these antibiotics (Knapp et al. 1997). Likewise, streptococci and staphylococci species that cause life-threatening infections are becoming increasingly resistant to antibiotics.

Summary

Actinomycetes are prokaryotes that look like fungi. There are pathogenic actinomycetes, but most are benign soil saprophytes. *Frankia* are beneficial actinomycetes because they fix nitrogen. Actinomycetes are the major source of diverse organic compounds with antimicrobial properties called antibiotics. Antibiotics are produced during secondary metabolism. Antibiotics are discovered at a rate of 300 per year and 75% of those discovered are produced by actinomycetes, with *Streptomyces* species producing most of the antibiotics. Most antibiotics are useless to humans because they are extremely toxic. Many antibiotics have become useless because the target microorganisms have developed resistance to them.

It is not clear why actinomycetes make antibiotics. They may give actinomycetes a competitive advantage in soil (Thomashow et al. 1990). Antibiotic production may force actinomycetes into stationary growth or signal them to begin sporulating. Antibiotics are rarely detected in soil and it's difficult to prove that they play a role there.

Sample Questions

1. How can you distinguish actinomycetes from fungi?
2. What are two environmental conditions favorable for actinomycete growth?
3. What carbohydrate is almost universally degraded by actinomycetes and where can it be found?
4. How do actinomycetes avoid being killed by their own antibiotics?
5. What would be the immediate effect on antibiotic production if a *Streptomyces* culture in stationary phase was amended with a readily available growth substrate?
6. Why might it be misleading if actinomycetes appear to make up 75% of the soil microbial population when the soil is sampled during a drought?
7. Actinomycetes suffer from some of the same problems in dilution plate counting that fungi do. Explain why.
8. In the graph below, provide a suitable explanation for the growth of the *Bacillus* that is being co-cultured with *Streptomyces griseus*.
9. Draw a graph that shows the relationship of actinomycete populations and soil depth.

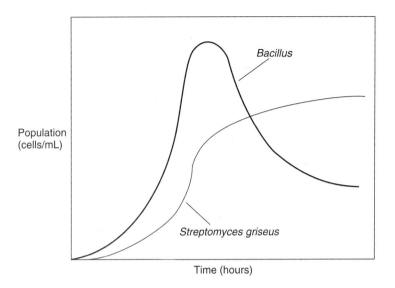

Time (hours)

10. Illustrate how pH can be used to control actinomycetes in soil.
11. What would be the biovolume and biomass of 1 mm of an actinomycete mycelium if it had an average diameter of 2 μm and there were 150 ng per μm³?
12. What is the approximate population of actinomycetes in a 10 g soil sample that has been diluted 100,000 times and yields 200 colonies when 1 mL of the last dilution is plated on media?

Thought Question

A long-standing practice in animal production has been to incorporate antibiotics into feed, since it seemed to improve weight gain. Why do you think that practice is now discouraged?

Additional Reading

Biotechnology: A Textbook of Industrial Microbiology by Wulf and Anneliese Crueger (1982, Sinauer Associates, Inc. Sunderland, MA) has a nice chapter on antibiotics that gives an industrial perspective of their manufacture and use.

References

Alexander, M. 1977. *Introduction to soil microbiology.* New York: John Wiley & Sons.

Baldry, P. E. 1965. *The battle against bacteria.* Cambridge, England: Cambridge University Press.

Knapp, J. S., K. K. Fox, D. L. Trees, and W. L. Whittington. 1997. Fluoroquinone resistance in *Neisseria gonorrhoeae. Emerging Infectious Diseases* 3:33–39.

Piddock, L. J. V. 1990. Techniques used for the determination of antimicrobial resistance and sensitivity in bacteria. *Journal of Applied Bacteriology* 68:307–18.

Thomashow, L. S., D. M. Weller, R. F. Bonsall, and L. S. Pierson. 1990. Production of the antibiotic phenazine-1-carboxylic acid by fluorescent *Pseudomonas* species in the rhizosphere of wheat. *Applied and Environmental Microbiology* 56:908–12.

Wellington, E. M. H., and I. K. Toth. 1994. Actinomycetes. In *Methods of soil analysis, part 2: Microbiological and biochemical properties,* R. W. Weaver et al. (eds.), 269–90. Madison, WI: Soil Science Society of America.

10

The Rest of the Prokaryotic World

Overview

After you have studied this chapter, you should be able to:

- Describe new taxonomic methods that have divided prokaryotes into two domains, bacteria and archaea, based on RNA sequences.
- Identify characteristics that describe those domains.
- Explain some basic microbiology concepts such as the basic forms and structure of prokaryotes.
- List—and appreciate—the advantages of being small.
- Define some ecological terms applying to prokaryotes.
- Identify a few of the dominant prokaryotes isolated from soil.

INTRODUCTION

Prokaryotes are the most ancient and successful form of life—if we count success as persistence and their spread around the Earth. Fossils show that bacteria are at least 2–3 billion years old. They have almost universal distribution. Bacteria were present at the dawn of life on Earth and bacteria will almost certainly be around at the dusk of life on Earth.

What bacteria do and how they do it are the principal emphasis of this text. Bacteria play crucial roles in the environment. Some scientists have suggested that bacteria are responsible for creating and regulating our environment (Lovelock 1991). Bacteria are metabolically diverse. Bacteria decompose organic and inorganic material—both natural and synthetic. Bacteria are involved in the formation, deposition, solubilization, and removal of chalk, elemental sulfur, and metallic sulfide ore, to name but a few compounds. Bacteria fix nitrogen. Bacteria are both predator and prey; pathogen and parasite.

TAXONOMY

Where do bacteria fit in the scheme of life? The classical grouping of kingdoms in nature follows this pattern (Last 1988):

Monera = bacteria and prokaryotic algae
Protista = protozoans, algae, and slime molds
Fungi = lichens, molds, and yeasts
Plantae = higher plants
Animalia = higher animals

New methods of bacterial taxonomy classify bacteria using molecular biology techniques (Woese 1987). They are based on comparing RNA (ribonucleic acid) sequences. RNA is so critical to cell function that mutations in it are frequently lethal. In theory, as time passes and species evolve, RNA sequences will reflect divergence between different species at a genetic level.

When RNA sequence analysis was first used, it became clear that there are two distinct classes of prokaryotic microorganisms that are as different from one another as they are from eukaryotes. One theory is that eukaryotes probably evolved as the assimilation of prokaryotes that later became cell organs: aerobic bacteria became mitochondria, spirilla became flagella, and photosynthetic bacteria became chloroplasts.

One prokaryotic domain, Bacteria, predominate in soils (Table 10-1). The other domain, Archaea, are frequently observed in extreme environments. Extreme halophiles, for example, grow in high-salinity environments such as salt mines. Methanogens, which convert carbon dioxide to methane, grow in sediments and sludges, insect guts, human and animal bowels, and the animal rumen—anywhere that's oxygen-free. Extreme thermophiles like *Sulfolobus* and *Thermoproteales* grow in acidic, hot waters and soils. However, as microbiologists look closer, they note that Archaea aren't limited to extreme environments.

Table 10-1 The basic groups of prokaryotes.

Prokaryotes		Eukaryotes
Bacteria	**Archaea**	
Thermotoga	Methanogens	Animals
Flavobacteria	Extreme halophiles	Plants
Cyanobacteria	Extreme thermophiles	Ciliates
Purple bacteria	Acidophiles	Flagellates
Gram-negative bacteria		Fungi
Green nonsulfur bacteria		Microsporidia
Gram-positive bacteria		
Photosynthetic bacteria		

(From Woese 1987; Pace 1996)

Bacteria and Archaea are morphologically indistinguishable. For our purposes, to describe Bacteria and Archaea in the soil, we refer to members of both prokaryotic domains as bacteria, the name they shared before phylogenetic techniques of analysis.

MORPHOLOGY

Bacteria have many shapes. The basic forms are spherical (cocci), rod shaped (bacilli), and spiral shaped. Vibrios form short spirals and look like commas. Spirilla form longer, fixed spirals. Spirochaetes form very long, unfixed spirals. Bacteria may also be pleomorphic, meaning they have no defined shape. At least some halophylic bacteria have square shapes. The individual bacterial cells may be associated in chains, sheets, packets, clusters, palisades, sheaths, and budding filaments. You are much less likely to find these associations in soil than in pure culture. Figure 10-1 illustrates some of these shapes.

Bacterial cells remain associated for mechanical and chemical reasons. They may be encrusted in an enclosing sheath or bound together by slime. They may associate because of hydrogen bonding, ionic attraction, or hydrophobic interactions. Whichever way they are associated, bacteria survive better in a crowd than singly.

WHY ARE BACTERIA SMALL?

Bacteria survive, in part, because they are small; exceptions to this rule exist, but aren't significant in soil. The typical bacterium (if one exists) is 0.15 to 4.0 μm in diameter, 0.2 to 50 μm in length, and has a volume of 0.1 to 5.0 μm³ (1 mL, or 1 cm³, contains 10^{12} μm³). Bacterial size varies widely depending on whether the cells are growing;

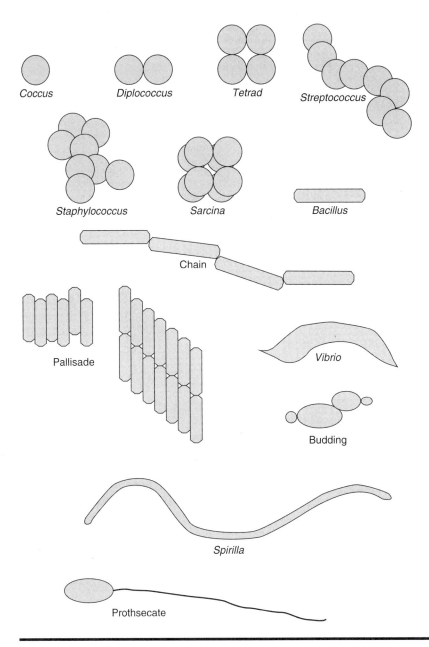

Figure 10-1   Typical shapes of bacterial cells.

nongrowing cells tend to be small in comparison with growing cells (Figure 10-2). Being small in the soil environment helps bacteria escape into pores too narrow for predators.

Metabolism is inversely proportional to size. Bacteria, which are small, have a high metabolism and can generally outgrow their predators. Bacteria also have a high sur-

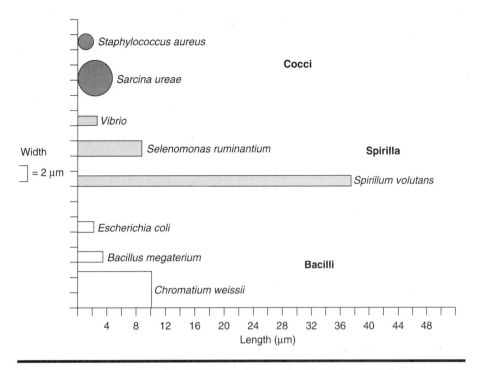

Figure 10-2   A comparison of bacterial sizes. (Adapted from Doetsch and Cook 1973)

Table 10-2 Effect of changes in bacterial diameter on the surface area and volume of a cocci.

Diameter (μm)	Surface Area (μm²)	Volume (μm³)	Surface Area /Volume Ratio
1	3.1	0.5	6.2
2	12.6	4.2	3.0
4	50.3	33.5	1.5
8	201.1	268.1	0.8
16	804.2	2,144.0	0.4

Note: The surface area of a sphere is calculated as $4\pi r^2$, while the volume of a sphere is calculated as $4/3\ \pi r^3$.

face area/volume ratio. For example, 10^{12} bacterial cells, which is a population that can easily be contained in 1 mL of media, may have a surface area of 3 m². Nutrient uptake is optimized by a high surface area/volume ratio. If the cell diameter increases, the volume of the bacteria increases faster than its surface area and this high ratio is diminished (Table 10-2). Unfortunately, diffusion is a relatively slow process. If the bacteria are too big, nutrients may not diffuse through the cell fast enough to be available where they're needed.

The maximum size of bacteria is partly restricted by diffusion limitations and the need to replicate genetic information. There are also restrictions governing how small bacteria can be. A minimum number of enzymes (1,000) has to be packed into a tiny space. As we'll see, viruses can be much smaller than bacteria, but by being so small, they lose many of the metabolic capabilities that bacteria possess, particularly that of independent reproduction.

CELL STRUCTURE

An idealized view of bacterial structure is shown in Figure 10-3. Bacteria can be thought of as bags of enzymes. Most of these enzymes are in the cytoplasm of the bacterial cell, which is a viscous material crammed with proteins, RNA, DNA, and soluble inorganic compounds. Enclosing the cytoplasm is the cell membrane, a differentially permeable lipid bilayer. Differential permeability is the ability to take in some things from the environment while excluding others. An analogy is a screen door—it is differentially permeable (based on size), because it lets in air and keeps out insects. Outside the cell membrane is the periplasmic space, which contains more inorganic solutes and proteins. Outside the periplasmic space is the cell wall. The cell wall gives bacteria rigidity and helps them resist the pressure of water entering the cell. Think of the bacteria as a balloon in a box. If you blew up the balloon outside the box, you could over pressurize and break it. If you blew up the balloon inside the box, it would be much more difficult to burst because the box would resist the balloon's expansion. Cell walls in the domain Bacteria are made of peptidoglycan, a polymer of alternating sugar subunits N-acetylglucosamine (NAG) and N-acetylmuramic acid (NAM), that are crosslinked by peptides. This creates an organic fabric of amazing strength—the bacterial equivalent of chain mail. Another name for peptidoglycan is murein. Cell walls in Archaea are composed of a variety of polymers and do not contain any peptidoglycan.

Gram staining characterizes bacteria on the basis of cell wall thickness because the cell wall's thickness affects whether it traps different stains (Figure 10-4). Gram-positive bacteria have a cell wall 40 times thicker than the cell walls of gram-negative bacteria (80 nm thick vs. 2 nm thick), and in the staining process appear to be blue. Gram-negative bacteria appear to be red because the thin cell walls can't retain the blue stain.

Some bacteria have an outer membrane. It gives antigenic properties to the bacterial cell (it causes animals to raise antibodies against it) and it can also serve as a bacterial toxin (the outer membrane components of coliform bacteria are what cause food poisoning). The outer membrane, like the cell membrane, is differentially permeable. The next structure, which also is not found in all bacteria, is the capsule. The capsule is a polymeric carbohydrate slime layer. It is synthesized by the bacteria and helps the cells prevent virus infection, trap nutrients, resist desiccation, and adhere to solid surfaces.

Some bacteria have additional structures that are permanent or temporary features. Pili, for example, are hollow protein tubes that assist pathogens in binding to their host cells and in the transfer of genetic information. Flagella, which are also composed of protein, are thinner and less rigid than pili. A cell may have from 1 to 100 flagella located at the end of the cell (polar location), close to the end of the cell (subpolar

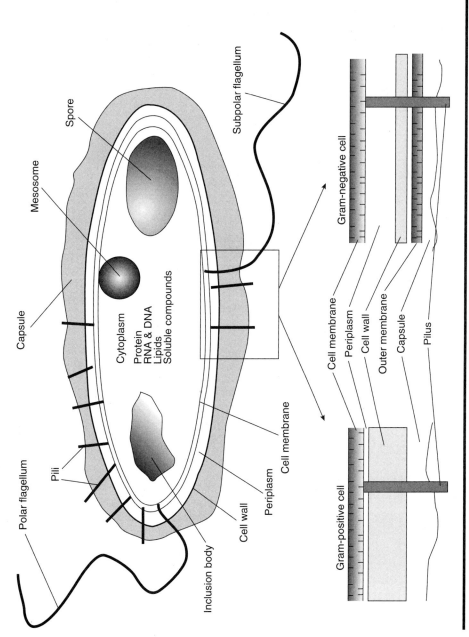

Figure 10-3   An idealized diagram of bacterial cell structure (Adapted from Ingraham et al. 1983).

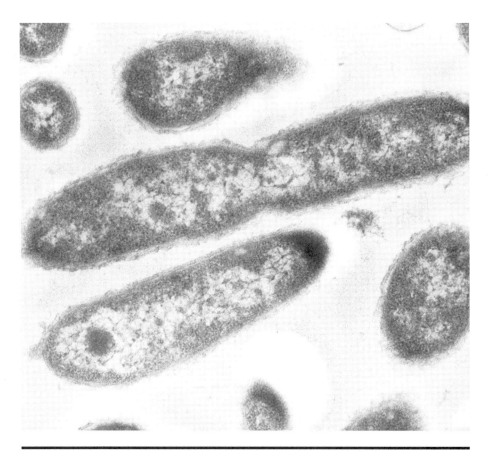

Figure 10-4   Transverse and longitudinal views of bacteria in an electron micrograph (magnification 48,450×). Note the density of material within the cell and the bacteria caught in the middle of cell division. (Photograph courtesy of M. S. Coyne)

location), or peritrichously (all over the cell) (Figure 10-5). You must use special stains to see flagella under a microscope. The flagella take the form of a helix and spin like a corkscrew. This allows the bacteria to be mobile (or motile). A 2 μm-bacteria can move 50 μm per second. This is the equivalent of a 6-ft-tall (2 meter) person swimming 100 mph (156 kph). Nearly all rods and spirilla, but few cocci, have flagella.

Bacteria use flagella to move in response to signals from their environment. Movement is known as taxis. In positive taxis, the bacteria move toward something. In negative taxis, the bacteria move away from something. Bacteria exhibit taxis in response to many different stimuli. Chemotaxis is response to a chemical gradient. Phototaxis is response to a light gradient. Aerotaxis is response to an oxygen gradient. Magnetotaxis is response to a magnetic field.

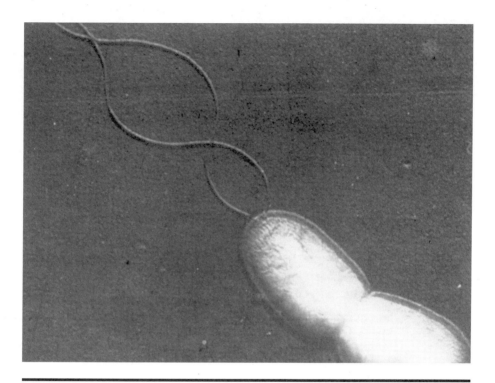

Figure 10-5   Flagellated *Bacillus*. Note polar location of flagella. (Photograph courtesy of the Soil Science Society of America)

ECOLOGY AND DISTRIBUTION OF BACTERIA

Bacteria can live where most other organisms cannot because of their metabolic diversity. Bacteria are usually more numerous in soil than all other organisms combined, with the exception of viruses. There are over 200 identified bacterial genera and a single soil sample may have over 4,000 genetically distinct bacteria (Torsvik et al. 1990), but it is estimated that less than 1% of bacterial species are culturable, so the actual diversity of soil bacteria is undoubtedly much greater.

Indigenous bacteria (also called "autochthonous" bacteria) can be considered to be the true, permanent residents of a given soil. Nonresidents, or foreign bacteria (also called "allochthonous" bacteria), can be considered as invaders or transients. They enter soil by precipitation, diseased tissue, manures, or sludges. They may persist and grow, but they rarely contribute to significant biological activity.

Other characterizations of bacteria have been used. Autochthonous bacteria are regarded as the slow-growing bacteria in soil. "Zymogenous" bacteria, which have bursts of activity depending on whether food is present, can be considered the fast-growing bacteria in soil. An "r-selected" bacteria is the equivalent of a zymogenous

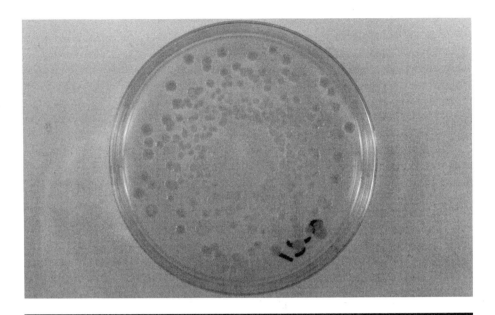

Figure 10-6   A pure culture of bacteria plated onto growth media. (Photograph courtesy of M. S. Coyne)

bacteria. The r-selected bacteria are adapted to conditions of bountiful energy that one might find in the rhizosphere (they go through cycles of feast and famine). In contrast, "k-selected" bacteria (equivalent to autochthonous bacteria) are adapted to uncrowded but physically and nutritionally restrictive environments like those found in bulk soil. The k-selected bacteria put a premium on high growth rate per unit of food (they deliver more bang for the buck).

Counting bacteria in soil is difficult because no single culture media is adequate for all groups. About 10^8 to 10^{10} bacteria per gram of soil can be seen by direct observation. Growing bacteria on solid media usually gives 1% to 10% of the direct counts (Figure 10-6). Part of the difficulty in counting bacteria occurs because minor deviations in the environment may have drastic influences on bacterial populations at points only centimeters apart.

There are typically more bacteria in pastures than in cropland; more bacteria in cultivated soil than in fallow soil; more bacteria in warm soil than in cool soil, and more bacteria in moist soil than in dry soil. The average bacterial cell may weigh only 1×10^{-12} g (0.001 ng). Although numerous, bacteria represent less than 10% of the biomass in soil, approximately 300 to 3,000 kg per hectare depending on the environment.

IMPORTANT SOIL BACTERIA

Arthrobacter, Bacillus, and *Pseudomonas* are three of the most commonly isolated soil bacteria. *Arthrobacter* is classified among the actinomycetes. It is pleomorphic, slow growing, slightly motile, and grows on diverse substrates, which is one reason why it is frequently isolated from soil. *Bacillus* species represent 7% to 67% of the soil isolates. They also have diverse metabolism. *Bacillus* species, like actinomycetes, are spore formers. *Bacillus* survive a pH range from 2 to 8 and a temperature range from –5° to 75°C. Some *Bacillus* species can fix nitrogen. *Pseudomonas* species represent 3% to 15% of soil isolates. They are noted for their diverse metabolism, particularly their ability to degrade organic chemicals like pesticides. Some *Pseudomonas* species are pathogenic.

> "It may be assumed that there is a reason for every activity because nature is too parsimonious to allow for random or purposeless functions."

> Doetsch and Cook 1973

It is not always clear why bacteria have the capacity to carry out the metabolic activity they perform. Unfortunately, microbiologists (and other scientists) often fall into a trap called teleological reasoning—giving human motives to biological processes. The reason why bacteria may carry out some activity may have little to do with what we perceive that activity to represent.

Summary

Prokaryotes are the most ancient and successful form of life; they are nearly 3 billion years old. Taxonomy and classification are constantly changing and newer classification methods rely on RNA sequencing to group prokaryotes. These methods show that there are two distinct prokaryotic groups: Archaea and Bacteria. In addition to differences in RNA, these prokaryotes have different cell walls, cell membranes, and habitats. Bacteria are most frequently associated with soil environments while Archaea are most frequently associated with extreme environments.

Prokaryotes (bacteria) are very small, which enables them to absorb nutrients from their environment quickly. Zymogenous bacteria grow rapidly in response to nutrients in their habitat; autochthonous bacteria are slower growing. Archaea and Bacteria are morphologically indistinct. Cells appear as spheres (cocci), rods (bacilli), and spirals (spirilla). All bacteria have some common structural elements such as a cell membrane. Other features such as the cell wall, flagella, and capsule may differ or be absent entirely. Flagella allow bacteria to move in response to environmental signals. The most common bacteria isolated from soil are *Arthrobacter, Bacillus,* and *Pseudomonas.* However, there is enormous diversity of bacterial form and function in the environment and less than 10% of the bacteria in an environment can actually be cultured.

Sample Questions

1. What are the principal differences between Bacteria and Archaea?
2. Draw a diagram of a typical bacteria showing its major structures.

3. What are the advantages of being small?

4. One analogy that could be used to describe r-selected bacteria is "feast or famine." Why do you think this analogy is used?

5. What's the difference between an "autochthonous" and an "allochthonous" bacteria, and where would you expect to find them?

6. Why do plate counts and direct counts of bacteria differ so greatly?

7. What characteristics of bacteria set them apart from other living organisms?

8. Calculate the volume of a cocci 3 μm in diameter and a bacilli 4 μm in length and 2 μm in diameter.

9. If you had a solution of *Pseudomonas fluorescens* (a rod-shaped bacteria) that had a concentration of 10^9 cells per mL, what would be the total surface area of the bacteria in 10 mL of culture if each cell had a diameter of 1 μm and a length of 5 μm? (Assume that the best approximation of the shape of this bacterium is a cylinder.)

10. What is the total bacterial population in 100 g of soil if 10 g of soil diluted to 10^6 times gives 150 colony-forming units on an agar plate when 1 mL is plated?

11. Assuming that each individual cell weighs 3×10^{-12} g, what is the total bacterial biomass in your sample?

12. Complete the following table that shows the change in surface area and volume of a cocci as its diameter changes. The first one is done for you.

Diameter (μm)	Surface Area (μm²) $4\pi r^2$	Volume (μm³) $4/3\pi r^3$	Surface Area /Volume
1	3.1	0.5	6.2
3			
5			
7			
9			

Thought Question

If you had to choose what kind of prokaryote you wanted to be, what would you pick and why?

Additional Reading

Two books by John Postgate that you have to read if you have any interest in bacteria are *Microbes and Man* (1992), which examines the way humans and microorganisms have adapted to one another, and *The Outer Reaches of Life* (1994), which explores some of the extreme environments to which bacteria have adapted. Both books are literate, witty, and a pleasure to read. Along the same lines, for those with a more clinical outlook, is a book by Wayne Biddle called *A Field Guide To Germs* (1995, Anchor Books, New York).

References

Doetsch, R. N. and T. M. Cook. 1973. *Introduction to bacteria and their ecobiology.* Baltimore, MD: University Park Press.

Ingraham, J. L., O. Maaløe, and F. C. Neidhardt. 1983. *Growth of the bacterial cell.* Sunderland, MA: Sinauer Associates Inc.

Last, G. A. 1988. Musings on bacterial systematics: How many kingdoms of life. *ASM News* 54(1):22–27.

Lovelock, J. E. 1991. *Gaia: A new look at life on earth.* Oxford, England: Oxford University Press.

Pace, N. R. 1996. New perspectives on the natural microbial world: Molecular microbial ecology. *ASM News* 62:463–70.

Torsvik, V., J. Goksoyr, and F. L. Daae. 1990. High diversity in DNA of soil bacteria. *Applied and Environmental Microbiology* 56:782–87.

Woese, C. R. 1987. Bacterial evolution. *Microbiology Reviews* 51:221–71.

Chapter 11

Mycoplasmas, Viruses, Viroids, and Prions—The Rest of the Microbiota

Overview

After you have studied this chapter, you should be able to:

- Describe viruses, the smallest microorganisms in soil.
- Discuss what distinguishes mycoplasmas from other bacteria.
- Give a definition for viruses and describe some basic viral shapes.
- Explain the life cycle of viruses in soil.
- Describe two entities that are at the fringes of life and have important pathogenic consequences—viroids and prions.

INTRODUCTION

As microorganisms in nature get smaller, there are fewer environments in which they can live or processes they can perform, simply because they don't have enough genetic information. Microbiologists have argued that genetically simple microorganisms represent steps down the evolutionary ladder in terms of sophistication. They have also argued that genetically simple microorganisms represent steps up the ladder to the ultimate in parasitism. Because mycoplasmas, viruses, viroids, and prions are so small, they have very little effect on biological processes in soil. However, they significantly affect the organisms with which they are associated, including those organisms we've already discussed (humans included), so no soil microbiology text would be complete without mentioning them.

MYCOPLASMAS

Mycoplasmas are bacteria (in the Domain Bacteria) that lack a cell wall. They are pleomorphic, meaning they do not have a defined shape (i.e., they're the boneless chickens of the bacterial world). Mycoplasmas are the smallest independently replicating prokaryotes. How small are mycoplasmas? Very small—about 0.1 to 0.3 μm in diameter. That means that mycoplasmas are small enough to pass through membranes that trap bacteria, and this plays havoc with supposedly filter sterile solutions. Mycoplasmas also have unusual trilayer cell membranes that contain sterol instead of lipid. Mycoplasmas make no spores and have no flagella.

You can find mycoplasmas in soil, sewage, insects, humans, and animals. Mycoplasmas have no known role in the major soil processes, which is why we discuss them in such little detail. Mycoplasmas have no cell wall, so they are osmotically sensitive. This means that if the water in their environment becomes too saline or too dilute, it can kill them. Mycoplasmas are also generally heat sensitive. The optimum pH for most mycoplasma growth is 7.0, although the pH range is from 5.5 to 10.0. *Thermoplasma,* isolated from smoldering coal mine tailings, is one exception and is both thermophylic and acidophylic. There are aerobic and anaerobic mycoplasmas. Mycoplasmas are also responsible for some plant diseases.

VIRUSES

There are obvious problems trying to attribute biological phenomenon to agents you cannot see. Viruses suffer from this problem. The first hints that living entities much smaller than bacteria existed came from the work of Ivanovski in Russia (1892) and Beijerinck in Delft (1898) (van Helvoort 1991). Both scientists independently observed that something in tobacco suffering from tobacco mosaic disease could be passed though a filter that excluded bacteria, and still cause tobacco mosaic disease in healthy tobacco leaves. If you recall, this satisfies one of the critical elements of Koch's postulates. It wasn't until the 1920s and 1930s that the largest viruses were made visible with improved microscopy techniques.

Once techniques were developed to examine viruses apart from the host, and filterability was abandoned as a criterion to define viruses (since they range in size from

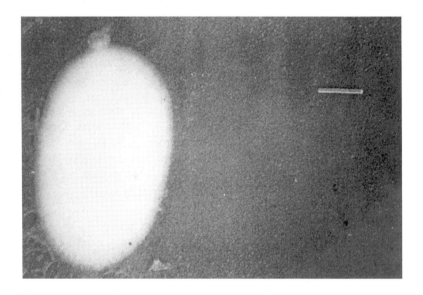

Figure 11-1   The size relationship of virus to bacteria. (Photograph courtesy of the Soil Science Society of America)

slightly larger than large proteins to slightly smaller than the smallest bacterium), their position as unique microorganisms became more firmly established (van Helvoort 1996). By 1957, Andre Lwoff at the Pasteur Institute had devised a good working definition of viruses that holds to this day. Viruses are "infectious agents made up of nucleic acids and proteins but unable to grow independently or reproduce by binary fission" (van Helvoort 1996).

Viruses are the smallest microorganisms, much smaller than bacteria, as Figure 11-1 demonstrates. Most viruses are only 20 to 30 nm or smaller in size. We're talking about really, really small organisms here, considering that there are 1,000 nm in a μm and 1,000 μm in a mm, and a mm is about as wide as the space between your thumb and index finger when you try to put them as close to touching as possible. Consequently, it takes an electron microscope to see viruses well.

Viruses are so small that only 10 to 200 genes are coded by the genetic information they contain, which is not enough information to carry out metabolic activity. As far as we know, therefore, viruses are inert (seemingly lifeless) and do not carry out biosynthetic functions like growth and reproduction apart from their hosts. Some scientists consider viruses to be nonliving, but within a favorable environment viruses behave as living cells. In other words, viruses are obligate intracellular parasites. However, viruses can survive outside their host and remain infectious.

PHYSIOLOGY AND MORPHOLOGY

Viruses are acellular organisms, which means they lack a cell membrane. They contain either RNA or DNA (which can be double or single stranded). Viruses containing RNA (such as the AIDS virus HIV) specify an enzyme called "reverse transcriptase" to transcribe DNA from the RNA. There are two structural components in viruses: a protein coat (the capsid) and a nucleic acid core (the virion). A few viruses also have a lipid envelope.

Like other microorganisms, viruses come in many shapes. Some are helical, such as TMV (tobacco mosaic virus) and most other plant viruses. Some are cubical, based on an icosahedral pattern (a 20-sided shape), such as the herpes virus. Some are binal, which means they have a cubical head with a helical tail. Bacteriophages such as T phage (phage is another term for virus) are a good example. Figure 11-2 shows some typical shapes of viruses that infect bacteria.

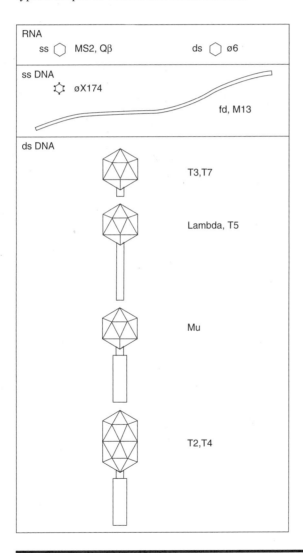

Figure 11-2   Schematic representation of some of the main types of viruses infecting bacteria. Sizes are to approximate scale. (Adapted from Brock and Madigan 1991)

How to Make Your Own Icosahedron

Nothing beats modeling to get a sense of three-dimensional items. You don't often encounter icosahedrons outside the microbial world, so it's probably hard to visualize what they actually look like. To help out, make a copy of Figure 11-3 and fit it together for yourself. When you're done, you'll have an icosahedron.

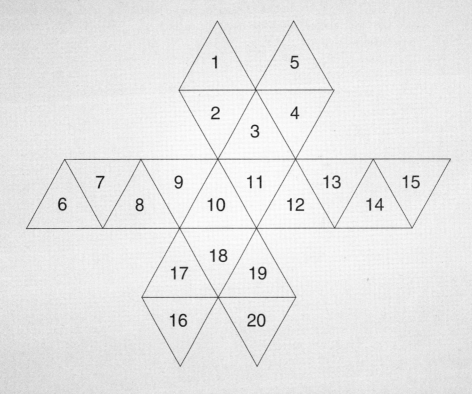

Figure 11-3   Form for constructing your own icosahedron. Fold at every line and tape the adjoining faces together at the tabs. (Adapted from Brock and Madigan 1991)

There are plant, animal, bacterial, fungal, algal, and protozoal viruses. Hardly a species of bacteria has been studied that does not have a bacteriophage that infects it. Viruses are extremely host specific, so people aren't infected by plant and bacterial viruses, and vice versa. Specific relationships (recognition) are determined by biochemical structures on cell surfaces such as polysaccharides, proteins, flagella, and lipoproteins. For example, the maltose receptor protein in the cell membrane of *Escherichia coli* (*E. coli,* for short) is the entryway for the lambda phage that infects it.

Lambda phage is a bacteriophage (a bacterial virus) with binal structure (Figure 11-4). Bacteriophages with binal structure have a head and a tail and are usually 0.5 to 1.0 nm in diameter.

In the mid-1930s there were reports of "fatigued" alfalfa fields in France (cited in Angle 1994). The poor growth in long-term stands was attributed to a virus infection of the bacteria in soil (rhizobia) that allowed the alfalfa to fix nitrogen symbiotically. Bacteriophages for any given bacterial species are not numerous, but the bacteriophage can be enhanced by enhancing bacterial growth. A good illustration of this phenomenon is in Table 11-1.

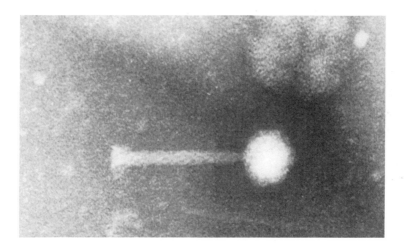

Figure 11-4 Binal bacteriophage. (Photograph courtesy of the Soil Science Society of America)

Table 11-1 Presence in various soils of the bacteriophage-infecting *Rhizobium meliloti,* the rhizobia species infecting alfalfa.

Soil	pH	Age of Stand (years)	Presence of Bacteriophage
Forest soil	4.9	NS	−
Forest soil	5.0	NS	−
Pasture	5.3	NS	−
Alfalfa field	5.3	6	+
Alfalfa field	5.9	3	+
Alfalfa field	6.1	8	+
Wheat field	6.1	NS	+
Pasture	6.2	NS	−
Alfalfa field	7.4	1	+

NS = not stated (From Katznelson and Wilson 1941)

What does the data tell us? Forest soils don't have any bacteriophage-infecting rhizobia. But the data are arranged in order of increasing pH, so maybe pH affects bacteriophage presence. However, pasture soils don't have any bacteriophage-infecting rhizobia either, even though one of the pasture soils has a soil pH lower than the alfalfa field with bacteriophage specific for rhizobia. So it seems that the presence and absence of bacteriophage is directly linked to alfalfa fields. Alfalfa is host to the root-infecting bacterium *Rhizobium meliloti,* which, in turn, is the host of the bacteriophage. The bacteriophages are present whether the alfalfa stand is young (1 year) or old (8 years); alkaline (pH 7.4) or acidic (pH 5.3).

Why are there bacteriophages specific for rhizobia in a wheat field? In the absence of their host, virus populations decline. We don't have any specifics about this site; it's possible that the wheat was planted in a converted alfalfa field. Even though their host is declining, bacteriophages persisted in soil long enough to be detected later.

VIRUS LIFE CYCLES

Viruses have two life cycles—the virulent/lytic cycle and the temperate/lysogenic cycle. Both cycles are illustrated in Figure 11-5. The lytic cycle (lytic meaning to lyse or dissolve) can be broken down into the following stages:

Virulent/lytic cycle
1. Infection
2. Synthesis of viral DNA/RNA
3. Synthesis of viral proteins for coat and lysozymes
4. Capsids (heads) filled
5. Capsids closed; tails added
6. Phage liberated
7. More infection; cycle continues

The lysogenic cycle differs in that the virus does not infect and immediately start replicating within its host. Rather, like HIV or hepatitis C virus, it remains latent until it begins a replication cycle at a later date. The lysogenic cycle can be broken down into these stages:

Temperate/lysogenic cycle
1. Infection
2. Incorporation into DNA
3. Division in synchrony with cells (at which time the virus is often called a prophage)
4. Immunity to further infection by similar phage (synthesis of a cytoplasmic repressor protein)
5. Triggering of lytic cycle by environmental shock
6. Excision of DNA
7. Initiation of lytic cycle

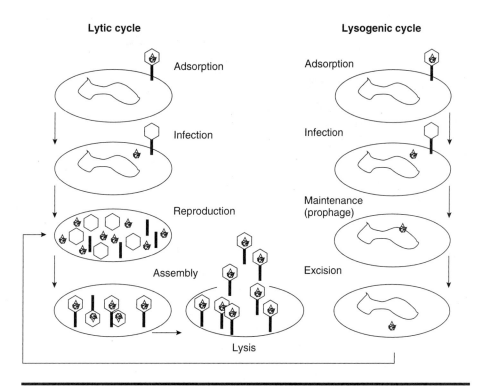

Figure 11-5   Schematic illustration of the lytic and lysogenic viral life cycles.

VIRUS ECOLOGY

Little is known about the field ecology of viruses that infect soil microorganisms, except that they persist in soil as dormant particles that retain parasitic ability. Plant viruses rarely survive in soil, while some insect viruses remain infective for years. Many soilborne viruses require transmission by nematodes or fungi. Viruses are an environmental concern when human and animal viruses survive in soil and move into the water table (Powelson et al. 1990). As a rule, viruses are readily removed by passage through soil. They are adsorbed by organic residues, humus, and clays (Gerba 1984). Unfortunately, adsorbed viruses can desorb (Bales et al. 1997). Virus persistence is also increased by moist conditions.

VIROIDS

What are viroids? Viroids are partially folded, infectious RNA of low molecular weight. They may just be the simplest form of life. At least 12 plant diseases have been linked to viroids. Very little else is known about them, particularly whether they play any role in soil, either as an active participant in soil biology or as a menace waiting for suitable human prey. So, we won't talk about them anymore.

How Do You Study Viruses in Soil?

It is difficult to study soil viruses. The number of viruses infecting soil microorganisms is low, either because they are in a lysogenic state or because the methods for counting them are poor (Angle 1994). Remember, if the wrong host is present on which a virus may grow, nothing will happen.

The initial steps in the procedure to study viruses in soil are to dilute a soil sample by a factor of at least 1 billion because there may be that many viruses per gram of soil. Next, each soil dilution is amended so that it contains 0.5% chloroform. This kills any microorganisms other than viruses, and liberates any viruses from cells that are in the midst of a lytic cycle but are not yet lysed. Chloroform is toxic, so an alternative is to filter the solutions through a 0.2 µm filter. Unfortunately, filtration tends to cause a lot of nonspecific virus adsorption to the filter, so it lowers the countable numbers.

The liberated viruses in each dilution bottle are assayed using the soft agar technique. A base of agar media is prepared. Then a thin

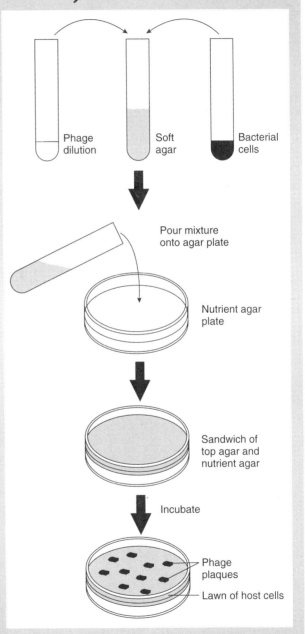

Phage dilution

Soft agar

Bacterial cells

Pour mixture onto agar plate

Nutrient agar plate

Sandwich of top agar and nutrient agar

Incubate

Phage plaques

Lawn of host cells

Figure 11-6a

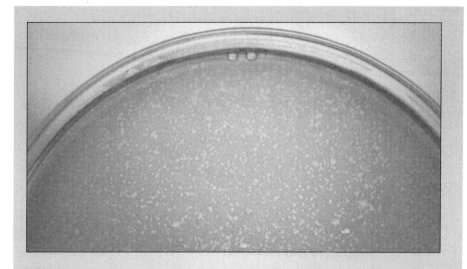

Figure 11-6b The soft agar technique for enumerating bacteriophages and a picture of plaque-forming units. (Adapted from Brock and Madigan 1991; photograph courtesy of M. S. Coyne)

layer of soft agar that contains bacteria and the viruses from your dilution samples (which can infect the bacteria) is poured over the surface. For example, *Escherichia coli* and coliphage (the viruses that infect *E. coli*) are mixed together in soft agar and spread on the surface of the agar plate.

As the bacteria grow, they become infected by the viruses from the diluted soil samples, and each infected cell becomes the center for infection of all the cells around it. So, instead of looking at a nice, uniform lawn of bacterial cells in the soft agar, when looking for phage, you observe a lawn spotted with cleared zones where all the bacteria have been infected and killed (Figure 11-6a, b).

Each cleared zone is called a "plaque" and the number of viruses per plate are called the plaque-forming units (PFU). Back-calculating from the volume of sample added to each plate, the dilution from which the sample came, and the amount of soil originally diluted, you can estimate the viruses in your soil sample that were capable of infecting the bacterial host you used.

PRIONS

The limbs that erstwhile charmed your sight
  Are now a savage's delight;
The ear that heard your whispered vow
  Is one of many entrées now;
Broiled are the arms in which you clung,
  And deviled is the angelic tongue.

O love, O loveliest and best,
  Natives this body may digest;
Whole, and still yours, my soul shall dwell,
  Uneaten, safe, incoctible.

"My Darling, I Was Captured By Cannibals, Who Poached the Eyes
You Loved So Well" (Rupert Brooke, 1887–1915)

The ill-fated poet in Rupert Brooke's work may yet have had the last laugh on the cannibal hosts who invited him to dinner. In recent years, another strange self-replicating entity has been discovered that leaves us wondering what the minimum definition for life is. Prions are self-replicating units of protein. They, like viroids, are associated with disease. Cannibals in New Guinea, for example, suffer from a strange disease known as "Kuru," which is a progressive degeneration of the central nervous system. Similar diseases are "scrapie" in sheep, and Creutzfeldt-Jakob disease in humans (Gibbs and Asher 1996).

Prions cause a disease known as spongiform encephalopathy, a condition in which the infected individual's brain develops large, open gaps in the tissue (like a sponge) and ceases to function properly. Prions raises no host defense mechanisms. Prions don't cause an immune response because the fibrils aren't foreign protein—they come from the host. The disease is always fatal; it has no cure. Prions are in the news because they are believed to cause "mad cow" disease or BSE (bovine spongiform encephalopathy). It is thought that the cattle were infected because they consumed feed supplements that included waste parts of sheep that were infected with scrapie.

If prions are transmitted only via infected tissue that is either consumed or transplanted, there would probably be little concern with their role in soil microbiology. However, they do affect how food is processed and what types of supplements that regulatory agencies will now allow into animal feed. If you're a soil microbiologist, prions raise at least two interesting questions: 1) How long do prions persist in soil before they decompose? and 2) Do macrofauna consuming diseased animal tissue transmit prions up the food chain?

Summary

The four smallest microorganisms in soil are mycoplasmas, viruses, viroids, and prions. Mycoplasmas are bacteria without cell walls. They play a role as human and plant pathogens. Viruses are acellular. They contain nucleic acids surrounded by a protein covering. All organisms are infected by viruses, which are obligate intracellular parasites that cannot reproduce themselves apart from a host.

Viruses come in many shapes. Those that infect bacteria are called bacteriophages and typically have an icosahedral head with a helical tail. Viruses can persist in soil without a host to infect, but they will eventually decline. When viruses infect a host they have two life cycles—lytic or lysogenic. In the lytic cycle, the virus immediately begins to replicate itself using the host's biochemical machinery and kills the host when it is liberated. In the lysogenic cycle, the virus DNA incorporates into the host DNA for an unspecified period until an environmental signal causes lytic stages of replication to occur. Lysogenic bacteria can transfer pieces of host DNA in a process called transduction.

Viroids are small pieces of self-replicating RNA. Prions are small self-replicating pieces of protein. Neither may be living things depending on your definition of life. Neither plays an important role in soil processes. Both, however, are pathogenic and prions cause several incurable diseases in mammals.

Sample Questions

1. Is it possible for viruses to infect other viruses?
2. Why are viruses obligate parasites?
3. What are the unique characteristics of mycoplasmas?
4. What are the basic structural characteristics of a virus?
5. Draw a diagram that shows the change in a lytic virus's population if it is cultured on a host bacterium.
6. Are viruses highly advanced microorganisms or extremely primitive microorganisms? Explain your reasoning.
7. Based on the criteria we previously established for living things, are prions and viroids alive? Explain your reasoning.
8. How do you explain the delay between the first observance of viruses in culture? Why does virus replication ever slow down?
9. How many viruses that are 50 nm in diameter will it take to span the head of a pin (about 1 mm)?
10. How many icosahedral viruses that are 50 nm in diameter will it take to fill up the entire volume of a *Bacillus* 1 μm in diameter and 5 μm long? For the sake of simplicity, assume that the icosahedron is approximately circular and that the *Bacillus* is approximately shaped like a cylinder.
11. Assuming that the *Bacillus* in question 10 weighs approximately 1.5×10^{-13} g to begin with, and that an equal weight of virus particles are synthesized, what is the weight of each virus in question 10?
12. If 1 mL of a soil sample diluted 10^9 times yields 250 plaque-forming units, how many viruses were in the soil sample to begin with?

13. Would you change your answer for question 12 if the original soil sample was briefly incubated in nutrient media before diluting it? Explain your answer.

Thought Questions

1. Assuming that prions can persist in diseased tissue in the environment, how might they find their way up the food chain?

2. What are some consequences of human development on the spread of viral diseases?

Additional Reading

It doesn't have much to do with soil microbiology, but you'll want to read *The Hot Zone* by Richard Preston (1994, Random House, New York). Not only is it a fun book to read (for a microbiologist), but many of the early and late chapters focus on trying to find out where deadly tropical viruses live, and why humans are increasingly coming in contact with them, which is good ecological food for thought. A more general introduction to virology is *The Invisible Invaders* by Peter Radetsky (1991, Little Brown and Company, Boston, MA).

References

Angle, J. S. 1994. Viruses. In *Methods of soil analysis, part 2: Microbiological and biochemical properties,* R. W. Weaver et al. (eds.), 107–18. Madison, WI: Soil Science Society of America.

Bales, R. C., S. Li, T.-C. Jim Yeh, M. E. Lenczewski, and C. P. Gerba. 1997. Bacteriophage and microsphere transport in saturated porous media: Forced-gradient experiment at Borden, Ontario. *Water Research* 33:639–48.

Brock, T. D., and M. T. Madigan. 1991. *Biology of microorganisms.* 6th ed. Englewood Cliffs, NJ: Prentice-Hall.

Gerba, C. P. 1984. Applied and theoretical aspects of virus adsorption to surfaces. *Advances in Applied Microbiology* 30:133-68.

Gibbs, C. J., and D. M. Asher. 1996. Subacute spongiform unconventional virus encephalopathies. In *Medical microbiology,* S. Baron (ed.), 865–76. Galveston, TX: The University of Texas Medical Branch at Galveston.

Katznelson, H., and J. K. Wilson. 1941. Occurrence of *Rhizobium meliloti* bacteriophage in soils. *Soil Science* 51:59–63.

Powelson, D. K., J. R. Simpson, and C. P. Gerba. 1990. Virus transport and survival in saturated and unsaturated flow through soil columns. *Journal of Environmental Quality* 19:396–401.

van Helvoort, T. 1991. What is a virus? The case of tobacco mosaic disease. *Studies in the History and Philosophy of Science* 22:557–88.

van Helvoort, T. 1996. When did virology start? *ASM News* 62:142–45.

3

THE SOIL AS A MICROBIAL ENVIRONMENT

*T*hink of the soil as a house or apartment, and the soil organisms as its
occupants. In Section 2 we introduced the occupants. Now we're going to tour
the facilities. In this section, we look at the soil as a unique environment in which
microorganisms live. We examine how the soil is put together, the important
environmental factors that make each soil unique, and the factors that influence
the activity of all soil organisms. Part of this section will be a review; part will be
a new way of looking at life in soil—getting away from studying individual
organisms and looking instead at how the organisms are collectively involved in
carrying out soil processes.

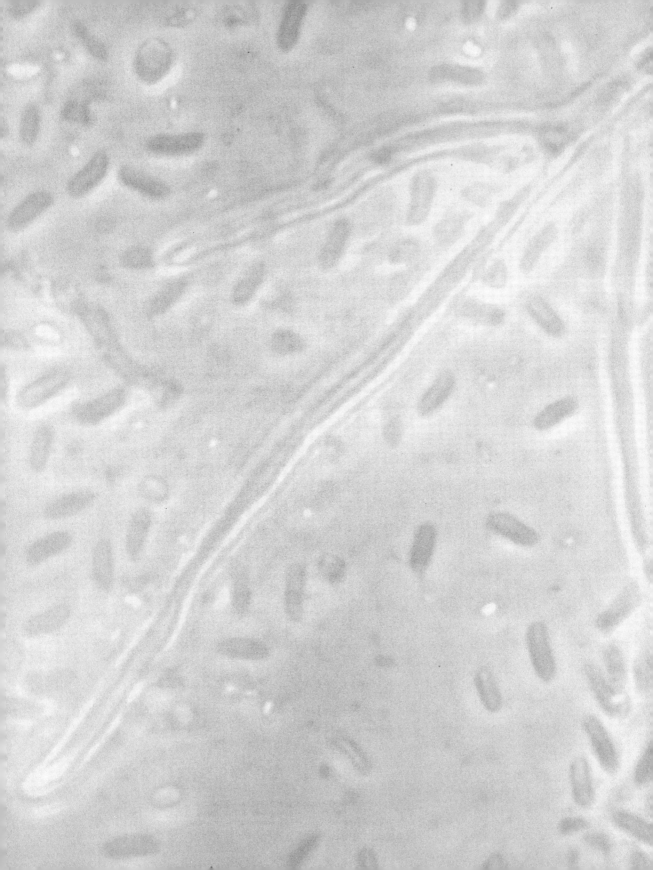

Chapter 12

Soil as a Microbial Habitat

Overview

After you have studied this chapter, you should be able to:

■ Summarize the rudiments of soil classification, soil texture, and soil composition.

■ Define "soil quality" and tell how it relates to soil biology.

■ Explain how microorganisms attach to soil, and state how the arrangement of soil into aggregates and pores influences microbial populations.

■ Discuss why soil atmospheric gases fluctuate and explain how that process is influenced by soil structure.

■ Describe how changes in extrinsic and intrinsic factors affect the distribution of soil microorganisms and control their populations.

■ Explain how soil management can affect soil microorganisms and state why clay has such an important effect on microbial activity in soil.

INTRODUCTION

Soil is a living organism—a consortia (mixture) of living cells in an organic/mineral matrix. Neither the living cells nor the composition of the organic/mineral matrix are constant; both vary with time and with location. Soils are rarely manipulated for microbial growth. Yet, when the soil is altered to affect plant growth, which is the essence of agriculture, these manipulations also affect microorganisms. The microorganisms, in turn, affect plant growth. Soil is not an inert potting medium. In fact, there are many and varied interactions among the soil, microorganisms, and higher plants that greatly influence plant growth and development. We can alter the soil environment to regulate these activities. Tilling a soil, draining a soil, adding organic waste to soil, fumigating soil, and fertilizing soil are all activities that alter the soil environment.

Factors that influence microbial distribution in soil may be intrinsic or extrinsic. Intrinsic factors arise from the structure and function of the microorganisms themselves. These include: 1) persistence mechanisms (spores are an example); 2) size; 3) motility; 4) structural features (stalks, holdfasts, filaments); and 5) and biochemical capacities.

Extrinsic factors are factors arising from the soil and the environment. These are the gross characteristics of the physical environment, and they include such factors as: 1) soil structure; 2) soil atmosphere; 3) precipitation and soil water; 4) soil pH; 5) soil and atmospheric temperature; 6) soil reduction-oxidation (redox) potential; 7) solar radiation; and 8) wind and relative humidity.

In this chapter, and in subsequent chapters, we combine what we know about organisms in soil with what we know about the composition and character of the soil itself. Our goal is to see how physical and chemical characteristics of the soil environment, either occurring naturally or as the result of human activity, influence the distribution of soil microorganisms and help dictate what and where various processes will occur.

SOIL FORMATION, MORPHOLOGY, AND CLASSIFICATION

What is a typical soil? There is none. Soils, like people, are unique. Hans Jenny, a soil scientist at the University of California in Berkeley, described soil formation as a function of climate, topography, parent material, time, and biota. Soils, like twins raised separately, may share common parent material but develop quite unique personalities depending on their environment. Nevertheless, soils have some shared characteristics.

A three-dimensional body of soil is called a pedon. The minimum size of pedon that encompasses enough characteristics to be classified as a soil is 1 to 10 m² by 1.5 m deep (Singer and Munns 1996). The pedon contains the solum—what we call the soil. The solum lies atop the parent material, the mineral material from which the soil originated (Figure 12-1). Sometimes windblown soil can be deposited on previously existing soil.

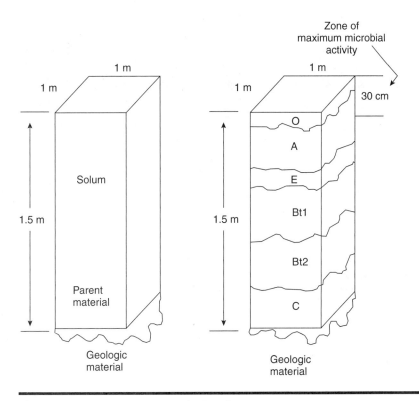

Figure 12-1   Gross composition of soil pedons.

The pedon is not uniform with depth, but can be divided into identifiable horizons (Figure 12-2a,b). Each horizon is given a designated symbol (Table 12-1). For example, an E horizon is an eluvial horizon, a zone from which many organic and mineral constituents have been removed. A zone that accumulates soil constituents is called an illuvial horizon (think of it as the difference between emigration and immigration—emigrants leave a country; immigrants enter a country). Not every soil has every horizon. Young soils (in geologic terms) may simply have A and C horizons. Mature soils have more numerous and more distinct horizons.

Soils are grouped together, based on broad morphological or visible features, into soil orders. In the system of soil classification used in the United States, there are 12 soil orders (Brady and Weil 1999). Table 12-2 lists the orders and describes some of their fundamental characteristics. The 12th and newest soil order is Gelisol. Gelisols are soils forming in environments with permafrost, such as the Alaskan tundra.

a   b

Figure 12-2a,b   Example of a soil showing well-developed horizons. (Photographs courtesy of J. A. Thompson, Dep. of Agronomy, Univ. of Kentucky)

Table 12-1   Symbols used in the characterization of soil horizons.

Horizon Symbol	Horizon Property or Characteristic
O	Surface layers dominated by organic debris
A	Mineral horizon formed at the soil surface or just below the O horizon; contains accumulating decomposed organic matter (humus)
E	Mineral horizon from which clays, iron, and aluminum have been lost, leaving mostly sand and silt particles
B	Horizons dominated by the destruction of the original rock structure and containing illuviated materials from upper horizons
C	Horizons, excluding bedrock, that are affected little by soil genesis
R	Bedrock such as basalt, granite, sandstone, or limestone

Table 12-2   The twelve orders in the U. S. soil classification system. Orders are listed in terms of their approximate degree of maturity (Brady and Weil, 1999).

Soil Order	Features
Entisols	Soils with no profile development except a shallow A horizon. They occur in places such as recent floodplains, and severely eroded areas. Sandy areas typically have Entisols.
Histosols	Organic soils containing > 20% to 30% organic matter in a layer > 30–46 cm thick that formed from the accumulated plant debris in bogs, marshes, and swamps.
Gelisols	Soils forming in environments with permafrost. Often have frost churning (cryoturbation). The defining element is a permafrost layer within 200 cm of the soil surface.
Inceptisols	Inceptisols are found in humid climates and have a weak to moderate horizon development. Horizon development may have been delayed because of cold, water logging, or insufficient time for soil development.
Andisols	Soils having greater than 60% volcanic ash, cinders, pumice, and basalt. Andisols have a dark A horizon and high absorption of phosphorus but relatively weak profile development.
Aridisols	Soils forming in arid climates. Some aridisols have horizons where lime, gypsum, or salt accumulates.
Vertisols	Soils with a high clay content that swell when wet and shrink when dry, leading to crack development. Vertisols have characteristic self-mixing A horizons and develop in temperate and tropical climates with distinct wet and dry seasons.
Alfisols	Soils developing in the forests of humid and subhumid climates where some leaching occurs. Clay accumulates in the B horizon (Argillic horizon). Alfisols are slightly to moderately acidic.
Mollisols	Soils that develop mostly in grasslands. Mollisols have deep, dark A horizons (mollic epipedon).
Ultisols	Strongly acidic and highly weathered soils of tropical and subtropical climates. Clay accumulates in the B horizon.
Spodosols	Leached, sandy soils of cool coniferous forests. Spodosols usually have an E horizon and a strongly acidic profile. The distinguishing feature of spodosols is a B horizon with accumulated organic matter plus iron and aluminum oxides (a spodic horizon).
Oxisols	Excessively weathered (oxic horizon), low fertility, acidic soils over 3 m deep. Predominantly composed of iron and aluminum oxide clays. Oxisols are found in tropical and subtropical climates.

If we made a pie diagram of the upper, A, horizon of a Mollisol we might come up with something that consisted of: 1) 50% void space (containing varying amounts of air and water); 2) 45% mineral material (composed of varying amounts of sand, silt, and clay); and 3) 5% organic matter (dead and alive) (Figure 12-3). The subsoil and the litter layer clearly won't have the same composition. Soil composition also differs among

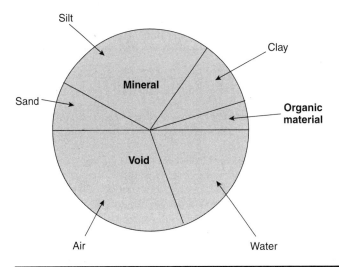

Figure 12-3 Gross soil composition in a Mollisol.

How Do You Remember the Twelve Soil Orders?

Here's a simple memory device that will help you to remember the names of the 12 soil orders in the U. S. system of soil classification, in the unlikely event that you'll ever have to.

Gee, IOU Alf's HAM & VASE

The capital letters stand for Gelisol, Inceptisol, Oxisol, Ultisol, Alfisol, Histosol, Andisol, Mollisol, Vertisol, Aridisol, Spodosol, and Entisol, respectively.

different soils. Furthermore, the proportions themselves, especially air and water, vary over time in all soils.

AGGREGATES

The mineral material of soil is composed of three fractions: sand, silt, and clay. Particles between 2 mm and 50 µm in diameter are sand-sized. Particles between 2 and 50 µm in diameter are silt-sized. Particles smaller than 2 µm in diameter are clay-sized. The proportion of each fraction determines a soil's texture. Figure 12-4 shows a textural triangle used for converting the various proportions to a textural class.

Sand, silt, and clay stick together and form aggregates (natural clumps of soil). Aggregates don't have a uniform shape. They range in size from the very small to the

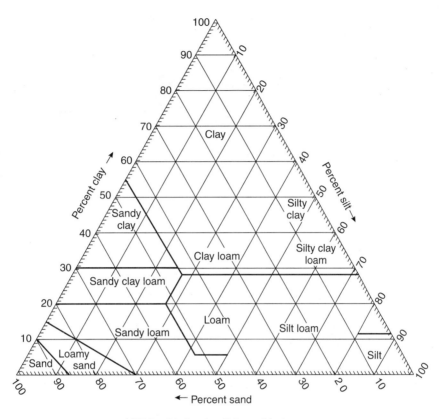

USDA guide for classifying soil textures

Size Limits of Soil Separates as Defined
by the U.S. Department of Agriculture

Fraction	Soil separate	Size, mm
Sand	Very coarse sand	2.00–1.00
	Coarse sand	1.00–0.50
	Medium sand	0.50–0.25
	Fine sand	0.25–0.10
	Very fine sand	0.10–0.05
Silk	Silk	0.05–0.002
Clay	Clay	<0.002

Figure 12-4   The USDA guide for classifying soil textures.

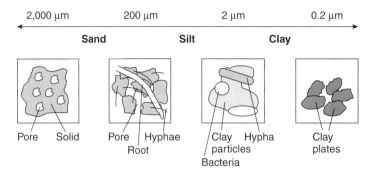

Figure 12-5   Schematic diagram of aggregate formation. (Adapted from Tisdale and Oades 1982)

very large. Microaggregates are less than 250 μm in diameter; macroaggregates are greater than 250 μm in diameter. Microaggregates are older, more stable, and less metabolically active than macroaggregates. The glue that holds microaggregates together is composed of precipitated inorganic minerals or humic substances (very resistant fractions of organic matter in soil, which we discuss in Chapter 25). Macroaggregates have larger pores, more aeration, water movement, diffusion, and microbial occupancy. They tend to be held together by roots, hyphae, and polysaccharides. Macroaggregates have elevated metabolic activity compared to microaggregates.

Aggregate formation starts when microflora and roots produce filaments and polysaccharides that combine with soil particles such as clays to form organic matter-mineral complexes (Figure 12-5). Aggregate shapes vary as they are molded by physical forces such as: 1) drying and wetting; 2) shrinking and swelling; 3) freezing and thawing; 4) root growth; 5) animal movement; 6) compaction; and 7) tillage activities in agricultural soils.

PORES

Aggregates aren't entirely solid. They contain open spaces called pores. Pore size is important because it controls entry and colonization of aggregates by microorganisms. Pore size also influences the diffusion of water and air through aggregates. Pore diameter in microaggregates ranges from 0.2 to 2.5 μm. Pore diameter range in macroaggregates is from 25 to 100 μm. Figure 12-6 shows an example of a large aggregate with equally large pores, probably formed by earthworms.

The microbial population in aggregates is regulated by soil water and pore size. Microorganisms usually occupy between 0.2% to 0.4% of the pore space within aggregates; it is more common to find them on the outside. The microorganisms present on the inside of aggregates were probably trapped there when the aggregate formed, and will stay there until the aggregate disperses. Microorganisms in or on aggregates may or may not be metabolically active depending on the availability of carbon. Fungi are usually on the outside of aggregates.

Figure 12-6   Large macropores in a soil aggregate. (Photograph courtesy of M. S. Coyne)

SOIL ATMOSPHERES

Pores are filled with water and gas. The major gases in the atmosphere are also found in soil. In well-aerated soil, this means that the three most important gases are nitrogen (N_2)(79%), oxygen (O_2)(18%–20%), and carbon dioxide (CO_2)(1%–10%). Carbon dioxide, you notice, is at a significantly higher concentration in soil than it is in the atmosphere (where its concentration is 0.03%, or 330 ppm).

The concentrations of O_2 and CO_2 vary with respect to one another. The lowest O_2 concentration typically corresponds with the highest CO_2 concentration. Oxygen decreases because it is consumed by biological activity. Carbon dioxide concentrations rise because of respiration. In a soil with a high clay content or with considerable organic matter and microbial respiration, CO_2 can form up to 10% of the soil atmosphere. Carbon dioxide disappears mostly through atmospheric diffusion, although some becomes dissolved in soil water and the resulting carbonic acid (H_2CO_3) contributes to the gradual dissolution of soil minerals. Other gases that form during microbial metabolic activity are methane (CH_4), nitric oxide (NO), nitrous oxide (N_2O), ethylene (C_2H_2), and hydrogen gas (H_2).

There are physical restraints on the movement of gases in soil, particularly gas dissolved in the soil water. The solubility of gas in water depends on: 1) gas type (ionizable gases such as CO_2, NH_3, H_2S are more soluble than gases such as N_2); 2) temperature (solubility decreases as temperature increases); 3) salinity; and 4) gas concentration in the atmosphere. Gas diffusion through water is much slower than it is through air. For example, O_2 diffuses 10,000 times slower through water than through air. The reason people drown is not because the water lacks oxygen, but because the oxygen in water can't diffuse fast enough into the lung tissue to allow life. An air-filled pore space of less than 10% indicates inadequate aeration for O_2-requiring organisms such as plants, and a change in soil from aerobic to anaerobic conditions occurs at about 1% air-filled pore space.

Some aquatic plants are adapted to waterlogged conditions because they contain air-filled stem chambers called aerenchyma. The aerenchyma transport oxygen from the leaves to the plant roots. Oxidized regions around roots are caused by facilitated diffusion of oxygen within soil by these plants. For microorganisms, they become microsites of oxygen availability.

The exchange of the soil atmosphere with the surface air is a function of: 1) moisture content; 2) the degree of aggregation of soil particles and pore size; 3) temperature gradients; 4) climatic conditions such as wind (turbulence increases diffusion); and 5) the type of plant cover. There is also biological control of soil atmospheric gas through root respiration. Of the total respiration that occurs in soils, 20% to 40% is due to roots; the rest is microbial in origin.

Gases associated with anaerobic metabolism are sometimes observed in a seemingly well-aerated soil. This suggests that there are aerobic sources of these gases, or that the soil is not entirely well-aerated. Both conditions are likely to be true. The average composition of the soil atmosphere (what you can measure) does not indicate the environment in microsites (localized sites in soil of very small diameter). For example, water-saturated crumbs less than 3 mm in diameter can be anaerobic in the center because metabolic activity depletes available O_2. This is the concept of anaerobic microsites. Figure 12-7 shows how the concentration of O_2 can change rapidly as one traverses from the surface to the interior of a water-saturated soil aggregate. Anaerobic zones are

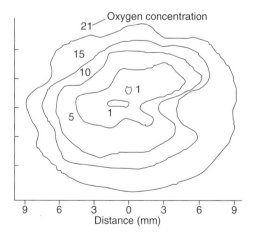

Figure 12-7   Contour map of O_2 concentration within aggregates from a cultivated soil. Lines indicate zones of the same O_2 concentration (% by volume). Distance is centimeters from the midpoint of the aggregate. (Adapted from Sexstone et al. 1985)

a function of C availability, O_2 consumption and diffusion rates, and gas diffusion path characteristics.

Anaerobic bacteria populations can be 10 times higher in the upper surface layers of soil, which are supposedly well-aerated, than they are at lower depths. How can this be? One theory is that these are metabolically inactive bacteria that are revived only when they are cultivated in anaerobic conditions during enumeration. Another theory is that they are living and growing in anaerobic microsites.

SOIL SURFACES

Microorganisms are not usually found evenly distributed over the surface of soil minerals. Microorganisms are typically found in microcolonies that may later develop into biofilms, primarily in aquatic environments. Figure 12-8 illustrates how microbial attachment begins and Figure 12-9 shows the scale of this attachment. Microbial distribution is spotty because, from the perspective of microorganisms, the soil is like a vast desert. Microorganisms tend to cluster and be metabolically active only in oasis-like sites that are favorable for growth; sites where C is available, for example.

Solids have other effects on microorganisms besides providing them with a surface on which to attach. These effects can be both positive and negative. Indirect effects of solids on microorganisms are pH buffering, and protection against desiccation, viruses, protozoa, chlorination, and radiation.

CLAY AND MICROORGANISMS IN SOIL

Clay is the dominant mineral fraction influencing microorganisms in most soils. Histosols, which are mostly organic, are an exception to this rule. Clay particles are 2 μm or smaller in diameter—microbial-sized or smaller. They have high dispersion and surface area per unit mass and a net negative charge (Figure 12-10). Clays are very chemically reactive. Because clay particles are small, they sometimes pack together tightly, limiting pore space.

Clays influence microbial activity by modifying the physical and chemical characteristics of the microbial habitat. Clays also have direct surface interaction with microorganisms. We know this for several reasons. First, people have tried, and failed, to wash microorganisms and viruses out of soil columns. This lack of movement, despite water flow, indicates that microorganisms must be interacting with something. Second, you can increase microbial release by sonication and by adding surfactants (detergents) to soil. Presumably this is because you're shaking microorganisms off the surface of clay particles or releasing encapsulated microorganisms. Third, you can see clay particles attached to the surface of microorganisms if you look at them with a microscope.

There are conflicting reports on how clays affect microbial metabolism. It depends on the conditions of the experiment. The effect of a particular clay on a particular organism might be masked because the combined metabolism of all microorganisms in soil is usually measured. Water adsorbed to clay surfaces is more viscous and freezes at lower temperatures compared to free water. This makes adsorbed water less available for metabolism by microorganisms. Microorganisms may, in fact, grow at some

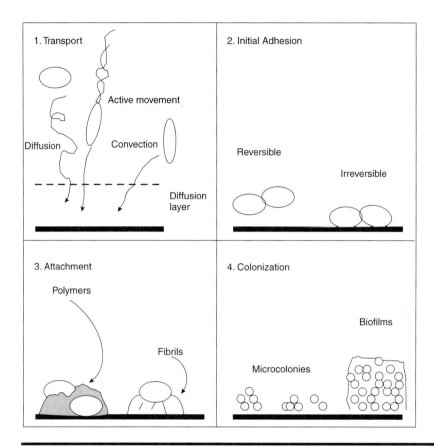

Figure 12-8   Steps in the colonization of soil surfaces by microorganisms. (Adapted from van Loosdrecht et al 1990)

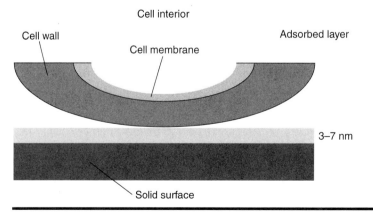

Figure 12-9   Representation of the zone of interaction between a microorganism and a surface. (Adapted from van Loosdrecht et al. 1990)

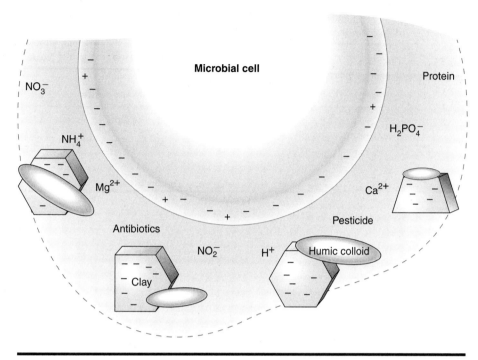

Figure 12-10   Interaction of microbial cells with clay.

distance from clay surfaces where water is more physiologically available. We do know that clays have an effect on N-fixation, the conversion of ammonium (NH_4^+) to nitrate (NO_3^-) (nitrification) and on the availability of essential mineral elements. They also offer some protection from desiccation. Clays are a cyclical source of nutrients. Inorganic compounds that microorganisms need for growth, such as K^+ and NH_4^+, adsorb and desorb from clay constantly. Organic compounds that are firmly bound to clays are not available as nutrients.

DIVERSITY AND DISTRIBUTION OF MICROORGANISMS

A growing topic of study is the concept of "soil health" or "soil quality." The number, diversity, and distribution of microorganisms reflects overall soil productivity. Microorganisms participate in the genesis of their environment. They form part of the biota in Jenny's equation for the factors of soil formation. What are the characteristics of a fertile, productive soil? In terms of soil quality, they are "the capacity of a soil to function within ecosystem boundaries to sustain biological productivity, maintain environmental quality, and promote plant and animal health" (Sims et al. 1997). Soil is unlike a monoculture in which the same variety of the same plant species is used to ensure that each plant is virtually a clone. If only one microbial species dominates a soil, you would expect that some environmental condition has affected or is affecting biological

Table 12-3 Distribution of microorganisms in various horizons of the soil profile of a Spodosol.

	Thousands of Organisms per g^{-1} of Soil				
Depth (cm)	Aerobic Bacteria	Anaerobic Bacteria	Actinomycetes	Fungi	Algae
3–8	7,800	1,950	2,080	119	25
20–25	1,800	379	245	50	5
35–40	472	98	49	14	<1
65–75	10	1	5	6	<1
135–145	1	<1	ND	3	ND

ND = Nondetectable
(From Richards 1987)

activity, or that a special ecological niche exists. What's an ecological niche? "The functional role of an organism within an ecosystem and its interaction with other organisms in that environment" (Atlas and Bartha 1993). The key point to remember is that diversity of microorganisms is the norm in soil.

We've discussed the idea of the soil being a desertlike environment as far as microorganisms are concerned—not necessarily dry, but barren of resources on which the microorganisms can grow. On the microbial scale, large inhospitable distances surround oases favorable for growth. You can also look at the distribution of microorganisms with soil depth (Table 12-3). Microorganisms are less numerous at lower depths mainly because there is less organic material for them to grow on. It's also possible that they simply can't be enumerated as well. Microorganisms are typically more numerous in soil horizons rich in silt and clay than in intervening sandy layers. The finer-textured materials usually hold more food for the microorganisms to grow on. Microorganisms are also abundant just above water tables. An example are the rhizobia that infect mesquite, a leguminous desert shrub. Instead of being at the hot, dry surface of desert soils, they are located several meters down the soil profile, near the water table.

The Rhizosphere

Microbial populations just centimeters apart can be dramatically different because of plant roots. Microbial populations are affected by plants just as plants are affected by microorganisms. Microorganisms cover 5% to 10% of the root surface, and most terrestrial plants are infected by fungi called endo- and ectomycorrhizae (more about them later in Chapter 29). There is a steep decrease in microbial populations that occurs with small distances (within 5 mm) from plant roots.

The effect of the rhizosphere is characterized by dividing soil into three fractions: 1) bulk soil—not affected by plants > 5 mm from the root; 2) rhizosphere soil—the area around plant roots affected by plant metabolic activity (respiration, exudates); usually the soil that adheres to roots; and 3) rhizoplane—the area at the plant–soil interface. The endorhizosphere is colonized root tissue (also called the histosphere or cortosphere). In soil microbiology, we're concerned with which microorganisms are affected by roots. The R/S ratio refers to the ratio of microorganisms in the rhizosphere compared to the microorganisms in the bulk soil. Every microbial species has its own R/S ratio, which varies depending on the plant and the soil environment. The R/S ratio typically ranges from 10 to 50, which means that there are 10 to 50 times as many microorganisms right around a plant root compared to the surrounding soil.

Microorganisms in the rhizosphere are characterized by saprophytic and pathogenic organisms. Other terms for these microorganisms are "allochthonous" or "r-selected" organisms. Remember that r-selected microorganisms maximize growth rate at the expense of survival when food sources are scarce. Plant pathogens are clearly rhizosphere-type microorganisms. Microorganisms in the bulk soil are characterized by autotrophic and lithotrophic growth (growth without using organic carbon as an energy source). Other terms for these microorganisms are "autochthonous" or "k-selected" microorganisms. The k-selected microorganisms, remember, get more "bang for the buck" from the nutrients they consume.

The Phyllosphere

Our discussion wouldn't be complete without at least mentioning the abundant number of microorganisms on the surface of plant leaves—the phyllosphere. Phyllosphere microorganisms are influenced by climate and plant type. Most plants also contain endophytes, bacteria and fungi that live in the intracellular spaces of the stems, petioles, roots, and leaves (Carroll 1988).

MANAGEMENT EFFECTS

Soil management affects the distribution and types of microbial populations. After tilling soil, there is an increase in microbial activity at all depths in the plow layer, the layer of soil that is directly affected by tillage. This is probably due to the disruption of aggregates, better aeration, and additional organic matter that is plowed into the soil. No-tillage is a soil management practice used to control erosion. As the name implies, planting operations are done with minimal disturbance to the soil. Consequently, microbial activity and populations are stratified near the surface in no-tillage management systems.

In general, microbial populations and activity in no-tillage soils are greater than in conventional tillage soils. This may be because no-tillage soils are generally more moist than conventional tillage soils and have more organic C. As one proceeds through the soil profile, however, the situation reverses. You can see this phenomenon by comparing the ratio of microbial numbers from no-tillage and tilled soil as soil depth

Table 12-4 Effect of tillage on microbial populations in an Alfisol after 23 years of continuous corn.

Microbial Group	Ratio of Microorganisms in No-tillage vs. Tilled Soil	
	0 to 7.5 cm	7.5 to 15 cm
Total aerobic bacteria	5.6	0.4
Actinomycetes	5.5	1.1
Fungi	1.2	0.7

(From Handayani 1996)

increases (Table 12-4). Microorganisms that don't require organic C for growth are less influenced by tillage practice.

Compacting soil has adverse effects on soil microorganisms. Organic C and total N decrease in compacted soils. Total microbial populations and microbial biomass decrease in compacted soils. Enzyme activity also decreases in compacted soil. This may not be as apparent in the surface soil as it is in the subsurface. Compacting soil exerts its adverse effects partly by reducing soil porosity and aeration.

Biocides and chemicals affect microbial populations, at least temporarily. Herbicides and foliar insecticides are usually not applied at a high enough concentration to reach soil and kill microorganisms. One exception is a group of bacteria called nitrifiers, which are unusually susceptible to chemicals. Fungicides and fumigants have a greater effect on microbial populations because they are much more microbially specific compounds. After fumigation or fungicide application, respiration in treated soil initially decreases. Ultimately, though, more CO_2 is respired in fumigated than unfumigated soil. Why? Because fungi represent the largest fraction of the soil biomass. By killing them, you release considerable organic material for surviving organisms to consume.

Fertilizer banding kills the microbial population in the localized region of soil receiving the fertilizer. The high chemical concentration in the fertilizer band leads to salt stress and chemical toxicity. It can also drastically alter pH (NH_3 is released, for example, after banding of anhydrous ammonia). As NH_3 increases, nitrification decreases in the immediate vicinity of the fertilizer band.

Clear-cutting of forests affects microbial populations and activities. When a forest is clear-cut, the roots die and root decomposition accelerates. The debris from leaves and branches on the soil surface that works into the disturbed soil, serves as food for microbial growth. Soil respiration increases and nutrients such as nitrogen that are found in the organic matter can be released. Unfortunately, this can also lead to nitrogen loss in streams because, for a brief period, there are no plants around to take up the nutrients that are being released by the microorganisms.

Summary

Intrinsic factors affecting microbial growth are properties of the cells and extrinsic properties come from the environment. Extrinsic properties are soil structure, atmosphere, water, pH, temperature, redox potential, solar radiation, and the temperature, humidity, and wind of the atmosphere.

The soil type is a function of climate, topography, parent material, time, and the biota, which includes soil microorganisms. Each soil is unique. There are 12 soil orders based on general soil characteristics. Individual soils are characterized at the level of a pedon, which contains the solum and parent material.

The solum has distinct horizons—zones of soil with different chemical and physical properties. A typical soil is composed of solids (minerals and organic matter) and void space (air and water). The minerals are combined into aggregates of different size, and are filled with pores of different size. The diameter of pores restricts entry and allows some microorganisms to inhabit the inside of aggregates.

The air in soils is enriched in CO_2. It also has N_2 and O_2. The O_2 can be depleted by consumption, and also because it diffuses slowly in water-filled pores. Small O_2-depleted zones in soil are called anaerobic microsites.

Soil surfaces are colonized sporadically by microorganisms because they bind only where conditions are favorable for growth. Soil surfaces have many effects on soil microorganisms—some promoting growth; some retarding it. The most important soil surface is represented by clay particles, which have an extremely high specific surface or surface area per unit mass compared to other particles.

We can hypothesize that soil quality is best when microbial numbers and diversity are highest. The microorganisms are generally found in the upper 15 cm of soil, but this can be affected by soil management and other environmental factors. Microorganisms are also found close to root surfaces. The R/S ratio reflects the population of microorganisms around roots compared to the bulk soil. The R/S ratio is usually 10 to 50. The phyllosphere is another microbial habitat that occurs on plant leaves.

Sample Questions

1. There are several types of evidence showing that organic matter holds aggregates together. What are they? Design an experiment to test one of those types of evidence.
2. How are microaggregates stabilized differently than macroaggregates?
3. Draw a diagram that shows two ways an anaerobic microsite might develop in soil.
4. Why is microbial distribution in soil compared to oases in a desert?
5. Draw a graph that shows the relationship between total anaerobic bacteria in soil and soil depth for a soil that has a perched water table on a fragipan (an impermeable layer) at a depth of 70 cm.

6. In enumerating bacteria from the rhizosphere of a plant, you observe 1×10^6 CFU g^{-1} of soil. Five cm away from the root, the bacterial population is 0.5×10^5 CFU g^{-1} of soil. What is the R/S ratio for this sample and what does it tell you?

7. Draw a graph showing the population of bacteria in relationship to distance from a plant root in soil.

8. Calculate the population of fungi in the rhizosphere of a corn root from which a 10 g dry soil sample yields 250 colonies after plating 0.1 mL of a 1/100,000 dilution.

9. Draw a graph that shows a comparison of CO_2 evolution with time from two treatments—one fumigated and the other unfumigated.

10. Interpret the data in the following table.

Effect of tillage on microbial populations

Microbial Group	Ratio of Microorganisms in Untilled vs. Tilled Soil at Two Depths	
	0 to 7.5 cm	7.5 to 15 cm
Aerobic bacteria	>1	<1
Anaerobes	>1	>1
Actinomycetes	>1	<1
Nitrifiers	>1	<1
Fungi	>1	>1

Thought Questions

1. What physical, chemical, or biological criteria might prevent soil yielding 300 bushels of corn (*Zea mays* L.) per acre (18,813 kg per hectare) from being quality soil?

2. The steady-state soil ecosystem is sometimes compared to the daily conditions that occur at the corner of any street in a large city. Discuss why this comparison is made.

3. Based on the environmental conditions existing around your home, what soil order do you predict it is on?

Additional Reading

For those of you with a greater interest in how microorganisms and soil interact, a good place to start is *Interactions of Soil Minerals with Natural Organics and Microorganisms* (P. M. Huang and M. Schnitzer, eds., 1986, Soil Science Society of America, Inc., Madison, WI). Refer to N. C. Brady, *The Nature and Properties of Soils, 10th ed.* (1990, Macmillan Publishing Co., New York) or other introductory soil texts for more information about soil genesis, soil classification, and soil mineralogy.

References

Atlas, R. M., and R. Bartha. 1993. *Microbial ecology: Fundamentals and applications.* 3d ed. Redwood City, CA: Benjamin/Cummings.

Brady, N. C. and R. R. Weil. 1999. *The nature and property of soils.* 12th ed. Prentice Hall Inc. Upper Saddle River, NJ.

Carroll, G. C. 1988. Fungal endophytes in stems and leaves: From latent pathogen to mutualistic symbiont. *Ecology* 69:2–9.

Handayani, I. 1996. Soil carbon and nitrogen pools and transformations after 23 years of no tillage and conventional tillage. Ph.D. diss., University of Kentucky, Lexington.

Richards, B. N. 1987. *The microbiology of terrestrial ecosystems.* Harlow, England: Longman Scientific & Technical.

Sexstone, A. J., N. P. Revsbech, T. B. Parkin, and J. M. Tiedje. 1985. Direct measurement of oxygen profiles and denitrification rates in soil aggregates. *Soil Science Society of America Journal* 49:645–51.

Sims, J. T., S. D. Cunningham, and M. E. Sumner. 1997. Assesssing soil quality for environmental purposes: Roles and challenges for soil scientists. *Journal of Environmental Quality* 26:20–25.

Singer, M. J., and D. N. Munns. 1996. *Soils: An introduction.* 3d ed. Upper Saddle River, NJ: Prentice-Hall.

Tisdale, J. M., and J. M. Oades. 1982. Organic matter and water stable aggregates in soil. *Journal of Soil Science* 32:141–63.

van Loosdrecht, M. C. M., J. Lyklema, W. Norde, and A. J. B. Zehnder. 1990. Influence of interfaces on microbial activity. *Microbiology Reviews* 54:75–87.

Chapter

13

Environmental Influences—Temperature, Reduction–Oxidation, and pH

Overview

After you have studied this chapter, you should be able to:

■ Describe three of the key environmental (extrinsic) factors affecting soil microorganisms: soil temperature, soil redox potential, and soil pH.

■ Explain how soil moisture affects soil temperature, and classify microorganisms based on their growth response to temperature.

■ Describe what a redox reaction is and write one.

■ List the elements in soil that have a major role in redox reactions.

■ Explain why excessively acidic or alkaline conditions impede microbial growth in soil.

■ Describe how soil microbiologists characterize microorganisms based on their growth response to pH.

INTRODUCTION

Microorganisms have little or no control over the temperature of their environment and, consequently, over the thermodynamics of their metabolism. Because temperature affects all chemical reaction rates, microorganisms have little control over how fast their own metabolic processes take place. Enzyme activity and structure are also affected by temperature; so is membrane fluidity. So, temperature is an extremely important environmental factor regulating microbial life.

The soil pH, or the measure of soil acidity and alkalinity, is a critical environmental feature. The pH of a microorganism's environment has a direct effect on whether the microorganism can survive and grow. More importantly, pH is one of the environmental factors in soil systems that is most often manipulated to promote plant growth.

Reduction–oxidation potential, redox potential for short, is a measure of how oxidized or reduced an environment is. Redox potential is a proxy measurement for assessing whether a soil environment is well-oxygenated; a well-oxygenated soil has a high redox potential. Redox potential is a critical environmental factor because it governs the chemical form of many inorganic compounds and their availability, and it also influences the products of microbial metabolism in soil.

TEMPERATURE RELATIONSHIPS

The range of temperature encountered in surface soil typically ranges from 0° to 60°C and is subject to daily (diurnal), weekly, and seasonal variation. Soil temperature is affected by the angle of solar irradiation, soil aspect (the direction a slope is facing), soil shading, soil color, soil depth, and soil moisture.

If sunlight strikes the Earth at an angle, whether because of the Earth's tilt or because of a slope's aspect, the same amount of solar radiation is distributed over a greater surface area, and less energy is available to heat the soil. Soil shading and cover affect soil temperature because plants and plant debris intercept solar radiation. Soil color affects soil temperature because of the albedo effect—the reflectance of materials. Dark soils have a low albedo and absorb more energy than do light soils, just as dark clothes on a sunny day make you feel hotter than do light-colored clothing.

In the Northern Hemisphere, soils are usually wetter and cooler in spring and fall, and drier and warmer in summer. This is because the evaporation of water carries off energy. So, the more water that is in soil, the more evaporation can occur, rather than soil warming. You can see this by working the example in the following section.

The Influence of Soil Water on Soil Temperature

Take 1 cm^3 of soil with a bulk density of 1.0 g cm^{-3} (bulk density is a common soil measurement; it is the mass of oven-dry soil per unit of volume).

The heat capacity (HC) of its components are:

Mineral solids—0.265 calorie/g°C
Water—1.0 calorie/g°C
Air—negligible heat capacity

Heat capacity is the calories needed to raise the temperature of 1 gram of material by 1 degree centigrade.

Table 13-1 Daily temperature fluctuations of a bare Alfisol with depth during June, 1994 in Lexington, KY.

Fluctuation	Depth
9° to 35°C	Air
19° to 35°C	10 cm
22° to 26°C	46 cm

(Univ. of KY Ag Weather Center 1994)

How much will the temperature of that cubic centimeter of soil rise for each increment of energy, if the total heat capacity equals:

$$(\text{Mass of minerals}) \times (HC_{minerals}) + (\text{Mass of water}) \times (HC_{water}) +$$
$$(\text{Mass of air}) \times (HC_{air})?$$

For a soil at 5% moisture:

$$HC_{total} = (1 \text{ g})(0.265 \text{ cal/g°C}) + (.05 \text{ g})(1.0 \text{ cal/g°C}) + (\text{negligible}) = 0.315 \text{ cal/°C}$$

It will take 0.315 calories of energy to raise the temperature by 1°C, or, for each calorie input per g, the soil temperature will rise 3.2°C.
For a soil at 30% moisture:

$$HC_{total} = (1 \text{ g})(0.265 \text{ cal/g°C}) + (0.3 \text{ g})(1.0 \text{ cal/g°C}) + (\text{negligible}) = 0.565 \text{ cal/°C}$$

It will take 0.565 calories of energy to raise the temperature by 1°C, or, for each calorie input per g, the soil temperature will rise only 1.8°C.

Soil temperature is depth dependent, as Table 13-1 demonstrates. The upper soil insulates the subsoil. Below 6 meters, temperature variation is $< 1°C$ on a yearly basis.

EFFECT OF TEMPERATURE ON MICROORGANISMS

Most microorganisms can grow over a temperature range of approximately 40°C. Some grow at greater than 90°C, and some grow at –10°C. Water normally turns into ice at 0°C, but on a microbial scale it can remain liquid below that temperature because of an increase in the solutes around and inside the microbial cells (freezing point depression). Microorganisms are grouped into three classes based on their growth response to temperature: thermophiles (heat-loving), mesophiles (temperate-loving), and cryophiles or psychrophiles (cold-loving).

For most microorganisms, high temperatures cause permanent denaturation of proteins (i.e., you can't unscramble an egg). High temperatures also alter the permeability of microbial cell membranes (Ingraham et al. 1983). Membranes must be in a semisolid state (like Jell-o) to function properly. If they are too liquid the cell contents diffuse away (like melting Jell-o). If they are too solid, the membranes fracture (like freezing Jell-o). Membranes become more fluid at elevated temperatures. From our previous analogy of a screen door as a cell membrane, it's as though at high temperature the screen has all the rigidity of cooked spaghetti. Membranes become more rigid in the cold. They become more brittle, as though the screen was now composed of ice strands.

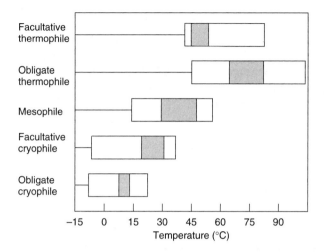

Figure 13-1 Classification of microorganisms based on optimum temperature regimes. The shaded region represents the temperature of optimum growth.

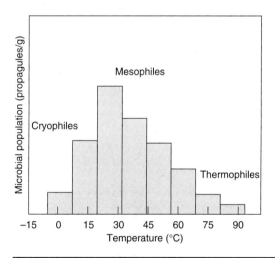

Figure 13-2 Dendogram of soil microbial population based on temperature.

In any given soil there is a diversity of microorganisms with a diversity of temperature limitations. Each microbial group functions over a range of temperatures in which its members grow, and an optimum temperature range in which they grow best, which Figure 13-1 illustrates.

The range of soil temperature across which microorganisms can grow forms a continuum. As a general rule, the largest microbial populations are found at mesophilic or moderate temperatures (Figure 13-2). However, that may simply reflect the way in which we isolate and grow microorganisms. We tend to incubate microorganisms at temperatures at which we feel comfortable.

A Little Microbe Music

"Cryophiles Want To" (to the tune "It's My Party and I'll Cry if I Want To")

I don't like to grow above 15 degrees,
At 20, well, I'm good as dead.
Try to refrigerate me,
It's just like I'm reading in bed.

Chorus

> Nobody grows here but
> a cryophile wants to,
> a cryophile wants to,
> a cryophile wants to.

You'd grow here too, if you didn't turn blue!

My lipids and membranes are fluid because
they're unsaturated or short.
My proteins have substitutions
that differ from the hotter sort.

Chorus

My cellular functions are so sensitive
to moderate changes in Temp,
I may roll up in a ball
or go filamentous and limp.

Chorus

I leave all my enzymes in pasteurized milk
or in cream or in butter or cheese.
My lipase breaks down their fat
to a flavor that's sure to displease.

Last chorus (repeated)

> It's really chilly for
> a cryophile microorganism,
> a cryophile microorganism,
> a cryophile microorganism,

> You would be, too, if you grew in the cold!

K. D. Young, University of North Dakota

REDUCTION-OXIDATION (REDOX) POTENTIAL

What is a redox reaction? A redox reaction is the transfer of electrons from one compound (A) to another compound (B). During this reaction, A is oxidized (loses electrons) and B is reduced (gains electrons). Microorganisms gain energy by oxidizing reduced material.

$$A_{reduced} \longrightarrow A_{oxidized} + electron\ (e^-)$$
$$\underline{electron\ (e^-) + B_{oxidized} \longrightarrow B_{reduced}}$$
$$A_{reduced} + B_{oxidized} \longrightarrow A_{oxidized} + B_{reduced}$$

The origin of reduced material (if it is organic) is usually plant photosynthesis. The elements commonly involved in redox reactions are H, C, N, O, S, Fe, and Mn.

Redox potential (E_h) describes the tendency of electrons to flow between compounds. Electrons flow from electron-rich, low-E_h, reduced compounds to electron-poor, high-E_h, oxidized compounds. Redox potential reflects an electrical potential—the potential to drive electrons (current; so E_h is measured in volts) from reduced to oxidized compounds. The more positive the E_h, the greater the oxidizing capacity. Microorganisms conserve or harness that electron flow to generate energy. The greater the difference in E_h, the greater the energy released to the microorganism. Use the analogy of a waterwheel in a river—the river is a stream of electrons that flows downhill from reduced to oxidized compounds. The greater the drop, the more rapid the flow and the more rapidly the waterwheel turns to generate energy.

The greater the difference between the E_h of two compounds, the greater the energy that can be generated in a redox reaction (Table 13-2). As a consequence of redox

Table 13-2 Biologically significant reduction reactions in nature.

Reduction Reaction		E° (mV)[1]	Electrons Transferred
CO_2	$\Rightarrow$ Glucose	−430	24
$2H^+$	$\Rightarrow H_2$	−420	2
CO_2	$\Rightarrow$ Methanol	−380	6
N_2	$\Rightarrow NH_4^+$	−350	8
CO_2	$\Rightarrow$ Acetate	−280	8
$S°$	$\Rightarrow H_2S$	−280	2
Pyruvate	$\Rightarrow$ Lactate	−190	2
Fumarate	$\Rightarrow$ Succinate	30	2
Ubiquinone$_{ox}$	$\Rightarrow$ Ubiquinone$_{red}$	110	2
Cytochrome c$_{ox}$	$\Rightarrow$ Cytochrome c$_{red}$	250	1
Cytochrome a$_{ox}$	$\Rightarrow$ Cytochrome a$_{red}$	390	1
NO_3^-	$\Rightarrow N_2$	740	2
Fe^{3+}	$\Rightarrow Fe^{2+}$	760	1
$1/2O_2$	$\Rightarrow H_2O$	820	2

(From Brock and Madigan 1991)

[1]Most tables report redox potential as E°, when the oxidized and reduced members of a redox reaction are both at 1 M concentrations. The E_h takes into account the concentration of oxidized and reduced compounds in the environment in determining the potential current.

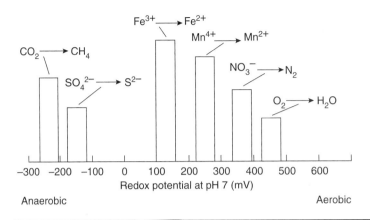

Figure 13-3 Microbially mediated redox reactions that occur as a soil changes from aerobic to anaerobic state.

reactions, oxidized compounds in reduced environments are electron acceptors and reduced compounds in oxidized environments are electron donors.

HOW DOES REDOX AFFECT SOIL REACTIONS?

Redox potential in soil is governed by the strongest oxidizing agent present. Oxygen is a very strong oxidizing agent; it has a tremendous affinity for electrons. As long as the soil is well-aerated, O_2 is at a high concentration and keeps the E_h high, which means that compounds will tend to be oxidized—to give their electrons to O_2. The ultimate electron acceptor is usually O_2. However, microorganisms also use NO_3^-, Fe^{3+}, Mn^{4+}, and SO_4^{2-} as electron acceptors. As the E_h decreases, synonymous with the disappearance of O_2, microorganisms switch to alternative electron acceptors if they have the metabolic capacity to do so (Figure 13-3).

What happens when the supply of O_2 is impeded? First, most of the O_2 in the system is consumed. In terms of E_h, O_2 accepts electrons and is reduced to H_2O. When O_2 is gone, the next strongest oxidizing agent is used, although there is some overlap of electron acceptor use. Figure 13-4 shows the likely sequence of reduced products that will appear in a flooded soil with time as the E_h declines and the soil becomes increasingly anaerobic and reduced. When O_2 returns, previously reduced electron acceptors simultaneously begin to reoxidize, though not necessarily speedily. N_2 and CH_4 do not spontaneously reoxidize, for example.

HOW CAN MICROORGANISMS AFFECT REDOX?

Any environment that has many microorganisms that are actively respiring and consuming O_2 can rapidly become anaerobic. This leads to anaerobic microsites in otherwise aerobic soils, as we discussed in Chapter 12. Gleization is an example of a process in which soil microorganisms help to create reduced compounds that influence the soil environment. Figure 13-5 shows a gleyed soil.

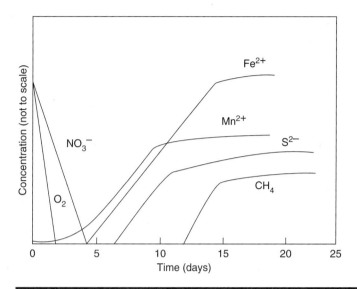

Figure 13-4   Sequence of reduced compounds forming in flooded soil. Note that the product of O_2 reduction is water (H_2O) and the usual product of NO_3^- reduction is N_2. Both products would be difficult to quantitatively measure.

Figure 13-5   A gleyed soil. (Photograph courtesy of A. Karathanasis)

In microsites where iron is present and oxygen is consumed, the following reaction occurs:

$$Fe^{3+}(oxidized \,/\, insoluble) \Rightarrow Fe^{2+}(reduced \,/\, soluble)$$
$$Red \qquad\qquad\qquad\qquad Gray$$

A mottled or gleyed soil is one in which fluctuating water tables cause iron to fluctuate between reduced (electron donor) and oxidized (electron acceptor) states.

SOIL pH

What is pH? pH equals $-\log [H^+]$. It is measured by an electrode and, as a result, its determination in soils is highly ambiguous. Most agricultural soil pH is usually between 4 and 8.5. Excessive acidity or alkalinity makes the soil inhospitable for plant and microbial growth. Microorganisms are, for the most part, metabolically intolerant of low pH. Microbial enzymes are pH dependent and they undergo folding and unfolding as pH changes, which reduces their activity. Aluminum (Al) and manganese (Mn) solubility increases in acidic soils and these elements become toxic to microorganisms. Plants and microorganisms also have to deal with P limitation in acidic soils. The P precipitates as Fe and Al phosphates and becomes unavailable for growth. In alkaline soil, the P precipitates as Ca_2PO_4. We say that P becomes limiting because without it, a plant or microorganism is limited in how much it can grow.

Microbial tolerance to acidity varies (Table 13-3). Fungi are more tolerant than are bacteria, which are, in turn, more tolerant of acidity than are actinomycetes. Exceptions always exist, and in the most extremely acidic conditions you routinely find bacteria rather than fungi. For example, *Thiobacillus desulfuricans* can grow in environments with a pH around 0.6. This environment exists around ore-leaching facilities or hot sulfur springs in volcanic environments.

Table 13-3   Tolerance of soil microorganisms to pH.

Organism	pH Allowing Growth		
	Minimum	**Optimum**	**Maximum**
Escherichia coli	4.4	6–7	9
Proteus vulgaris	4.4	6–7	8.4
Aerobacter aerogenes	4.4	6–7	9
Pseudomonas aeruginosa	5.6	6–7	8
Erwinia cartovora	5.6	7.1	9.6
Clostridium sporogenes	5–5.8	6–7.6	8.5–9
Nitrosomonas	7–7.6	8–8.8	9.4
Nitrobacter	6.6	7.6–8.6	10
Thiobacillus thiooxidans	1	2–2.8	4–6
Lactobacillus acidophilus	4–4.6	5.8–6.6	6.8

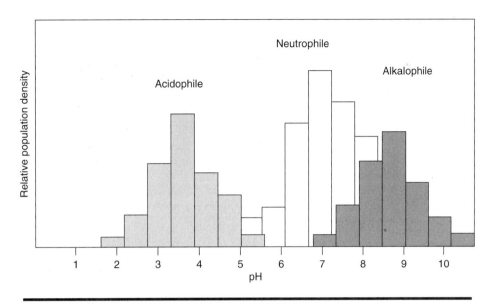

Figure 13-6 Range of pH growth among soil microorganisms.

We can classify microorganisms into three groups based on the pH at which optimum growth occurs: acidophiles (acid-loving), neutrophiles (best growth at neutral pH), and alkalophiles (alkaline-loving) (Figure 13-6). There is an optimum pH at which each microorganism will grow, but as we saw with temperature, the range of pH over which individual microorganisms grow can be wide. Consequently, although we classify the microorganisms into three basic groups, there is considerable overlap among them. Most microorganisms seem to grow best around pH 7 and this is reflected by the populations of microorganisms that are enumerated from soils of different pH.

SOIL pH VS. CULTURE pH

There are numerous observations of microorganisms that cannot grow in solution culture at a pH similar to that of the soil from which they were isolated. Nitrification, for example, is a pH-sensitive process. But nitrification in forests is observed at pH 4, while in the lab it requires a pH greater than 6. Some *Streptomyces* species from acidic soils fail to grow when cultured at a pH of less than 7.5. One explanation is similar to the microsite hypothesis for soil processes. Pockets of intense metabolic activity can raise the pH in microsites. Decomposition of protein-rich organic matter in microsites, for example, produces NH_3, which reacts with H_2O and raises the pH. Solubilization of $CaCO_3$ also occurs. The pH at the charged surfaces of clay may be several times more acidic than the adjacent soil solution. The highest $[H^+]$ (10 to 100 times greater than the bulk soil solution) or lowest pH is near the clay surface.

Summary

Soil temperature is important because soil microorganisms have no control over the thermodynamics of their environment. Thermophiles, mesophiles, and cryophiles represent three classifications of soil microorganisms based on whether they grow best when it is hot, moderate, or cold. Soil moisture is critical in controlling soil temperature because water has a high heat capacity. The upper and lower temperature limits of microbial life are primarily due to the unfolding of proteins and the fluidity of the cell membrane. At high temperatures the cell membrane becomes too fluid and at low temperatures it becomes too solid.

Redox reactions refer to the transfer of electrons from reduced to oxidized compounds. Soil microorganisms extract energy out of this process. The redox potential, E_h, is a measure of the potential of electrons to flow in an environment. When redox potential is very low, or reducing, electrons rapidly flow to any oxidized compound placed in that environment. When redox potential is very high, reduced compounds placed in that environment rapidly lose their electrons to oxidized compounds. When oxygen is present in most soils, it is usually the ultimate acceptor of any electrons that flow. When oxygen is depleted, other oxidized compounds can be used by soil microorganisms to accept electrons. Alternative electron acceptors are used sequentially depending on their affinity for electrons.

The pH is an important factor in soil. Most soils are neutral and most soil microorganisms are neutrophiles. Acidophiles grow best in acidic conditions, and alkalophiles grow best in alkaline conditions. Fungi are more tolerant of acidic pH than are bacteria, and actinomycetes are more tolerant of alkaline conditions than are other soil microorganisms.

Sample Questions

1. Why is soil temperature important?
2. What controls soil temperature?
3. How does soil moisture affect soil temperature?
4. What sets the upper and lower limits of microbial life?
5. Why is an acidic or alkaline pH bad for many soil microorganisms?
6. Write an example of a redox reaction.
7. What is redox potential and why is it important?
8. How many calories will it take to raise the temperature of 10 cm³ of soil by 10°C if the soil has a bulk density of 1.5 and a moisture content of 25%?
9. By how much will the temperature of 5 g of wet soil rise if it has 2.5 g of water and receives 10,000 calories of energy?
10. Both of the following reactions are written as reductions:

NO_3^- $\Rightarrow$ N_2 $E^o = 740$ mV
Pyruvate $\Rightarrow$ Lactate $E^o = -190$ mV

If they are coupled in a redox reaction, which one will you switch so that it is written as an oxidation? What is your reason?

11. Draw a graph that shows the relationship between the percent fungi in soil (as a fraction of the total microbial population) and soil pH.

12. If the molar concentration of H^+ in solution is 0.7×10^{-4}, what is the pH?

Thought Question

Louis Pasteur had an ongoing debate with a French physician named Colin that culminated in a public demonstration by Pasteur to show that chickens could get anthrax. Healthy chickens did not suffer from the disease. However, Pasteur arranged to have four chickens treated in the following manner: one chicken was injected with *Bacillus anthracis* and left at room temperature; two chickens were injected with *Bacillus anthracis* and kept in a cold water bath; one chicken was similarly kept in a cold water bath, but was not injected with *Bacillus anthracis*. In short order, the two chickens in the water bath that had been infected with *Bacillus anthracis* became deathly ill. The uninfected chicken in the water bath, though perturbed, was otherwise healthy. The infected chicken kept at room temperature, likewise, showed no signs of anthrax. Pasteur removed one of the infected chickens from the water bath and allowed its body temperature to rise back to normal (41.7°C, about 107°F). It soon recovered. The remaining infected chicken in the water bath soon died. Explain, in terms any soil microbiologist could understand, the likely explanation for Pasteur's results. In addition, identify what Pasteur's controls were, and why they were necessary.

Additional Reading

Let me refer you back to *The Outer Reaches of Life* by John Postgate (1994, Cambridge University Press, Cambridge, England). Postgate devotes individual chapters to each of the topics we addressed. More importantly, he gives some insight into the actual mechanisms by which microorganisms survive extremes of these environmental factors.

References

Brock, T. D., and M. T. Madigan. 1991. *Biology of microorganisms.* 6th ed. Englewood Cliffs, NJ: Prentice-Hall.

Ingraham, J. L., O. Maaløe, and F. C. Neidhardt. 1983. *Growth of the bacterial cell.* Sunderland, MA: Sinauer Associates, Inc.

Turner, F. T., and W. H. Patrick. 1968. Chemical changes in waterlogged soil as a result of oxygen depletion. *Transactions of the 9th International Congress of Soil Science* 4:53–65.

University of Kentucky Agricultural Weather Center. 1994. University of Kentucky, Lexington. (http://www.ca.uky.edu/agcollege/agweather/).

Chapter 14

Soil Water and Microbial Activity

Overview

After you have studied this chapter, you should be able to:

- Explain why soil water critically affects microbial activity.
- Discuss the mechanisms controlling nutrient diffusion and microbial movement in soil.
- Describe water availability, including the concept of water potential.
- List the components of water potential that are important to the soil microbiologist, and tell how microorganisms can be classified in terms of their response to water stress.
- Describe the strategies that microorganisms use to tolerate water stress.

INTRODUCTION

Liquid water is the one factor that most distinguishes Earth from other planets. Among our closest planetary neighbors, Venus is too hot and Mars is too cold for liquid water to exist. We take water for granted, but it's an unbelievably useful molecule. Its properties are due to a unique atomic structure. There is a 105° angle between H atoms that creates a dipole, a molecule that has both negative and positive poles. The dipole helps hydrogen bonding, and this property helps to make water a universal solvent. Water also has different characteristics depending on whether it is free or soil bound. Bound water can remain unfrozen at 0°C, but it may also be unavailable to microorganisms.

Soil water has critical effects on environmental factors (extrinsic factors) that influence microbial activity, such as: 1) soil temperature; 2) soil aeration; 3) the nature and solubility of compounds; 4) salinity; and 5) soil solution pH (generally only when it dilutes high chemical concentrations).

We've already examined how water can affect soil temperature because of its considerable heat capacity. We've also discussed how water can contribute to anaerobic microsites in soil by limiting oxygen diffusion. In this chapter, we discuss mathematical ways of expressing water availability and examine how microorganisms in soil behave when water becomes increasingly unavailable.

WATER AND MOVEMENT IN SOIL

How does water affect soil microorganisms? Let's start with the way water affects the movement of nutrients and the movement of the microorganisms themselves. Plants, for example, acquire nutrients by: 1) mass flow—movement of nutrients with bulk water flow; 2) diffusion—movement of nutrients across a concentration gradient; and 3) interception—movement of plant roots to nutrients.

Mass flow is important for plants but less important for microorganisms. Diffusion dominates at the microbial scale. An exception is in the rhizosphere where plant growth brings water (and nutrients) to microorganisms by mass flow, or carries root-bound microorganisms to nutrients. Water is also essential for microbial mobility (taxis— remember, taxis is movement, either positive or negative, in response to a gradient that can consist of light, air, chemicals, or magnetic field, etc.).

The dominant factors controlling nutrient diffusion and mobility in soil are the water-film thickness and the continuity of water films. As soils dry, water films quickly thin, and diffusion slows. This reduction in water-film thickness impedes bacterial and protozoal mobility. *Pseudomonas aeruginosa,* for example, needs water-filled pores that are 1 to 1.5 μm in diameter in which to move. Thin water films retard nutrient diffusion and availability. Soil particles are covered by a thin water film in which the microorganisms move; if that water film becomes discontinuous, then microorganisms are stuck or have to move in much more tortuous paths where water films remain.

DESCRIPTIONS OF WATER AVAILABILITY

How can we describe water availability in soil? Field capacity, gravitational water, capillary water, and hygroscopic water are all descriptive terms for water availability.

Gravitational water is water that can drain out of soil pores. Field capacity is the water held in soil once all of the larger pores have drained. Capillary water is held in pores and at mineral surfaces. Capillary water can be available or unavailable to plants. The point at which it becomes unavailable to plants is called the wilting point. Hygroscopic water is water that can be removed by oven-drying soil or is the water that remains in soil that is air-dried.

There are also quantitative descriptions of water that soil microbiologists use. They are useful for quantitatively describing the amount of water in different soils. Gravimetric water content (ω) describes the weight of water in a given mass of soil:

$$\frac{\text{Grams of water}}{\text{Grams of oven-dry soil}}$$

Volumetric water content (θ_v) is:

$$\frac{\text{Grams of water}}{\text{Volume of soil}}$$

Gravimetric water and volumetric water are usually expressed as a percent. You can find the volumetric water content by multiplying the gravimetric water content by the bulk density:

$$\text{Volumetric water content } (\theta_v) = \omega\rho_b$$
$$= (\text{Grams of water/grams soil})(\text{Grams soil/cm}^3 \text{ soil})$$

The percent water-filled pore space is:

$$\frac{100(\theta_v)}{\text{Total porosity}}$$

Total porosity (TP) = $(1 - \rho_b/\rho_p)100$. ρ_p is particle density and is usually given as 2.65 g/cm^3 in most soils dominated by silicate minerals.

All of these terms reflect the influence of soil structure and mineralogy on water availability. But as you already know, the soil structure and mineralogy of all soils are unique and affect the availability of water. For the same amount of water, less is available in clay soils than in sandy soils. So, some term is needed to reflect the actual availability of water in different soils. This term is "water potential."

WATER POTENTIAL (Ψ)

Water potential (ψ) is a mathematical description of water availability. Water flows to plant roots and microorganisms along free energy gradients. Water potential (ψ) measures the tendency of water to flow—the more likely it is to flow, the more available the water is. In more exact terms, water potential (ψ) is a measure of the potential energy of water at a point in a system relative to the potential energy of pure, unbound water (Papendick and Campbell 1981). Water flow is spontaneous from high to low water potential. Water potential (ψ) is almost always negative because the moment you add something to water, the water molecules begin interacting with it and they can do less work; they are less likely to move freely. The preferred units of water potential (ψ) are megapascals (MPa). A MPa is 1×10^6 pascals. One atmosphere = 1.013 bar = 101.3 $\times$ 10^3 Pa = 0.1013 MPa.

THE COMPONENTS OF WATER POTENTIAL

The important components of water potential (ψ) in soil can be described as:

$$\psi_{soil} = \Sigma\,\psi_{\pi} + \psi_{m} + \psi_{g}$$

That is, the total water potential in soil (ψ_{soil}) equals the sum of an osmotic potential (ψ_{π}), a matric potential (ψ_{m}), and a gravitational potential (ψ_{g}). High water potential (ψ_{soil} > –0.03 MPa or > –0.3 atm) means unstressed soil. Low water potential (ψ_{soil} < –3 MPa or < –30 atm) means a water-stressed soil.

Osmotic potential (ψ_{π}) is always negative and is due to solutes in the soil water. Osmotic potential becomes more negative as more solutes are added to a soil and the soil becomes more saline. Matric potential (ψ_{m}) is always negative and is due to adsorption by the soil solid phase. As the soil surface area increases, ψ_{m} becomes more negative. Gravitational potential (ψ_{g}) can be positive or negative depending on the position of water in a gravitational field relative to the reference level. Gravitational potential (ψ_{g}) is proportional to elevation differences from the reference level and is usually positive with respect to soil microorganisms.

Matric potential (ψ_{m}) is the most important component of water potential in unsaturated soil. At saturation, ψ_{m} = 0. At high matric potential (ψ_{m} = 0 to –0.1 MPa) the amount of water retained by soil is strongly influenced by soil structure. Microbial activity is reduced much more by reducing ψ_{m} than by reducing ψ_{π}. Reducing ψ_{m} affects substrate availability because the substrates become increasingly unable to diffuse. Diffusion is affected severely at ψ_{m} = –0.1 to –1.0 MPa. Osmotic potential becomes important in saline soils or fertilizer-amended soils.

Soil microorganisms have their own water potential. The components of water potential that are important in soil microorganisms are the osmotic potential (ψ_{π}) and pressure potential (ψ_{p}), or turgor pressure. Osmotic potential (ψ_{π}) in microorganisms is negative and is due to solutes in the cytoplasm such as inorganic nutrients, amino acids, carboxylic acids, and carbohydrates. Pressure potential (ψ_{p}) is positive.

Water potential in soil microorganisms ($\psi_{microorganisms}$) is in near equilibrium with the environment because the cell membrane is relatively permeable to water.

Let's say you have the system diagrammed in Figure 14-1. On one side of the chamber you have absolutely pure water. On the other side of the chamber you have a mixture of water and dissolved solutes. Between the two is a semipermeable membrane that lets water flow back and forth but doesn't let solutes move. Water moves into the more

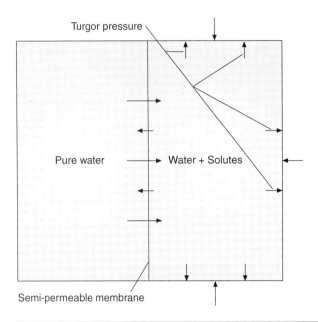

Figure 14-1   Schematic diagram of an experiment to show osmotic potential and turgor pressure. The direction of the arrows across the semipermeable membrane shows the overall direction of water flow and the size of the arrows indicates the magnitude of flow. The small arrows indicate the force of incoming water attempting to increase the volume of the chamber while the external arrows show the resistance of the chamber walls to expansion.

saline chamber because it has a negative ψ_π. This causes the chamber to expand because its volume has increased. The chamber resists expansion, however, like the walls of a balloon (the more the balloon expands, the more difficult it becomes to fill). This is equivalent to turgor pressure, or ψ_p.

Be careful when you refer to water potential because it can be confusing. Since water potential is almost always negative, as are the components that are most important to soil microorganisms, a decreasing water potential means greater water stress, and an increasing water potential means decreasing water stress. Remember that $\psi_{soil} = -0.01$ MPa $> \psi_{soil} = -0.3$ MPa.

High ψ_{soil} $\xrightarrow{\text{Decreasing } \psi_{soil}}$ Low ψ_{soil}

-0.03 MPa   -0.3 MPa   -3.0 MPa

$\xleftarrow{\text{Increasing } \psi_{soil}}$

RESPONSE OF MICROORGANISMS TO CHANGES IN WATER POTENTIAL (Ψ)

Water moves in and out of microorganisms in response to changes in the water potential gradients of their environment (Figure 14-2).

Potato Sticks and Water Potential

How can you show that changes in osmotic potential affect cells? Try the potato stick experiment. Take a fresh potato and cut it into long, thin sticks. Make sure each stick is exactly the same length. Put some of the sticks into a cup of distilled water so that they're completely covered. Place some of the sticks into a cup of salt water (the more salt, the better). Wrap some of the sticks in cellophane and store them in the refrigerator (these are your controls—you could incubate them in a cup of potato juice, but that would be a lot of work).

Wait a couple of hours and then compare the lengths of all your potato sticks. Those incubated in distilled water should be longer than your controls and those incubated in salt water should be shorter than your controls. Why? Because the osmotic potential of the potato sticks was lower than the osmotic potential of the distilled water. So water flowed into the sticks and caused them to expand to the point where turgor pressure stopped further water uptake. Conversely, the osmotic potential in the potato sticks was higher than the osmotic potential in the salt water, so water flowed out of the potato sticks and they got shorter.

Try it at home. It really works.

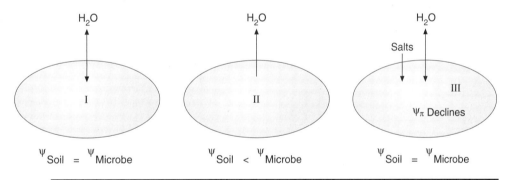

Figure 14-2   Schematic diagram of a microbial cell response to a decrease in the ψ_{soil} of its environment. The cell and its environment are initially in equilibrium (I). If the ψ_{soil} declines, the water in the microbial cell starts to move out of the cell in response to the new water potential gradient (II). The cell can take in salts to artificially lower its own water potential ($\psi_{microbe}$) (III) until it is back in water potential equilibrium with its environment.

Water loss increases the internal solute concentration in a microorganism and ψ_{π} decreases it. As water leaves the cell and salts enter the cell, the membrane distorts. Microorganisms could lose water and take in solutes until:

$$\psi_{microorganism} = \psi_{soil}$$

This will have harmful effects because salt toxicity can become a problem. High salt concentrations disrupt cell metabolism because cellular enzymes may not function well at high salt concentrations. This environment will also be energy-demanding because the microorganisms will be creating artificial concentration gradients. Too much water loss causes the cell membrane to pull away from the cell wall—a process called plasmolysis (Figure 14-3). You may recall that before refrigeration, salt-packing meat was one of the ways meat was preserved against microbial decay; the other was sun-drying (e.g., making beef or deer jerky).

Microorganisms have developed several strategies for surviving water stress. They accumulate stress solutes (Na^+, Cl^-, K^+). They form compatible solutes (amino acids such as proline and glutamic acid; quaternary amines such as glycine betaine; sugars such as trehalose; polyols such as glycerol). They constitutively accumulate solutes such as glutamate and arabitol. Constitutive solute accumulation means that the internal microbial ψ_π is kept very low. Consequently, the internal pressure potential (turgor pressure, ψ_p) must be high to balance it. The primary role of the cell wall is to protect the cell and provide the structural rigidity necessary to prevent cell bursting due to increased turgor pressure. Cell walls have limited selective permeability. Gram-positive cells have thick walls to provide this rigidity.

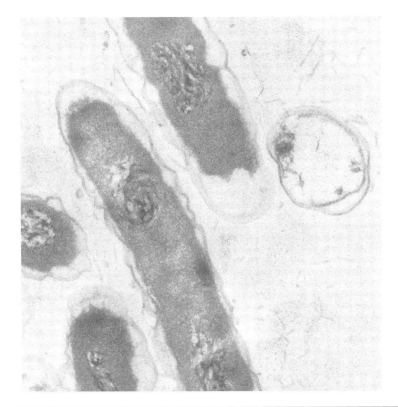

Figure 14-3 Plasmolysis of a bacterium. Note how cell contents have pulled away from the cell (Magnification 48,450×). (Photograph courtesy of M. S. Coyne).

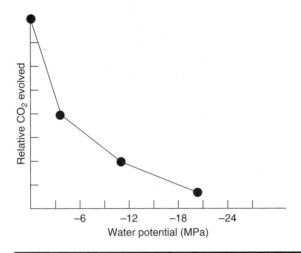

Figure 14-4   Effect of water potential on the decomposition of oat straw. (Adapted from Bartholomew and Norman 1946)

MICROBIAL ACTIVITY

Microbial activity is optimal at a water potential of –0.01 MPa (–0.1 atm, or field capacity). For example, oat straw decomposition declines as the straw dries and as the water potential declines (Figure 14-4). At higher water potentials, aeration becomes a problem because O_2 diffusion is impeded. The microorganisms least tolerant of water stress maximize growth and sacrifice xerotolerance (tolerance to low water) by not maintaining a low internal water potential. The microorganisms most tolerant of water stress sacrifice growth by maintaining high internal solute concentrations against a concentration gradient, but maximize tolerance to water stress. Table 14-1 shows the limits of some other microbial processes in soil to water potential stress.

Table 14-1   Microbial tolerance to matric-controlled water stress.

Maximum Tolerance (ψ MPa)	Water Film Thickness (μm)	Microbial Activity Affected
–0.03	4.0	Denitrification
–0.10	1.5	Movement of protozoa and bacteria
–0.5	0.5	
–1.5	30×10^{-4} (10 water molecules)	Nitrification Sulfur oxidation
–4.0	$< 30 \times 10^{-4}$	Bacterial and actinomycete growth
–10.0	$< 15 \times 10^{-4}$ (< 5 water molecules)	Fungal growth
–40.0	$< 9 \times 10^{-4}$ (< 3 water molecules)	Fungal growth

(Harris 1981)

Summary

Water is important to soil microorganisms. Although soil organisms can obtain nutrients by mass flow and interception, diffusion is the most important way of obtaining nutrients for soil microorganisms. As water films around soil particles become thinner, nutrients and microorganisms must move around more tortuous paths.

Soil water can be described descriptively, as in "gravitational water," and mathematically, as in "gravimetric water content." However, the concept that best describes water availability, and takes into account differences in soil type on water availability, is "water potential" (ψ). Water potential is the mathematical description of the potential for water to flow within soil. Water flows from high to low potential. Water potential is composed of osmotic, matric, pressure, and gravitational components, but only the first three components are of significance to soil microorganisms. Osmotic potential reflects the contribution of solutes, matric potential reflects the contribution of solids, and pressure potential reflects the contribution of cell membranes and cell walls to overall water potential.

Optimum microbial activity is at about –0.01 MPa and declines rapidly as the soil dries. In response to soil-drying, microorganisms have numerous strategies that include taking in external solutes and synthesizing internal solutes on demand or constitutively. The most resistant microorganisms to water stress are fungi, which can withstand water potentials as low as –40 MPa.

Sample Questions

1. What are six reasons why soil water critically affects microbial activity?
2. What does "taxis" refer to?
3. What are the dominant factors controlling diffusion and motility in soil?
4. What are some ways of describing soil water?
5. What are the components of water potential and which are most important?
6. What are some microbial responses to changes in water potential?
7. What are microbial strategies for surviving water stress?
8. Describe classes of microorganisms based on water stress response.
9. What are some examples of the limits of microbial tolerance to water stress?
10. Draw a graph that illustrates the relationship between microbial activity and soil and water potential.
11. For the following pairs of water potentials (ψ), draw an arrow that shows the direction of water flow:

(ψ_{soil})	($\psi_{microorganisms}$)
1.00 MPa	0 MPa
–1.50 MPa	–1.50 MPa
–0.01 MPa	0.50 MPa
–0.02 MPa	–1.00 Mpa

12. What is the mass of water in a 10 g soil sample that has a gravimetric moisture content of 33%? Remember, gravimetric moisture content is based on oven-dry soil.

13. How much water is in a soil sample that has a volume of 5 cm^3, a bulk density of 2 g cm^3, and a gravimetric water content of 10%?

Thought Questions

Here's a home remedy for open wounds that's crazy enough to work. Sprinkle granulated sugar on open wounds or skin ulcerations to help kill bacteria and speed healing. Why do you think it works?

Additional Reading

One of the best books you can get that specifically addresses the topic of water and soil microorganisms is *Water Potential Relations in Soil Microbiology* (Special Publication Number 9, 1981, Soil Science Society of America, Inc., Madison, WI). Some of it is very technical, but much is a clear exposition of the major topics we covered in this chapter.

References

Bartholomew, W. V., and A. G. Norman. 1946. The threshold moisture content for active decomposition of some mature plant materials. *Soil Science Society of America Proceedings* 11: 270–79.

Harris, R. F. 1981. Effect of water potential on microbial growth and activity. In *Water potential relations in soil microbiology*, J. F. Parr et al. (eds.), 23–95. Madison, WI: Soil Science Society of America.

Papendick, R. I., and G. S. Campbell. 1981. Theory and measurement of water potential. *Water potential relations in soil microbiology*. In J. F. Parr et al. (eds.), 1–22. Madison, WI: Soil Science Society of America.

SOIL MICROORGANISMS AND NUTRIENT CYCLES

*T*he nutrient cycles in soil are at the heart of soil microbiology. Indeed, without soil microorganisms, there would scarcely be any nutrient cycling at all and life as we know it could not exist. In this section we look at the cycling of the major inorganic nutrients that plants require for growth: sulfur, phosphorus, iron, and nitrogen. We devote considerable time to nitrogen because it is often the most limiting plant nutrient and because it undergoes many varied transformations. Likewise, we devote a lot of time to examining carbon cycling in soil because carbon, as a structural component of cells and as an energy source, and its cycling, drive almost all other nutrient cycles in the environment. By the end of this section you should have enough information to see how models of decomposition can be used to predict how much nutrient cycling will occur in the soil environment.

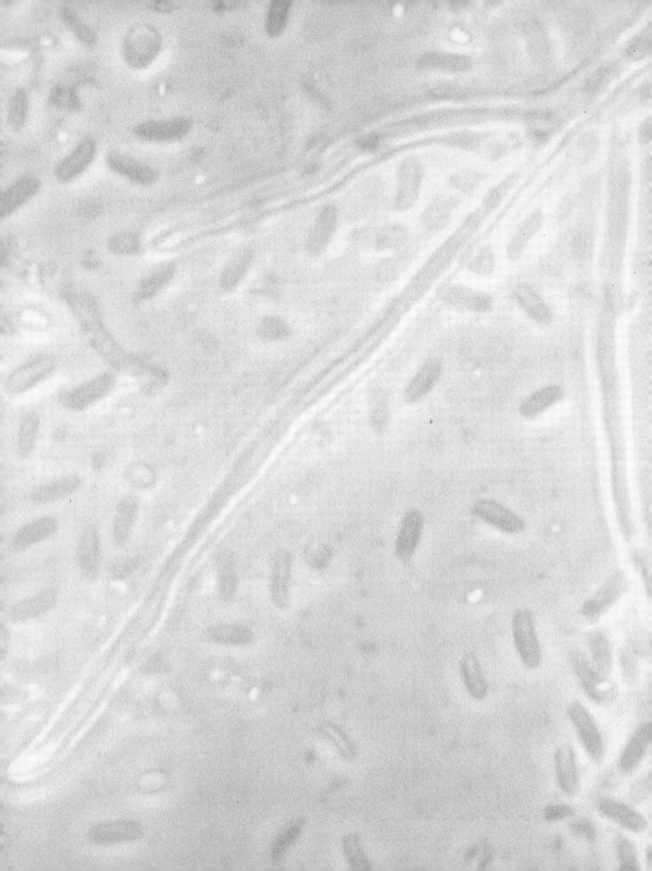

Chapter 15

The Sulfur Cycle

Overview

After you have studied this chapter, you should be able to:

- List the major inorganic, organic, and gaseous forms of S in the environment.
- Combine these S forms into a diagram that depicts the S cycle in nature.
- Describe the processes of S mineralization, assimilation, reduction, and oxidation.
- Explain some of the environmental consequences of microbial S transformations.
- Give examples of specific bacteria that carry out each type of S transformation.

INTRODUCTION

The sulfur cycle was first described by Martinus Beijerinck and Serge Winogradsky in the late 1880s. It's a fascinating topic, because S undergoes many transformations in the environment. One easy way to think of the sulfur cycle is that it is basically an oxidation–reduction cycle with organic and inorganic pools, as illustrated in Figure 15-1.

To simplify things, we'll focus on:

1. What it means for S to be immobilized/assimilated
2. What it means for S to be mineralized
3. What reactions S undergoes in soil under different redox conditions
4. What organisms carry out those reactions, and when
5. How this affects agriculture and our environment

SULFUR TRANSFORMATIONS

Inorganic Transformations

Anaerobic, reductive environments

$$SO_4^{2-} \xrightarrow{\hspace{5cm}} S^{2-}$$
$$\xleftarrow{\hspace{5cm}}$$

Anaerobic, oxidizing environments
S oxidation is energy yielding

Photosynthetic Transformations

Anaerobic, light environments
S is an electron donor

$$H_2S + CO_2 \xrightarrow{\hspace{5cm}} S^0 + (CH_2O)_n$$

Organic Transformations—Immobilization/Assimilation

Aerobic and anaerobic environments
S is converted from an inorganic form to an organic form
Many bacteria and fungi

$$SO_4^{2-} \longrightarrow R - OS + R - SH$$

R represents some larger carbon compound such as a sugar.

Mineralization

$$R - SH \Rightarrow H_2S + R$$
$$R - OS \Rightarrow SO_4^{2-} + R$$

ASSIMILATION/IMMOBILIZATION

Inorganic S forms typically found in soil are:

1. SO_4^{2-} = Sulfate (oxidation state +6)
2. SO_3^{2-} = Sulfite

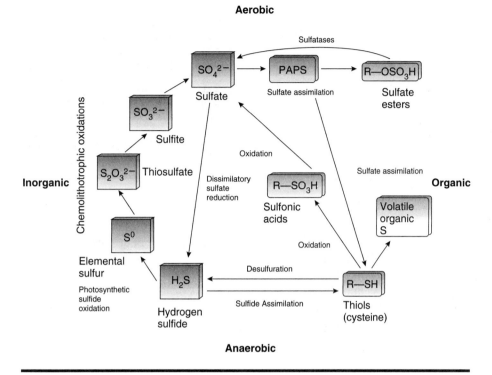

Aerobic

Anaerobic

Figure 15-1   The sulfur cycle in soil. (Adapted from Doetsch and Cook 1973)

3. $S_2O_3^{2-}$ = Thiosulfate
4. S^0 = Elemental sulfur
5. S^{2-} = Sulfide (oxidation state −2)

Sulfate is the dominant form of inorganic S in aerobic environments. The oceans are vast reservoirs of SO_4^{2-}. Elemental S is insoluble, but SO_4^{2-} is an anion and can be leached. This is also the inorganic S form that is taken up by microorganisms and plants. Microorganisms can also assimilate H_2S.

Sulfate assimilation is a reductive process. This means that the assimilated S has more electrons (is more reduced) than when it started. We say that it is immobilized because it is no longer available for reaction in the soil. About 35% of the assimilated S is contained in the C—SH bonds (sulfhydryl bonds) of amino acids and glutathione, for example. The remainder is found in a variety of N—S or —C—O—SO_3^- bonds (sulfate ester bonds). The sulfate ester bonds are easier to decompose (more labile) than sulfhydryl bonds. The principal organic S forms include amino acids such as methionine and cysteine. Most of the plant-available S in soils is tied up in soil organic matter where it forms anywhere from 0.02% to 5.0% of the mass. About 1% to 3% of the organic S becomes available each year through decomposition (mineralization) of the organic matter. Sulfur-deficient soils occur in the Pacific Northwest and Canada where the parent material is low in S, and SO_4^{2-} leaching occurs.

MINERALIZATION

Mineralization is the transformation of an unavailable compound into one that is available (i.e., organic S $\Rightarrow$ inorganic S).

$$R - O - SO_3^- + H_2O \xrightarrow{\text{Sulfatases}} ROH + H^+ + SO_4^{2-}$$
$$R - SH \Rightarrow RSO_2H + RSO_3H$$
$$R - SH \Rightarrow H_2S$$

Mineralization of organic S compounds, like other organic compounds in soil, can be modeled as a first-order reaction. That is, the amount mineralized depends on the substrate concentration. The equation that describes this is:

$$\delta S/\delta t = \ kS$$

The change in substrate concentration (S) per unit time (t) ($\delta s/\delta t$) is equal to a rate constant ($-k$) times the starting substrate concentration (S). The rate constant is negative because the substrate is disappearing. Because the substrate concentration is always changing, the value of $\delta S/\delta t$ does not remain constant. So, if you plotted the substrate concentration against time it would give you a curve, not a straight line (Figure 15-2).

Mathematically, the curve would be described by this equation:

$$S_t = S_0 e^{-kt}$$

The substrate concentration at any given time (S_t) is equal to the starting substrate concentration (S_0) times "e" (the natural, or Naperian base, which has a value approximately equal to 2.718) raised to the power of kt, where $-k$ is the rate constant and t is

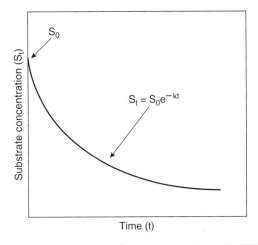

Time (t)

Figure 15-2 Plot of a first-order mineralization reaction.

Table 15-1 Volatile biogenic sulfur compounds.

Compound	Formula	Atmospheric Concentration	Amount Released (10^9 kg/year)
Hydrogen sulfide	H_2S	0.2–1 ppb	16.5–70.6
Sulfur dioxide	SO_2	0.2–5 ppb	15.0
Carbon disulfide CS_2	0.1–0.4 ppb	3.8–4.7	
Carbonyl sulfide COS	0.2–0.6 ppb	2.7–3.5	
Methyl mercaptan	CH_3SH		
Ethyl mercaptan	CH_3CH_2SH		
Dimethyl sulfide	CH_3	SCH$_3$ 58 ppt	39.6–45.4
Dimethyl disulfide	CH_3 SSCH$_3$		1.3–3.4

ppb = parts per billion; ppt = parts per thousand
(From Kelly and Smith 1990)

the elapsed time. It may look complicated, but it's really not, especially if you have a calculator with logarithmic functions.

Mineralization involves ester cleavage and cleavage of C—SH bonds to release products such as methionine. In anaerobic environments, such as swine lagoons, this produces malodorous compounds like methylmercaptan (CH_3SH). Dimethylsulfide (CH_3SCH_3) is a volatile organic S compound that is evolved from oceans. Volatilization of S from soils and plants is mostly in the form of hydrogen sulfide (H_2S). Natural sources of sulfur dioxide (SO_2) primarily come from burning biomass (clearing rain forests, etc.) and volcanoes (Table 15-1).

Once in the atmosphere, H_2S and CH_3SH are oxidized to SO_2 and precipitate as inorganic sulfur. Perhaps half of the S in the atmosphere is industrial and this has led to some serious environmental concerns (Schlesinger 1991). Burning coal produces SO_2, which eventually oxidizes to H_2SO_4. Buildings decay as a result of S in the atmosphere and acid smog. The effect of this is acid rain. Rainwater in equilibrium with CO_2 has a pH of 5.7, but just 1.5 ppm SO_2 lowers the pH to 4.0. Acid rain is very important where soils are not buffered, such as in the northeastern United States. Lichens are extremely sensitive to SO_2 and consequently make good bioindicators of SO_2 pollution.

REDUCTION

Once back in the inorganic pool, S is subject to oxidation and reduction. SO_4^{2-} can be used as a terminal electron acceptor by some microorganisms. Remember that the terminal electron acceptor in aerobically and anaerobically respiring microorganisms is the last compound to accept the electrons that are moving through the respiratory chain.

$$9H^+ + SO_4^{2-} + 8e^- \Rightarrow 4H_2O + HS^-$$

Some microorganisms use thiosulfate as an electron acceptor:

$$S_2O_3^{2-} + 4H_2 \Rightarrow 2HS^- + 3H_2O$$

Sulfate and other oxidized S compounds are used as electron acceptors when the redox potential, E_h, = –100 to –200 mV. The microorganisms that use SO_4^{2-} as an electron acceptor are anaerobic, chemoheterotrophic (chemoorganotrophic) bacteria. *Desulfovibrio desulfuricans* is typical of this group. It is a heterotrophic, gram-negative, curved rod found in soils and sediments. *Desulfovibrio desulfuricans* can also use H_2 as an energy source. Less than 10% of the S is assimilated, so we refer to these reactions as "dissimilative" S reductions. Other sulfate-reducing bacteria (Amann et al. 1992) include:

Desulfuromonas acetoxidans	*Desulfococcus multivorans*
Desulfobacter postgatei	*Desulfosarcina variabilis*
Desulfobacter curvatus	*Desulfovibrio gigas*
Desulfobacterium thermoautotrophicum	*Desulfovibrio vulgaris*
Desulfobacterium vacuolatum	*Desulfomonas pigra*

If you suspect a microorganism is a S reducer, look for the prefix "desulfo" in its name.

Sulfur reduction reduces the availability of S because S may be lost as a gas. Gaseous H_2S is toxic and corrosive—degrading stone and concrete. Hydrogen sulfide also has antimicrobial, antifungal, and antinematode activity. Sulfate reduction can be used to raise soil pH:

$$Na_2SO_4 + Fe(OH)_3 + 9H^+ \Rightarrow FeS + 2NaOH + 5H_2O$$

Sulfate reduction is also involved in pipe corrosion, which is a serious problem in sewage treatment plants. The process is known as cathodic depolarization, a spontaneous oxidation of iron-containing compounds that causes iron loss, and it is accelerated by SO_4^{2-}-reducing bacteria.

The overall reaction is shown below (Boopathy and Daniels, 1991). Sulfur reduction continuously precipitates the iron and removes the electrons from the system; this drives the overall reaction forward.

$$4H_2 + 4FeO + SO_4^{2-} \Rightarrow FeS + 3Fe(OH)_2 + OH^-$$

Sulfate reduction in sewage systems leads to the formation of sulfides that are corrosive, toxic, and smelly. One solution is to add NO_3^- to the system. Why does this work? Nitrate is a more readily available electron acceptor than is SO_4^{2-} (it has a higher E_h), so it is used preferentially by microorganisms to deposit their excess electrons and this inhibits S reduction. Sulfate reduction is retarded by any electron acceptor like NO_3^-, Fe^{3+}, or Mn^{4+} that has a higher E_h.

OXIDATION

Reduced S in anaerobic sediments and water can be oxidized by organisms for energy before it reaches aerobic zones. Photolithotropic H_2S use occurs in green and purple sulfur bacteria. These are cocci, vibrios, rods, spirilla, and budding and gliding bacte-

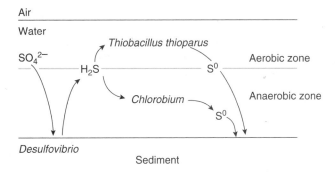

Figure 15-3   Stratification of sulfur transformations in sediment. Both *Thiobacillus* and *Chlorobium* oxidize sulfur. *Thiobacillus* does so aerobically and *Chlorobium* does so anaerobically and photosynthetically using the H_2S to supply electrons for CO_2 reduction. *Desulfovibrio*, which produces the H_2S, grows on dissolved organic carbon. (Adapted from Doetsch and Cook 1973).

ria found in mud and stagnant water containing H_2S and exposed to light. They carry out anaerobic, photosynthetic sulfur oxidation using S^0 and S^{2-} as a source of electrons.

$$CO_2 + 2H_2 S \xrightarrow{\quad Light \quad} (CH_2O)_n + 2S + S + H_2O$$

$$CO_2 + 2S + H_2O \xrightarrow{\qquad\qquad} 3(CH_2O)_n + SO_4^{2-}$$

Purple sulfur bacteria include Rhodospirillaceae and Chromatiaceae such as *Rhodospirillum* and *Rhodopseudomonas*. They have bluish violet, purple, deep red, and orange-brown coloration. The color is caused by photosynthetic pigments such as bacteriochlorophylls and carotenoids. These bacteria deposit S inside the cell and reutilize it when S concentrations are low in their environment.

Green sulfur bacteria are represented by Chlorobiaceae and Chloroflexiaceae such as *Chlorobium*. The green color is due to chlorophyll. These bacteria deposit elemental sulfur (S^0) outside the cell. They don't further oxidize the S^0 because they are acid sensitive. Both groups of photosynthetic S oxidizers reoxidize H_2S from more anaerobic depths (Figure 15-3).

OTHER ORGANISMS THAT OXIDIZE SULFUR

Mesophiles, moderate thermophiles, and thermophiles are the predominant S oxidizers in soil (Ahonen and Tuovinen 1992). Truly cryophilic S oxidizers are not known. The typical temperature range is 4° to 37°C with an optimum of around 28° to 37°C. Mesophiles (*Thiobacillus ferrooxidans, Leptospirillum ferrooxidans*) can be active in bioleaching piles at temperatures up to 40°C. Moderate thermophiles can grow at 40° to 60°C. *Sulfolobus* can thrive at 80°C. *Sulfolobus acidodurans* has an optimum pH of 0.5. It is a thermophilic S oxidizer with an optimum temperature of 60° to 80°C (range 21° to 90°C). It grows in hot, acidic, S soils (sulfatara soils) where the temperature may be 85° C and the pH 2.5 (it grows in a pH range from 0.7 to 5).

How Do You Make a Winogradsky Column?

A Winogradsky column is a self-contained microbial ecosystem designed to select and enrich photosynthetic sulfur-oxidizing bacteria. Put some rice grains and shredded paper in the bottom of a clear tube or bottle. Fill the tube about three-quarters full with a 1:1:1 mixture of sediment from the bottom of a stream, soil, and gypsum ($CaSO_4$). Try to avoid leaving any air pockets. Top off the tube with tap water and cover it with aluminum foil. Then put it in the light and wait.

After about a day, you may see potworms wriggling at the surface of your column. In another day or two you should start to see a black coloration developing at the bottom of the tubes and in isolated spots throughout the tube. Within a week you should start to see patches of green, brown, and purple developing wherever the column is exposed to light.

The black coloration is caused by H_2S precipitating with reduced metals at the bottom of the tube where *Desulfovibrio* and other SO_4^{2-} reducing bacteria are growing on the carbohydrates in the rice grains and paper. The colored patches are photosynthetic bacteria oxidizing the H_2S that gradually percolates up the column. If you cover half the column with paper to block the light, you can show that the black color still develops but the pigments appear only on the light side of the column. If you take off the aluminum foil cap and look at it, it should appear tarnished. That's also the work of the H_2S escaping into the atmosphere.

As S is oxidized, acidity is generated:

$$1\tfrac{1}{2}O_2 + H_2O + S^\circ \Rightarrow H_2SO_4 \text{ (pH declines)}$$

Sulfur oxidizers are mostly organotrophic. Filamentous S oxidizers include *Beggiatoa, Leucothrix,* and *Thiothrix.* They deposit S^0 granules intracellularly. *Beggiatoa* grows in the rhizosphere of rice and may protect the rice from H_2S in flooded sediment by oxidizing the H_2S to elemental S. *Sulfolobus acidodurans* is a chemolithotroph. It gets electrons (energy) from S oxidation and C from CO_2. *Thiomicrospora* and *Thermothrix* are two other S oxidizers. *Thermothrix* grows in neutral hot springs.

Many soil heterotrophs are able to oxidize S although they do not appear to gain useful energy from the oxidation. Examples of these microorganisms include *Arthrobacter, Bacillus, Micrococcus, Mycobacterium,* and *Pseudomonas.* The S-oxidizing soil heterotrophs are more numerous in soil than are chemolithotrophs or photolithotrophs and they are important because they are the primary oxidizers of S in neutral and alkaline soil until the pH falls low enough for thiobacilli to grow.

THIOBACILLI

No discussion of S oxidation would be complete without including *Thiobacillus* species; 22% of this genus are facultative and 13% are autotrophic S oxidizers.

Thiobacilli oxidize S°, H_2S; $S_2O_3^{2-}$ (thiosulfate), and FeS_2. They carry out the following reactions:

$$H_2O + S^\circ + 1\frac{1}{2}O_2 \Rightarrow H_2SO_4$$
$$Na_2S_2O_3 + 2O_2 + H_2O \Rightarrow 2NaHSO_4$$
$$FeS_2 + 3\frac{1}{2}O_2 + 2H_2O \Rightarrow FeSO_4 + H_2SO_4$$

Thiobacilli can be responsible for deposition of elemental sulfur (S°).

Normally soils have few thiobacilli, but if S is added, the population increases dramatically. Adding S° to soils acidifies them. As the thiobacilli increase, actinomycetes, which grow less well in acidic conditions, decrease. This is one method of controlling plant diseases caused by actinomycetes, such as potato scab. The consequences of oxidizing S in some environments are serious. Draining marine sediments and mangrove swamps that are high in FeS_2 and FeS leads to tremendous microsite S oxidation.

$$FeS_2 \text{ or } FeS + O_2 \Rightarrow H_2SO_4 \text{ pH } 4.0$$

Likewise, when coal mines are exposed to air, acid mine drainage can occur. The pyrite (FeS_2) intermixed with the coal deposits during their formation becomes a perfect substrate for S- and Fe-oxidizing bacteria, which grow and produce so much acidity that they kill streams.

Thiobacilli vary in their tolerance of acidity (Table 15-2). *Thiobacillus thioparus* and *Thiobacillus denitrificans* grow when pH is 5 to 8. *Thiobacillus thiooxidans* is an obligate aerobe that grows at pH 1.5–5.0. *Thiobacillus ferrooxidans* grows at pH 1.5 to 5. It oxidizes S and Fe^{2+} to Fe^{3+} for energy. Iron-oxidizing *Thiobacillus* has a temperature optima around 30°C but can be active at 2° to 4°C. *Thiobacillus intermedius* grows between pH 3 and 7. It is a facultative lithotroph.

While S is transformed throughout almost the entire range of conditions that permit life, the various bacteria have distinct conditions in which they must live, as Figure 15-4 illustrates.

Table 15-2 pH range of thiobacilli.

Thiobacillus Species	pH Range	Optimum pH	Energy Substrates
T. thiooxidans	1.5–5.0	2.0–3.5	S°, $S_2O_3^{2-}$, $S_4O_6^{2-}$
T. ferrooxidans	1.5–5.0	2.0–3.5	S°, $S_2O_3^{2-}$, Fe^{2+}
T. intermedius	3.0–7.0		
T. thioparus	5.0–8.0	7.0	S^{2-}, S°, $S_2O_3^{2-}$, $S_4O_6^{2-}$
T. novalis	5.0–8.0	7.0	$S_2O_3^{2-}$ Organic compounds
T. denitrificans	5.0–8.0	7.0	S^{2-}, S°, $S_2O_3^{2-}$, $S_4O_6^{2-}$

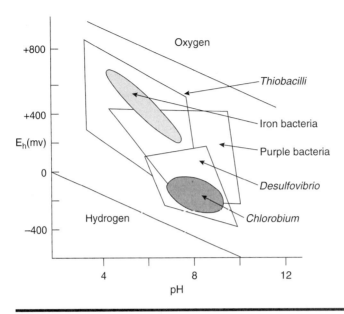

Figure 15-4   Distribution of different sulfur-transforming microbes in the environment based on their acceptable pH and E_h limits. (Adapted from Doetsch and Cook 1973)

Summary

The S cycle is best described as an oxidation–reduction cycle with organic, inorganic, and gaseous S pools. The major organic S compounds are the amino acids cysteine and methionine. Sulfate is the major inorganic S compound in aerobic soils. Sulfide is the major inorganic form of S in anaerobic soils. Sulfide is usually found precipitated as metal sulfides such as FeS or is a gas, hydrogen sulfide, H_2S. The H_2S can be oxidized photosynthetically and anaerobically by bacteria such as *Chlorobium.*

The primary microorganisms responsible for making H_2S are SO_4^{2-}-reducing bacteria such as *Desulfovibrio.* When reduced S compounds are exposed to aerobic conditions, they can be used to generate energy by S-oxidizing bacteria, best typified by the thiobacilli. When S is oxidized, it generates acidity, so many of the thiobacilli are acid tolerant. Acid mine drainage occurs when S-oxidizing bacteria oxidize the reduced S in pyrite (FeS_2). This has devastating environmental effects.

Sample Questions

1. Which microbial population would you expect to see dominating an acidic sulfate soil?
2. Based on what you know about the role of S oxidation in the formation of acid mine drainage, illustrate or describe some potential biological solutions that will either prevent acid mine drainage from occurring or control its products.

3. Outline the biologically mediated reactions of the S cycle. Under what conditions are each of these reactions important in soil? In which reactions are soil bacteria and fungi important?

4. Why is flour of sulfur (S^0) commonly used as a soil amendment for gardens? What is the mechanism of this effect?

5. Draw a simplified diagram of a Winogradsky column and write in the potential reactions that inorganic S compounds can undergo, and the physiological groups that carry out those reactions. Name specific examples.

6. Why does adding NO_3^- to sediment stop SO_4^{2-} reduction?

7. Draw a diagram of the inorganic S cycle showing the major S pools in nature and the environments in which they occur.

8. What are four consequences of SO_4^{2-} reduction in anaerobic environments?

9. Given the following reaction that is used to raise the pH of acid soils:
$$Na_2SO_4 + Fe(OH)_3 + 9H^+ \Rightarrow FeS + 2NaOH + 5H_2O$$
what conditions would you recommend creating to accelerate the process?

10. Draw a graph that shows what you think would happen with time to the pH of a marine sediment that is drained for agricultural purposes. To what mechanism(s) do you ascribe your prediction?

11. How can S reduction contribute to pitting of buried iron pipes?

12. What type of metabolism does oxidation of H_2S by *Thiobacillus* represent?

Thought Question

You are trapped on a deserted island with a sheep and you have just drained a mangrove swamp that developed in marine sediment. Unfortunately, you have a difficult time growing anything but the most acid-tolerant plants. What do you think is the cause of this problem, and how will you fix it?

Additional Reading

Introduction to Bacteria and Their Ecobiology by Doetsch and Cook (1973, University Park Press, Baltimore, MD) is a good reference for looking at the sulfur cycle in terms of its practical significance to environmental phenomenon. W. H. Schlesinger (1991, *Biogeochemistry: An analysis of global change,* Academic Press, San Diego, CA) puts the overall sulfur cycle in terms of a global perspective.

References

Ahonen, L., and O. H. Tuovinen. 1992. Bacterial oxidation of sulfide minerals in column leaching experiments at suboptimal temperatures. *Applied and Environmental Microbiology* 58(2):600–606.

Amann, R. I., J. Stromley, R. Devereux, R. Key, and D. A. Stahl. 1992. Molecular and microscopic identification of sulfate-reducing bacteria in multispecies biofilms. *Applied and Environmental Microbiology* 58(2):614–23.

Boopathy, R., and L. Daniels. 1991. Effect of pH on anaerobic mild steel corrosion by methanogenic bacteria. *Applied and Environmental Microbiology* 57:2104–08.

Doetsch, R. N. and T. M. Cook. 1973. *Introduction to bacteria and their ecobiology.* Baltimore, MD: University Park Press.

Kelly, D. P. and N. A. Smith. 1990. Organic sulfur compounds in the environment: Biogeochemistry, microbiology, and ecological aspects. *Advances in Microbial. Ecology* 11:345–85.

Schlesinger, W. H. 1991. *Biogeochemistry: An analysis of global change.* New York: Academic Press.

Chapter 16

The Phosphorus Cycle

Overview

After you have studied this chapter, you should be able to:

- Draw the soil P cycle.
- List the important inorganic and organic forms of P in soil.
- Describe how microbial activity makes P available for plant uptake.
- Explain how soil management can affect the distribution of P-containing compounds.
- Discuss the environmental consequences of excessive P availability.

INTRODUCTION

Phosphorus is second only to N as an inorganic nutrient needed by plants and microorganisms. It is an essential component of RNA, DNA, ATP, and phospholipids. Phosphorus is not an abundant component of the environment, and long-term cultivation without fertilization depletes soil P. Consequently, there are certain environments, notably freshwater lakes and streams, where P is probably the most limiting nutrient for plant and microbial growth.

Phosphates were once used in detergents because they precipitated Ca^{2+} and Mg^{2+} cations that interfered with how well detergents worked. However, phosphates were eventually banned from use for this purpose. When the wastewater that contained phosphates was discharged, it caused considerable environmental damage because the phosphates contributed to eutrophication in P-limited environments (eutrophication means excessive or abundant growth). In this chapter, we briefly discuss some of the forms of P in the environment, and examine how microorganisms are involved in biological transformations of P. We also look at some of the ecological ramifications of P use and abuse in the environment.

FORMS OF PHOSPHORUS

Phosphorus, unlike C, N, and S, does not have biological fluxes to and from the atmosphere, except in the most extreme conditions. Most P is in rocks and soil, and once it gets to the sea, it is permanently lost. Consequently, the largest reservoir of P is in ocean sediments. Phosphorus is present in the terrestrial environment in several forms and in several major pools that can be categorized as absorbed P (soluble), organic P, and mineral P (Figure 16-1). The absorbed form of P is the anion orthophosphate, PO_4^{3-}, although at the pH of most soils it is found as mono-($H_2PO_4^-$) and dibasic (HPO_4^{2-}) P. Orthophosphate precipitates with Ca^{2+}, Mg^{2+}, and Fe^{2+} at neutral and alkaline pH, although at high pH it once more begins to become available because it is associated with Na^+ ions. Unfortunately, this doesn't do plants any good, because the alkalinity inhibits their growth.

Much of the organic P in soil is in unidentified forms. The most common identified form is inositol phosphate (Figure 16-2). Phytin is inositol hexaphosphate (inositol with $6PO_4^{3-}$), one of the most common forms of plant-produced organic P. For unknown reasons, more inositol phosphate is found in forest soils than in grassland soils. Since inositol phosphate is not typically of microbial origin, this probably reflects the different storage forms for organic P that occur in vegetation. Inositol phosphate may be 15% to 30% of the total organic P in soil. Other identifiable organic forms of P in soil are nucleotides (2% to 5%) and phospholipids (1% to 2%).

There are over 200 mineral forms of P in soil. Some of the most common mineral forms are the apatites, which have the general formula:

$$M_{10}(PO_4)_6X_2$$

where:

$$M = Ca, Mg$$
$$X = F, Cl, OH, CO_3^{2-}$$

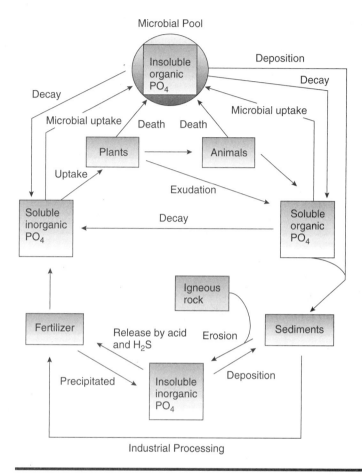

Figure 16-1   The terrestrial phosphorus cycle.

For example, $Ca_{10}(PO_4)_6F_2$ is fluoroapatite. Fluoroapatite is the same as enamel—the material covering your teeth. Cities fluoridate their water to prevent this important mineral, which fights tooth decay, from dissolving.

REDOX REACTIONS

Microbial transformations of P do not generally alter its oxidation state. Phosphorus is almost always in the +5 state (PO_4^{3-}) except under extreme reducing conditions. Neither plants nor microorganisms reduce P when they incorporate it into organic matter. Redox reactions occur and it is possible to reduce HPO_4^{2-} to HPO_3^{2-} and further, but this process is usually not important, and the role of microorganisms in the redox reactions is not clear.

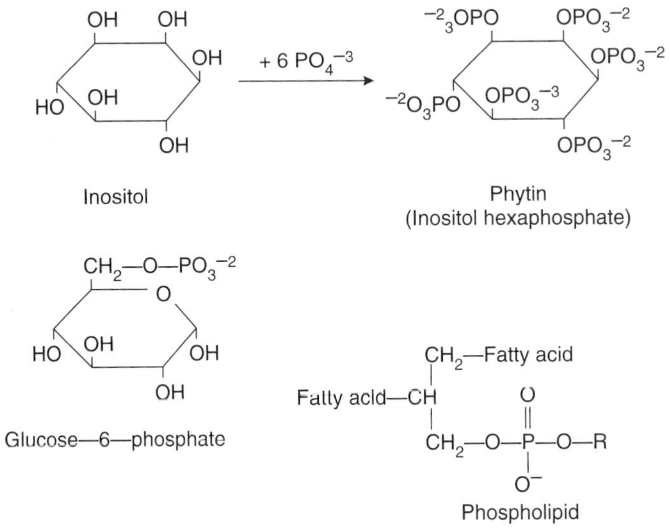

Figure 16-2 Common organic phosphorus compounds in soil.

$$+5 \qquad\qquad +3 \qquad\qquad +1 \qquad\qquad -3 \leftarrow \text{Oxidation state}$$

$$H_3PO_4 \Rightarrow \qquad H_3PO_3 \Rightarrow \qquad H_3PO_2 \Rightarrow \qquad PH_3$$

Phosphate   Phosphite   Hypophosphite   Phosphine

Some microorganisms apparently use PO_4^{3-} as a terminal electron acceptor. However, this is very unusual, especially if reduction occurs all the way to the level of PH_3. Phosphine is a volatile gas and spontaneously ignites with a greenish glow on exposure to O_2. It is produced only after SO_4^{2-} and CO_2 reduction, which means the redox potential has to be exceptionally low ($E_h = -345$ mV at pH 7.0).

Phosphine, Swamps, and Buried Treasure

Microbial processes in soil occasionally find their way into literature. Early in Bram Stoker's gothic novel *Dracula*, Stoker describes a scene in which Count Dracula, disguised as a coachman, drives Jonathan Harker to castle Dracula. To Harker's surprise, the Count periodically stops and wanders into the swamps surrounding their route and is seen driving stakes into glowing greenish spots in the distance. Spontaneous phosphine ignition apparently occurs in swamps where it ignites swamp gas (CH_4) and glows eerily at night. In eastern European folklore, the greenish glow marks the spot where the devil and his minions have buried treasure. The bravehearted drive a stake into the glowing spot to mark it (but only at night) and return to dig up the treasure the next day. It is very bad luck to be greedy and dig up the treasure that same night.

There is evidence that some organisms oxidize phosphite:

$$HPO_3^{2-} \Rightarrow HPO_4^{2-}$$

but there is no evidence that this oxidation yields energy useful for the metabolism of the microorganism.

MICROBIAL CYCLING—SOLUBILIZATION

Microbial cycling of P involves transforming P between inorganic and organic pools and insoluble and soluble forms. Microorganisms have a major role in solubilization, immobilization, and mineralization. The amount of dissolved P in soil at any time varies between 0.1 and 1 kg per hectare (Troeh and Thompson 1993). Since crops require 10 to 30 kg of P per hectare, the ability of microorganisms to solubilize and mineralize the available P pools in soil is vital.

Bacteria that actively solubilize P represent about 10% of the soil microbial population. They are primarily rhizosphere organisms such as *Bacillus, Micrococcus, Mycobacterium, Pseudomonas,* and some fungi. There are three basic mechanisms of solubilizing mineral P and making it more available: chelation, iron reduction, and acidification. All three methods destabilize the minerals in which P is found.

Organic compounds made by microorganisms, such as oxalic acid, can chelate (bind) Ca^{2+}, Mg^{2+}, and Fe^{3+}, thus destabilizing the phosphate mineral and making P soluble. Ferrous iron (Fe^{2+}) is more soluble than ferric iron (Fe^{3+}). Consequently, any mineral in which Fe^{2+} is found will slowly dissolve. Anaerobically, for example:

$$H_2S + Fe(PO_4)_2 \Rightarrow FeS + 2HPO_4^{2-}$$

Figure 16-3 is an interesting illustration of this point. The available phosphorus (on the *y*-axis) is greater in air-dried soils compared to waterlogged soils, with

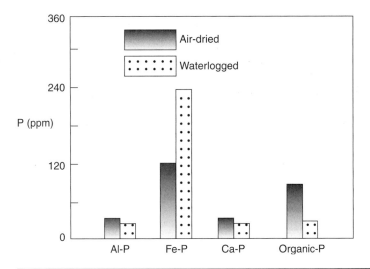

Figure 16-3   Effect of waterlogging conditions on phosphorus availability. (Adapted from Patrick and Khalid 1974)

one exception—when iron-phosphate compounds are examined. None of the other P-containing compounds can undergo the reduction reactions that Fe can. This suggests that the reason the P becomes more available when the soil is waterlogged is because the Fe becomes reduced and destabilizes the mineral (Patrick and Khalid 1974).

Acid production by microorganisms dissolves minerals. Thus, organic acids, nitric acid (produced by nitrifiers), sulfuric acid (produced by thiobacilli), and carbonic acid (H_2CO_3) all release P from mineral forms. For example, before P fertilizers were readily available, an effective means of solubilizing P was mixing soil + manure + S^o + rock phosphate.

$$S^o \Rightarrow H_2SO_4 \text{ } \textit{Thiobacillus}$$
$$Ca_3(PO_4)_2 \Rightarrow H_2PO_4^-$$

You can see visual evidence of microorganisms solubilizing phosphorus minerals by looking for cleared zones on agar plates (Figure 16-4).

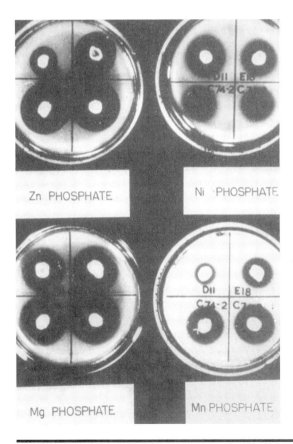

Figure 16-4   Solubilization of phosphorus minerals by microorganisms. (Photograph courtesy of the Soil Science Society of America)

IMMOBILIZATION

The concentration of P in soil solution is typically 0.1 to 1.0 ppm. Microbial concentrations of P are 10 times higher than they are in plants. Phosphorus may compose 0.5% to 1% of the fungal mycelium and 1% to 3% of bacterial biomass. The bulk of P in microorganisms is in RNA (30% to 50%). At low P concentrations, microorganisms accumulate P from inorganic or organic sources at the expense of plants. Bacteria also accumulate excessive P for their needs.

Phosphorus gets into cells after conversion to phosphate esters or ATP and is stored as polyphosphates. The end result is that bacteria can immobilize P and make it unavailable to plants. Immobilization depends on the growth demands of the microorganisms and the proportion of P in organic compounds. Phosphorus equal to 0.3% of the weight of an organic compound is required for the microbial community to develop to its full extent. Phosphorus deficiency can be caused by adding C in excess to soil.

MINERALIZATION

Organic P, which makes up 30% to 50% of the total P in soil, must be mineralized before it is available. Phospholipids and nucleic acids degrade rapidly, but inositol phosphate is slowly mineralized. Mineralization is favored by thermophilic temperatures, neutral-to-alkaline pH, and organic matter that is rich in P.

The principal enzymes involved in P mineralization are phytases and phosphatases (Figure 16-5). Although phytases are widespread among organisms, they are limited by the small amount of phytin in soil. Phytin is adsorbed to soil at acidic pH. Phytases remove one phosphate group at a time from phytic compounds. Inadequate supplies of P stimulate the production of phosphatase and mineralization of labile organic P. Phosphatases can be extracellular. Phosphate monoesterases (a phosphate monoester has the form $R-O-PO_3^{3-}$) have distinct acidic (pH 6.5) and alkaline (pH 11.0) pH optima. A single phosphatase may have multiple substrates. Phosphate diesters ($R-O-P-O-R$) require several enzymes to be completely mineralized. Soils with high levels of inorganic P inhibit phosphatase activity and mineralization of organic P (Dalal 1982).

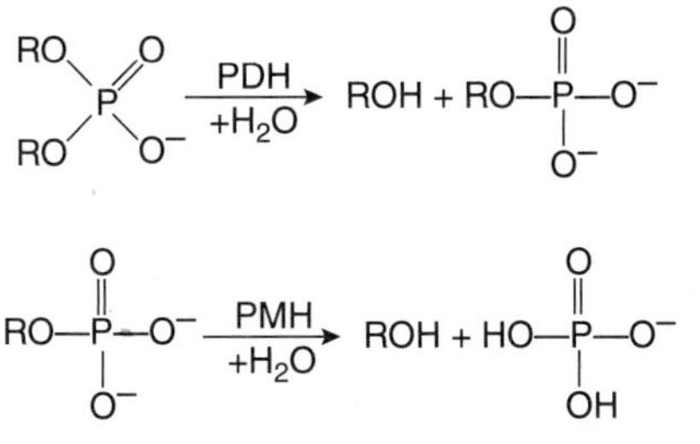

PDH = Phosphoric diester hydrolase
PMH = Phosphoric monoester hydrolase

Figure 16-5   Mineralization of a phosphate diester (a nucleic acid) by phosphatases.

How Do You Assay Phosphatase Activity in Soil?

The typical method for examining phosphatase activity in soil doesn't look for the production of phosphate, although that could be done. However, looking for phosphate directly requires some sensitive and expensive instruments. Instead, soil microbiologists use compounds that are substrates for phosphatases in soil, and they look for the formation of colored products as an indicator of phosphatase activity. The technique is simple, quick, and all you need is an inexpensive spectrophotometer.

The most common substrate is para nitrophenol-phosphate. It's a phosphate monoester that is mineralized by acid and alkaline phosphatases in soil. To assay either type of phosphatase, you simply adjust the pH of your incubation to either acidic or alkaline conditions (Tabatabai 1994).

The para nitrophenol-phosphate is colorless. Para nitrophenol, however, is yellow in alkaline conditions. So, the rate of yellow color production with time (as measured at 420 ηm with a spectrophotometer) is an indicator of the rate at which the substrate (para nitrophenol-phosphate) is being hydrolyzed by phosphatase once the reaction is stopped and the solution is made alkaline.

You can figure out exactly how much phosphate is being hydrolyzed by making a standard curve of various para nitrophenol concentrations. Since there's a one-to-one relationship between the para nitrophenol produced and the phosphate released, measuring the concentration of one is as good as measuring the concentration of the other.

MYCORRHIZAE

No discussion of P in soil would be complete without also discussing Mycorrhizae, although we're going to reserve our discussion of this important group until Chapter 29. Mycorrhizae are fungi that infect plant roots, sometimes in obligatory symbiotic relationships. Mycorrhizae contribute to the overall P nutrition of plants by four main mechanisms. They solubilize mineral P by producing organic acids and CO_2 production during respiration. Mycorrhizae expand the volume of soil from which P is adsorbed because they grow out from the plant roots extending the surface area available for contact with P-containing minerals. Phosphorus uptake may occur at lower soil P concentrations in mycorrhizae than in plant roots. Some mycorrhizae may release phosphatases that mineralize organic P in soil.

ENVIRONMENTAL ASPECTS OF PHOSPHORUS IN SOIL

Soil Management

Phosphorus concentrations and mineralization are affected by soil management. Phosphorus is relatively immobile in soil because it readily precipitates. Consequently, it will stratify in surface horizons where the soil is not cultivated, for example, in no-till soils. The extractable P concentration in the surface 7.5 cm of soil may triple when a soil is in no-tillage rather than conventional tillage (Kladivko et al. 1991). This stratification can take place in as little as 6 years (Eckert and Johnson 1985).

Phosphatase activity in soil ranges from 0.3 to 6.0 µmol PO_4^{3-} hydrolyzed per g soil per hour (Kirchner et al. 1993). Lower rates correspond to conventional tillage or unfertilized soil, and higher rates correspond to no-tillage or fertilized soil. Because inorganic P inhibits phosphatase, phosphatase activity declines and disappears when soluble P contents of soil reach 260 to 350 ppm (Speir and Ross 1978).

Like inorganic P, phosphatase activity is stratified at the surface of undisturbed soils. This makes perfect sense if you consider that both organic matter, which is a major source of available P, and soil microbial populations, the principal source of soil phosphatases, are also stratified at the surface of no-tillage soils compared to conventionally tilled soils. You can see this phenomenon in the data graphed in Figure 16-6. Acid

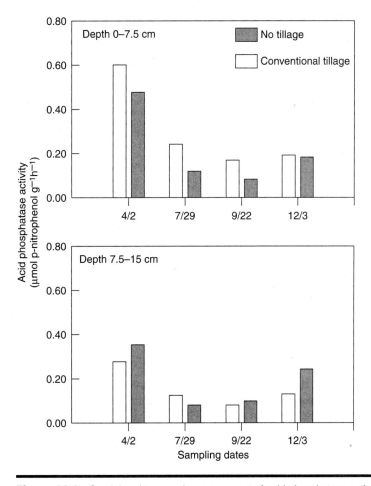

Figure 16-6 Spatial and temporal measurement of acid phosphatase activity in no-tillage (NT) and conventional tillage (CT) soils. (Adapted from Handayani 1996)

phosphatase was measured in soil samples from no-tillage and conventional tillage soils at two depths. At 0 to 7.5 cm, phosphatase activity is, on average, 37% greater in no-tillage soils than in conventional tillage soils. At lower depths, however, not only is phosphatase activity lower in general, but in most cases it is also higher in conventional tillage soils.

Note that there is also a distinct temporal aspect to phosphatase activity in soil. It is highest in spring and lowest in fall. These results were significantly correlated with both microbial biomass and water. As you would expect, the more microorganisms present, the more phosphatase that can be produced. Since microbial populations and activity are strongly influenced by the available water in soil (see Chapter 14), the more water present (to a point), the more microorganisms and phosphatase activity you have as well. Gravimetric moisture content in these soils in fall is half what it is in spring.

Eutrophication

Eutrophication refers to a condition in aquatic systems—ponds, lakes, and streams—where nutrients are so abundant that plants and algae grow uncontrollably. As the plants die, the resident microbial population decomposes the material and depletes the available O_2 during respiration. This ultimately kills other aquatic animals, particularly fish, that cannot tolerate low O_2 conditions. Eutrophication is a natural process that occurs as lakes age and fill with sediment, as deltas form, and as rivers seek new channels. When human activity accelerates the process and causes it to occur in previously clean but nutrient-poor water, the process is often referred to as "cultural eutrophication."

Cultural eutrophication is a major problem in watersheds and waterways that are surrounded by urban populations. The nutrients that cause eutrophication usually come from surface runoff of soil and fertilizer in agriculture, or from domestic and industrial wastes discharged into rivers and lakes. Phosphorus and N are two of the nutrients that most limit plant growth in water. When they are supplied, plant growth can explode, and eutrophication can occur. Eutrophication has been observed in surface waters where total P concentrations exceeded 0.05 to 0.15 ppm (Breeuwsma et al. 1995). However, the chemical, biological, and physical processes that occur in individual surface waters dictate that the waters will respond differently to the same P input and management (Combs and Bundy 1995). Some waters are exceptionally sensitive to P while others are not. Many states now restrict the total P that can be applied to land surrounding sensitive watersheds, such as the Hudson River Valley. Preventative action also requires sewage treatment facilities to chemically remove P from the water they discharge.

Summary

Phosphorus is an essential element in nucleic acids, phospholipids, and ATP. It is frequently one of the most limiting elements in soil. Soluble inorganic forms are the orthophosphate anions HPO_4^{2-} and $H_2PO_4^-$. The most common organic form is inositol phosphate. Apatite, $Ca_{10}(PO_4)_6(OH)_2$, is a common mineral form. Although P can change its valence state from +5 (PO_4^{3-}) to –3 (PH_3), it does not usually undergo redox reactions in soil nor is it lost as gaseous forms. Mineral P is solubilized by chelation, iron reduction, and acidification. Organic P is mineralized by enzymes such as phytases and phosphatases to release inorganic orthophosphate. Mycorrhizae are important to P nutrition in plants because they solubilize mineral P, expand the plant root area, and release phosphatases. Phosphorus in soil is relatively immobile and is typically stratified near the soil surface in undisturbed soil. When fertilization with P is used, it can sometimes lead to eutrophication as excess P runs off of fields and into adjacent watersheds where it stimulates aquatic plant growth.

Sample Questions

1. How might thermophilic temperatures help P availability?
2. What role do mycorrhizae play in P uptake by plants?
3. Discuss how soil microbial processes affect P availability. (Consider both free-living microorganisms and plant–microbe associations.)
4. What are the basic methods of microbial P mineralization in soil?
5. What mechanisms do microorganisms use to solubilize P and make it more available?
6. What biological and chemical conditions could lead to P unavailability in soil?
7. Illustrate three ways that microorganisms make P available from P-containing minerals.
8. Based on the following data, what is the effect of soil fertility on mycorrhizal infections?

Treatment	% of Wheat Root Infected	% of Root Length Infected
No fertilization	61	36
N + P + K + Mg	22	10

9. Based on the following data, what is the effect of mycorrhizal infection on *Pinus strobus* nutrition?

	Mycorrhizal	Nonmycorrhizal
Dry mass (mg)	405	303
% N	1.24	0.85
% P	0.20	0.07
% K	0.74	0.43

10. Describe an experiment to assess biomass P.

11. For the following data, what are the approximate values of the important kinetic parameters, K_m and V_{max}?

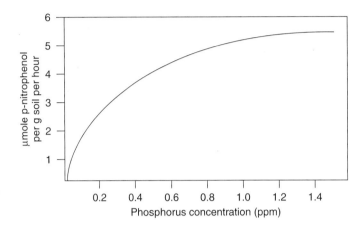

12. For this enzyme assay (what enzyme is being assayed?) what would the velocity of the reaction be if the P concentration was 1.0 ppm? Would this be a concentration in the range of first-order or zero-order kinetics?

Thought Question

You are stranded on a deserted, volcanic island with a sheep and a cargo of rock phosphate $(Ca_3(PO_4)_2)$. How can you use the sheep, the rock phosphate, and a supply of elemental sulfur from a nearby volcano to make phosphorus available for palm growth?

Additional Reading

A recent textbook that has several chapters related to eutrophication is *Pollution Science* (1996, edited by I. Pepper et al., Academic Press, San Diego, CA).

References

Breeuwsma, A., J. Reijerink, and O. Schoumans. 1995. Impact of manure on accumulation and leaching of phosphate in areas of intensive livestock farming. In *Animal waste and the land-water interface*, K. Steele (ed.), 239–50. Boca Raton, FL: Lewis Publishers.

Combs, S. M. and L. G. Bundy. 1995. Waste-amended soils: Methods of analysis and considerations in interpreting analytical results. In *Animal waste and the land-water interface*, K. Steele (ed.), 15–26. Boca Raton, FL: Lewis Publishers.

Dalal, R. C. 1982. Effect of plant growth and addition of plant residues on the phosphatase activity in soil. *Plant and Soil* 66:265–69.

Eckert, D. J. and J. W. Johnson. 1985. Phosphorus fertilization in no-tillage production. *Agronomy Journal* 77:789–92.

Handayani, I. 1996. Soil carbon and nitrogen pools and transformations after 23 years of no tillage and conventional tillage. Ph.D. diss. University of Kentucky, Lexington.

Kirchner, M. J., A. G. Wollum, and L. D. King. 1993. Soil microbial populations and activities in reduced chemical input agroecosystems. *Soil Science Society of America Journal* 57:1289–95.

Kladivko, E. J., G. E. van Scoyoc, E. J. Monke, K. M. Oates, and W. Pask. 1991. Pesticide and nutrient movement into subsurface tile drains on a silt loam soil in Indiana. *Journal of Environmental Quality* 20:264–70.

Patrick, W. H. and R. A. Khalid. 1974. Phosphate release and sorption by soils and sediments: Effect of aerobic and anaerobic conditions. *Science* 186:53.

Speir, T. W., and D. J. Ross. 1978. Soil phosphatase and sulphatase. In *Soil enzymes,* R. G. Burns (ed.), 197–250. New York: Academic Press.

Tabatabai, M. A. 1994. Soil enzymes. In *Methods of soil analysis, part 2: Microbiological and biochemical properties,* R. W. Weaver et al. (eds.), 775–833. Madison, WI: Soil Science Society of America.

Troeh, F. R., and L. M. Thompson. 1993. *Soils and soil fertility.* 5th ed. New York: Oxford University Press.

Chapter 17

Iron and Manganese Transformations

Overview

After you have studied this chapter, you should be able to:

■ Draw a diagram of the iron (Fe) and manganese (Mn) soil cycles.
■ Identify some examples of microbial groups that transform Fe and Mn.
■ Explain why redox transformations in Fe and Mn cycling are so important.
■ Discuss some important environmental consequences of Fe and Mn transformations that affect agriculture.

The sword of Charlemagne The Just is ferric oxide, known as rust.
Anonymous

INTRODUCTION

We continue our discussion of nutrient cycles by examining metal transformation and cycling. Microbial cycling of metals is one of the clearest indications that soils are not inert. Without these cycles, metal transformations would be almost impossibly slow. Microbial metal transformations are essential to soil formation and production of metallic ores. They are important in extracting metals from low-grade ores, acidifying mine wastewater, and polluting water supplies.

Mercury (Hg), arsenic (As), and selenium (Se) are also metals, but they are part of a group termed "heavy metals" that are important environmental contaminants. We focus on iron (Fe) and manganese (Mn) transformations in this chapter because we know more about Fe and Mn than we know about any other metals in soil. Iron and Mn transformations are typically oxidations in which the metals are an energy source, and reductions in which the metals are an electron acceptor. Iron and Mn are also transformed to organic forms (assimilated/immobilized) and the organic forms are transformed back into inorganic forms (mineralization).

IRON (Fe)

Iron is the fourth most abundant element in the Earth's crust and the most abundant metal. Precambrian oxidized deposits known as BIFS (biochemically immobilized ferric substances) consisting of up to 28% iron are associated with the period when oxygen (O_2) first appeared in the Earth's atmosphere. However, BIFS represent O_2 scavenging by purely chemical reactions that occurred as the Earth's atmosphere was changing from an anaerobic (no O_2) to aerobic (O_2 present) state.

Microbial Fe transformations involve Fe scavenging and uptake, Fe oxidation and precipitation, and Fe reduction and solubilization. In aerobic environments, microbial Fe oxidation dominates in acidic environments while chelation dominates in neutral environments. In anaerobic environments, the Fe cycle is dominated by Fe reduction and precipitation of iron sulfides.

Iron-containing compounds are strongly influenced by microbial activity. Oxidation and reduction reactions lead to goethite ($Fe_2O_3 \cdot H_2O$), hematite (Fe_2O_3), and pyrite (FeS_2) formation. Iron hydroxide ($Fe(OH)_3$) oxidation produces goethite. Pyrite comes from the interaction of Fe^{2+} and H_2S (H_2S is a product of SO_4^{2-} reduction). Bog iron comes about from the oxidation of Fe^{2+} that has been bound to organic matter. An overall summary of Fe cycling in soil is illustrated in Figure 17-1.

Iron exists as metallic iron (Fe^0), ferrous iron (Fe^{2+}), and ferric iron (Fe^{3+}). Metallic iron (Fe^0) spontaneously oxidizes in acidic conditions.

$$Fe^0 \Rightarrow Fe^{2+} \text{ at pH} < 5$$

Ferrous iron spontaneously oxidizes to Fe^{3+} when the pH is above 5, but the oxidation is not necessarily rapid; spontaneous reactions are slow from a microbial perspective.

$$Fe^{2+} \Rightarrow Fe^{3+} \text{ at pH} > 5$$

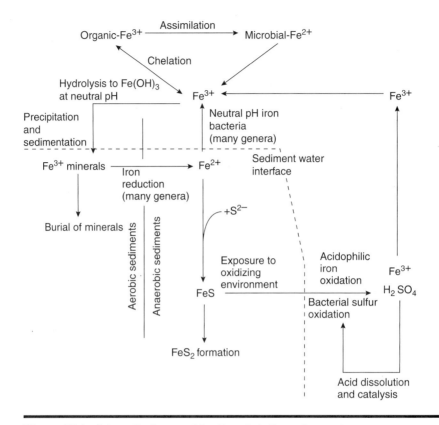

Figure 17.1   Schematic diagram of the Fe cycle in the environment.

Iron Oxidation

In aerated conditions, bacteria obtain energy from oxidizing Fe^{2+}. *Thiobacillus ferrooxidans, Leptospirillum ferrooxidans,* and *Sulfolobus acidocaldarius* are all Fe-oxidizing bacteria.

$$2Fe^{2+} + 1/2O_2 + 2H^+ \rightarrow 2Fe^{3+} + H_2O$$

Thiobacillus ferrooxidans is a chemolithotrophic, motile bacillus that we previously observed in the S cycle. Pyrite (FeS_2) oxidation to SO_4^{2-} and Fe^{3+} by *Thiobacillus ferrooxidans* is responsible for high acidity in drainage from coal mines. *Thiobacillus ferrooxidans* oxidizes the iron in ferrous sulfate to produce ferric sulfate:

$$4FeSO_4 + 2H_2SO_4 + O_2 \rightarrow 2Fe_2(SO_4)_3 + 2H_2O$$

The ferric sulfate contributes to acidity after it hydrolyzes to form ferric hydroxide:

$$2Fe_2(SO_4)_3 + 12H_2O \rightarrow 4Fe(OH)_3 + 6H_2SO_4$$

Thiobacillus ferrooxidans grows at about pH 2 to 4.5 and can grow at a pH as low as 0.8.

Ideally, 7 kcal/mole are needed to generate one ATP. About one to two Fe^{2+} ions are consumed per ATP generated and about 50 to 100 moles of Fe^{2+} are required to turn 1 mole of C into biomass (Okereke and Stevens 1991). Since it takes approximately 120 kcal to assimilate 1 mole of C, this is a conversion efficiency of only 3% to 21% and means a lot of Fe^{2+} is oxidized for growth. Ferrous iron oxidation follows zero-order kinetics because the concentration of Fe^{2+} in the environment is usually much greater than a microorganism's K_m for Fe^{2+} uptake.

What are the benefits of Fe^{2+} oxidation? It allows microorganisms to gain energy. The insoluble Fe^{3+} that is produced physically protects the microorganisms, and when it acts as a cementing agent, it stabilizes microcolonies on solid surfaces.

Iron oxidation is not always associated with energy production. Many bacteria such as *Crenothrix, Gallionella, Hyphomicrobium, Leptothrix, Metallogenium, Pedomicrobium, Planctomyces, Seliberia,* and *Sphaerotilus* oxidize Fe^{2+} without gaining any apparent energy. The cell walls act as catalytic surfaces for metal precipitation and ferrous iron oxidation leads to precipitated Fe^{3+} around filamentous sheath-forming bacteria. *Leptothrix* species, for example, include heterotrophic freshwater bacteria that oxidize Fe and Mn (Boogerd and de Vrind 1987). *Leptothrix discophora* is found in well waters and iron pipes. It forms sheath-covered cells, strings, and empty sheaths encrusted with ferromanganous oxides. *Leptothrix* gains no energy from the oxidation; you can assume this because you can't grow *Leptothrix* using reduced Fe and Mn as the sole energy source, which is the basis of chemolithotrophic growth. *Leptothrix* appears to excrete proteins with Fe-oxidizing or Mn-oxidizing properties.

Crenothrix creates $Fe(OH)_3$ precipitates (iron hydroxides). It requires low concentrations of Fe^{2+} and a little O_2. *Crenothrix* forms gelatinous masses that clog pipes. When the organic mass decomposes, it has a terrible odor. *Gallionella* is a neutrophilic chemolithotroph, a 0.5×2 µm-long, bean-shaped cell, growing in flat 200 to 300 µm long-mucilaginous ribbons encrusted with $Fe(OH)_3$. *Gallionella* is also responsible for precipitates in pipes. *Metallogenium, Pedomicrobium,* and *Seliberia* deposit $Fe(OH)_3$ on the surface of filaments. They also cause problems in drains. *Sphaerotilus natans* has only Fe-oxidizing ability.

Some classes of Fe-oxidizing microorganisms are listed in Table 17-1.

Iron Reduction

Many bacteria and fungi reduce Fe^{3+}; among them are *Alternaria, Bacillus, Clostridium, Fusarium, Klebsiella, Pseudomonas,* and *Serratia.* Iron and Mn are good electron acceptors. You can prove this by adding a concentrated $FeCl_3$ solution to soil with a lot of organic matter and watching the bubbles form as the Fe^{3+} oxidizes the reduced C in soil to CO_2. Unfortunately, solid substrates such as Fe^{3+} pose a problem because they have low solubility. Recent reports indicate that Fe and Mn reduction can be coupled exclusively to anaerobic respiratory growth by organisms such as *Shewanella putrefaciens* and *Geobacter metallireducens,* but this property appears to be rare (Lovely 1991).

Table 17.1 Iron-oxidizing and depositing bacteria.

Acidophilic Iron-Oxidizing Bacteria

Ferrobacillus sulfooxidans	*Leptospirillum*	*Sulfolobus*
Ferrobacillus ferrooxidans	*Metallogenium*	*Thiobacillus ferrooxidans*

Neutral pH Iron-Oxidizing Bacteria

1. Deposition directly on cell surfaces

Siderococcus	*Actinomyces* species	*Ochrobium tactum*
Planctomyces	*Acholeplasma*	*Pedomicrobium*
Peloploca	*Caueococcus*	*Metallogenium*
Hyphomicrobium	*Naumaniella*	

2. Deposition in polymer layers

Arthrobacter	*Leptothrix*	*Clonothrix*
Thiopedia	*Sphaerotilus*	*Crenothrix*

3. Deposition on stalks

   Planctomyces
   Gallionella
   Toxothrix

(Uren and Leeper 1978)

Redox potential and pH determine whether Fe is in an oxidized or reduced state. Both O_2 and NO_3^- suppress Fe^{3+} reduction because they keep (poise) the redox potential at a level too high for reduction to occur. At an E_h of 200 to 300 mV, Fe^{2+} oxidation and Fe^{3+} reduction begin to occur, depending on whether the redox potential is rising or falling. This appears as mottles in soil and indicates a fluctuating water table. You saw an example of a mottled soil in Chapter 12.

Gleying refers to a condition in which the Fe in soil is reduced and has a grayish-green color. It simply reflects the microbial use of iron as an electron acceptor for oxidizing C substrates. Under reducing conditions, organic C substrates such as glucose are fermented by some members of the microbial population:

$$C_6H_{12}O_6 \rightarrow CH_3COOH + CH_3CH_2COOH + CO_2 + H_2$$
$$\text{Glucose} \qquad \text{Acetate} \qquad \text{Pyruvate}$$

The H_2 they produce is oxidized by other members of the microbial population using Fe as a terminal electron acceptor (Lovely and Phillips 1989).

$$2Fe^{3+} + H_2 \rightarrow 2Fe^{2+} + 2H^+$$

Iron Availability and Assimilation

Ferric iron is generally insoluble, but it can be solubilized by acidification and complexation with organic material. An example of this in soil is called podzolization. Fer-

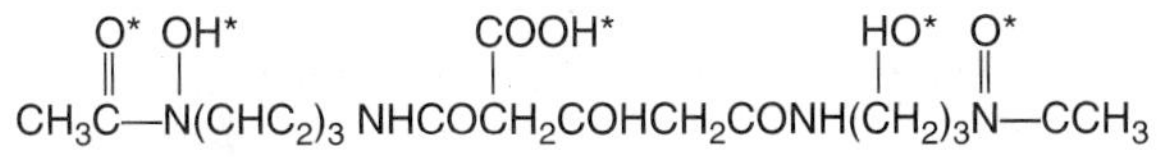

$$\begin{array}{cccccc}
O^* & OH^* & & COOH^* & & HO^* \; O^* \\
\parallel & \mid & & \mid & & \mid \quad \parallel
\end{array}$$

CH$_3$C—N(CHC$_2$)$_3$ NHCOCH$_2$COHCH$_2$CONH(CH$_2$)$_3$N—CCH$_3$

Schizokinen (* represents Fe binding site)

Figure 17.2   Structure of schizokinen, an iron-chelating compound.

ric iron combines with organic acids in forest soils, becomes more soluble, and percolates through the soil profile. Eventually the Fe^{3+} precipitates in the B horizon where it forms a distinct layer.

Ferric iron is unavailable compared to Fe^{2+} because it is less soluble. Iron solubility is poor in alkaline soils. One consequence is that soybeans grown in calcareous soils, relatively alkaline soils with high concentrations of $CaCO_3$, suffer from a condition called iron chlorosis. Iron chlorosis is an Fe deficiency that causes leaf yellowing because chlorophyll synthesis is impaired. The same condition appears in maple trees growing in neutral to alkaline soil.

Biological Fe consists of cytochromes, enzymes, ferredoxins, and FeS proteins. There are typically low Fe concentrations in water—0.1 ppb to 0.7 ppm—so microorganisms may be Fe-limited at times. Iron is often scavenged from the environment by organic chelating compounds. Nonspecific Fe chelators include citrate, oxalate, dicarboxylic acids, humic acids, and tannins. Specific Fe chelators include hemes, transferins, ferritin (iron-storage compounds), and siderophores. Fluorescent pseudomonads, are a group of bacteria that secrete soluble fluorescent Fe-chelating siderophores. One example of this group is *Pseudomonas fluorescens*. Siderophores are not taken up, instead they pass the Fe to receptor proteins in the cell membrane. Bacterial chelation of available Fe may inhibit fungal root pathogens because it deprives the pathogens of the Fe they need to grow. Hydroxamic acids, produced in streptomycetes and other bacteria, are another type of chelator. Schizokinen is an example of a hydroxamate (Figure 17-2).

MANGANESE (MN)

Manganese is one of the most abundant metals in the Earth's crust. Only aluminum (Al), iron (Fe), magnesium (Mg), and titanium (Ti) are more abundant metals. Manganese is an essential trace element for plants but is toxic at high concentrations. It is found in the enzyme superoxide dismutase, and also in Photosystem II. Manganese is important in the Mn peroxidases of lignin-degrading fungi. Some of the inorganic forms of Mn include oxide/hydroxides such as manganite (MnOOH) and pyrolusite (MnO$_2$), carbonates (MnCO$_3$), and sulfides (MnS) (Marshall 1979).

In surface waters, Mn ranges from 100 to 1000 ppm, while in groundwater it may be 1 to 10 ppm. Usually only Mn^{2+} (soluble) and Mn^{4+} (insoluble) are found. No volatile forms of Mn exist. Indirect microbial activity that affects Mn includes changes of redox potential, pH, and aeration. Manganese cycling is dependent on C inputs, which

A Short Biography of Serge Winogradsky

Soil microbiologists don't live forever, but it almost seems as though Serge Winogradsky (1856–1953) did. His life spanned an era in microbiology stretching from Pasteur in the 19th century to Waksman in the 20th century. Winogradsky's own contributions to fundamental concepts in soil microbiology are so great that he is commonly regarded as the "Father of Soil Microbiology." Winogradsky laid the foundations for studying sulfur-oxidizing, iron-oxidizing, and nitrifying bacteria. He introduced the concept of chemoautotrophic growth. He discovered the first anaerobic, nitrogen-fixing bacterium. He helped develop the technique of enrichment culture (Doetsch 1960).

Serge Winogradsky was born in Kiev, Russia on September 1, 1856. He was uncertain about his career when he grew up, and his studies at Kiev and St. Petersburg included law, natural science, and music. Winogradsky eventually graduated from the University of St. Petersburg in 1881 and became a microbiologist, first at the University of Strasbourg (1885), where he investigated sulfur and iron bacteria, then in Zurich (1889–1891), where he definitively isolated the bacteria responsible for nitrification in soil. In 1891, Winogradsky returned to St. Petersburg to work at the Institute of Experimental Medicine, but his stay there was brief, and he retired to his estates in the Ukraine in 1905.

For 17 years Winogradsky managed his country estates, but the Russian Revolution of 1917 and its aftermath made his continuation as a member of the landed gentry impossible. So, from 1922 until his death in 1953, Winogradsky found himself an expatriate at the Pasteur Institute in Paris. During this period, he performed fundamental research on nitrogen fixation and cellulose decomposition in soil.

influence redox potential and microbial activity, and Mn is cycled between oxidized and reduced states by microorganisms in much the same way as Fe.

We can simplify the microbial cycling of Mn into two basic compartments:

$$pH < 5.5$$
$$\text{Poor aeration, low } E_h$$
$$\text{Mn reducers}$$
$$\text{High organic matter content}$$

$$\longleftarrow$$

$$Mn^{2+} \text{ (immobilized)} \leftrightarrow Mn^{2+} \text{ (soluble)} \qquad\qquad Mn^{4+} \text{ (precipitated)}$$

$$\longrightarrow$$

$$pH > 8$$
$$\text{Aerated, high } E_h$$
$$\text{High organic matter content}$$
$$\text{Mn oxidizers}$$

Microbial Mn^{2+} oxidation occurs in soils and sediments. Chemical Mn^{2+} oxidation occurs only at a pH > 8. Manganese is also used as a terminal electron acceptor by bacteria when the redox potential is around 100 to 200 mV.

Manganese Oxidation

Manganese is oxidized by various soil and aquatic bacteria and fungi. About 5% to 10% of the microbial population are Mn oxidizers. Examples of bacterial oxidizers include *Arthrobacter* and *Leptothrix,* which oxidizes Mn^{2+} to Mn^{4+} (MnO_2) and precipitates it as a sheath. Other microorganisms that oxidize manganese are *Bacillus, Cladosporium, Corynebacterium, Curvularia, Gallionella, Klebsiella, Metallogenium, Pedomicrobium, Pseudomonas,* and *Sphaerotilus.*

There is little evidence that energy is obtained from Mn oxidation by most of these microorganisms (Table 17-2). The energy that is generated is modest, so considerable oxidation of manganese must occur to support microbial growth (Nealson et al. 1988).

$$Mn^{2+} + 1/2O_2 + H_2O \Rightarrow MnO_2 + 2H^+$$
$$\Delta G = -7 \text{ kcal/mole}$$

Extracellular proteins oxidize manganese, but it is not known whether they are coupled to energy or metabolic processes.

Manganese oxidation forms manganate (MnO_2), which is insoluble in water. Plants assimilate Mn^{2+} (soluble) rather than Mn^{4+} (insoluble), so Mn oxidation can cause Mn deficiency in plants. Rhizosphere bacteria precipitate MnO_2 on plant roots. You can demonstrate that Mn-oxidizing bacteria are present in rhizosphere soil by adding soil crumbs to $MnCO_3$-containing agar plates. Brown spots (MnO_2) should develop if Mn oxidizers are present.

Table 17-2 Manganese-oxidizing microorganisms.

Enzymatic Oxidation of Soluble Mn^{2+}

1. Obtain energy
    Hyphomicrobium manganoxidans
    Pseudomonas
2. Do not obtain useful energy
    Arthrobacter
    Leptothrix
    Metallogenium
3. Not known if they obtain energy

Arthrobacter ictreus	*Hyphomicrobium*
Arthrobacter globiformis	*Pedomicrobium*
Arthrobacter simplex	*Pseudomonas*
Citrobacter freundii	

Nonenzymatic Oxidation of Mn^{2+}

Pseudomonas manganoxidans
Streptomyces
Bacillus

(Uren and Leeper 1978)

Mn Availability and Plants

Manganese deficiencies are common in soils that are rich in organic matter and are between pH 6.5 to 8.0. Manganese deficiency in oats goes by the name "gray speck disease." For example, in a controlled experiment, the following evidence was observed:

1. Oats + nutrient solution $\Rightarrow$ No disease
2. Oats + nutrient solution + soil from a diseased site $\Rightarrow$ Diseased
3. Oats + nutrient solution + sterilized soil from a diseased site $\Rightarrow$ No disease
4. Oats + sterile soil + soil from a diseased site $\Rightarrow$ Diseased

The symptoms could be reduced by adding a bactericide and increased by adding straw mulch. The evidence pointed to Mn-oxidizing bacteria as being responsible for the deficiency syndrome, particularly when evidence for Mn-oxidizing bacteria was looked for in the rhizosphere of oats (Table 17-3).

What do the data tell you? The nonrhizosphere soil indicates that Mn-oxidizing bacteria make up about 6% of the total bacterial population. Their population doesn't change much when you amend the soil with manure and $MnSO_4$, even though the overall bacterial population nearly doubles. A susceptible oat strain has a higher total bacterial population in the rhizosphere, as you would expect. There isn't as much benefit to adding manure because the rhizosphere is already rich in available carbon for bacteria to grow on. But look at the population of Mn-oxidizing bacteria. Their population increases at least 20-fold, and in the untreated soil they become at least 42% of the total bacterial population. Adding $MnSO_4$ or manure seems to affect the Mn oxidizers since their population in the rhizosphere declines with this treatment. In contrast to the susceptible oat strain, the resistant oat strain has fewer Mn oxidizers in its rhizosphere to start and they form a smaller percentage of the total bacterial population.

Table 17-3 Microorganisms in the rhizosphere of oats that are resistant and susceptible to Mn deficiency (gray speck disease).

Oat Variety	Soil Treatment	Total Bacteria ($\times 10^3$ g^{-1})	Mn-oxidizing Bacteria ($\times 10^3$ g^{-1})	Mn Oxidizers (as % of total)
		Nonrhizosphere Soil		
Bare	None	129,000	8,000	6.2
Soil	Manure + $MnSO_4$	224,000	9,400	4.2
		Rhizosphere Soil		
Susceptible	None	564,000	225,000	42.2
(R.L. 909)	Manure + $MnSO_4$	638,000	181,000	28.2
Resistant	None	266,000	41,800	15.7
(Acton)	Manure + $MnSO_4$	509,000	42,200	8.3

(From Timonin 1946)

Different oat varieties have different tolerances to gray speck disease because they have selective microbial populations in their rhizospheres. As Table 17-3 shows, this selection may have something to do with the exudates released by the oat roots because the resistant oat variety had a much lower rhizosphere population than the susceptible variety and this rhizosphere population responded more readily to manure addition.

Manganese deficiency can be relieved temporarily by adding $MnSO_4$ as a foliar spray, by acidifying soil, or by flooding soil, all of which promote the presence of soluble Mn^{2+} ions. On the other hand, Mn^{2+} is sometimes too readily available. Mycorrhizal fungi that precipitate Mn^{2+} before it reaches the plant can help to reduce Mn^{2+} availability and prevent Mn toxicity.

ENVIRONMENTAL CONSEQUENCES OF IRON AND MANGANESE TRANSFORMATIONS

Interaction of Iron and Manganese

Figure 17-3 shows schematic diagram of Fe and Mn redox transformations in the environment. In a SO_4^{2-} poor environment, Mn^{4+} and Fe^{3+} can be important electron acceptors. Glacial lake sediments are an example of one such environment.

Reduced compounds can react chemically with Mn^{4+} and reduce it to Mn^{2+}. This includes Fe^{2+} and S^{2-}. For this reason, Mn^{2+} often forms in anaerobic environments before Fe^{2+}, even though, based on relative redox potentials, the opposite should be true. Any organism capable of reducing Fe is indirectly capable of chemically reducing Mn (Figure 17-4). Reduced manganese can also be generated by acid production.

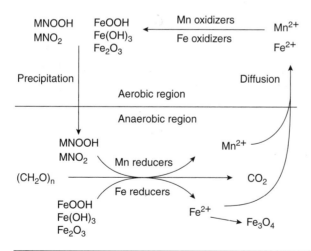

Figure 17-3   Schematic diagram of Fe and Mn transformations in natural environments. (Adapted from Nealson and Meyers 1992)

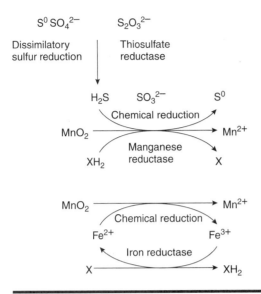

Figure 17-4   Chemical and biological reduction of Mn and Fe in the environment. (Adapted from Nealson and Meyers 1992).

Metal Sulfide Precipitation

One of the consequences of Fe and Mn reduction is metal sulfide formation, particularly in environments that also produce S^{2-}. Why is the Black Sea black? Because microbial metal reduction occurs in stratified environments, such as the Black Sea, in which Fe and Mn and sulfates are present. You can make your own model of the Black Sea by covering a sediment column with a solution of Fe^{3+}, Mn^{4+}, and SO_4^{2-} and adding a C source, as Figure 17-5 a,b illustrates. In this example, the C source is shredded newspaper. In a few days you should see a black precipitate forming at the sediment surface as Fe and Mn sulfides begin precipitating in the stratified environment (stratified in terms of O_2; remember, O_2 diffusion through water is very slow, so while there may be O_2 at the surface of the water, there is virtually none by the time you reach the sediment).

More Ways to Corrode Buried Pipes

Under certain conditions ($E_h < 400$ mV, low O_2 concentration, moderate temperatures, pH > 5.5, and the presence of SO_4^{2-}) Fe-containing pipes can corrode in soil. You saw this when we talked about SO_4^{2-} reduction in soil, but it's worth repeating here in the context of the Fe.

The iron pipe acts as an anode (oxidation occurs):

$$Fe^0 \Rightarrow Fe^{2+} + e^-$$

a   b

Figure 17-5   Stages in the development of metal sulfide precipitates. (a) At the start of incubation; (b) two weeks later. Note how a dark band has developed at the sediment–water interface. (Photographs courtesy of M. S. Coyne)

Simultaneously, H_2 gas forms at the cathode (reduction occurs):

$$2H^+ + 2e^- \Rightarrow H_2 \text{ gas}$$

Sulfate reducers contribute to this problem because they can oxidize H_2 and use SO_4^{2-} as a terminal electron acceptor. This produces S^{2-}, which precipitates the Fe^{2+}. The overall reaction is:

$$4Fe + SO_4^{2-} + 4H_2O \Rightarrow FeS + 3Fe(OH)_2 + 2OH^-$$

FeS precipitates, or H_2S, a corrosive gas, can form. As a result, 3 mm of pipe can be corroded in 5 to 7 years. This is a serious problem in the oil industry and in sewage treatment plants.

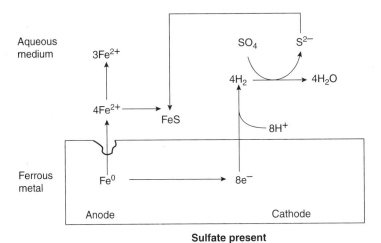

Sulfate present

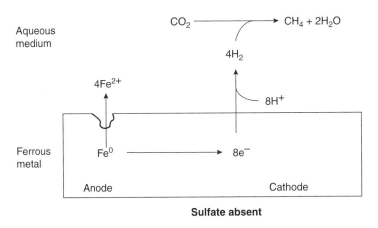

Sulfate absent

Figure 17-6   Two mechanisms of cathodic depolarization leading to the solubilization of Fe and the corrosion of buried iron pipes in soil. (Adapted from Boopathy and Daniels 1991)

Methanogens can also drive the reaction forward by consuming H_2 (Figure 17-6).

Biologically Influenced Iron Deposits

Several different origins of minerals depend on the way Fe was deposited. Bog iron is concretions, 10 to 15 cm in diameter, that occur at depths of less than 5 m in lakes, bogs, and taiga–podzolic soil regions (histosols and tundra). Iron in acidic environments tends to be reduced and leach since Fe^{2+} is relatively soluble compared to Fe^{3+}. Leaching is

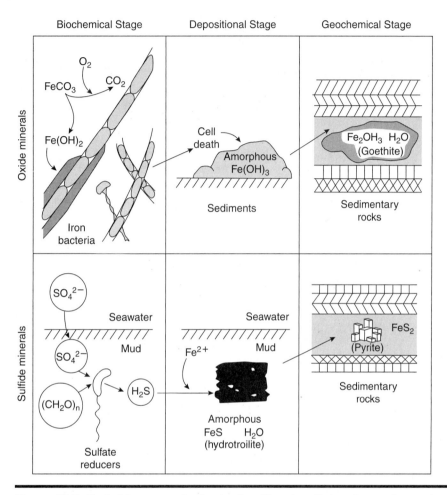

Figure 17-7   Bacterial processes leading to mineral iron deposits in various environments. (Adapted from Doetsch and Cook 1973)

promoted by binding to humic substances. The Fe may precipitate in a more alkaline environment or be oxidized elsewhere. Sulfide minerals, pyrite, for example, arise through the action of SO_4^{2-} reducing bacteria helping to precipitate Fe^{2+} as Fe sulfides. Iron-rich water with neutral to alkaline pH develops a characteristic flora with many ribbon-forming *Gallionella* and sheathed bacteria such as *Leptothrix*.

$$4FeCO_3 + O_2 + 6H_2O \rightarrow 4Fe(OH)_3 \text{ (siderite)} + 4CO_2$$

Oxides such as goethite ($Fe_2O_3 \cdot H_2O$) and hematite (Fe_2O_3) form from the deposition of $Fe(OH)_3$ around the bacteria. Thus, Fe deposits can reflect whether the Fe was deposited in anaerobic or aerobic environments and whether it likely formed in marine or freshwater environments (Figure 17-7).

Iron and Manganese and Biofouling

Ochre deposits are amorphous Fe or Mn oxide deposits in field drains (tile lines) that are used to improve soil drainage. Ochre deposits occur because the field drains represent aerobic sites in otherwise anaerobic or reducing environments. Iron- and Mn-oxidizing bacteria take up residence in the openings of these drains, and when reduced Fe or Mn in drainage water passes through, a portion of it is oxidized and precipitates. The deposition process can be relatively rapid. One consequence of this activity is that eventually the openings of the drains are rendered useless because they become clogged with a mass of precipitated Fe, Mn, and gelatinous microbial growth. This also occurs in groundwater supply wells. The screens surrounding the wells become clogged with metal-oxidizing bacteria that restrict water flow.

The same phenomenon also occurs in municipal water systems. Iron and Mn oxidation in water distribution lines by oligotrophic microorganisms (microorganisms growing in water with very little C) coat the inside of the water distribution lines with a fluctuating mass of precipitated Fe, Mn, and microbial cells (Sly et al. 1990; Tyler and Marshall 1967a,b). This coating restricts water flow and increases pumping costs. The coating also periodically sloughs and lowers the aesthetic quality of the water. Periodically, water mains are flushed to remove these deposits and for a brief period the potable water may have an unsightly, though harmless, brownish tinge.

Rock Varnish

Rock varnishes (devil's paintbrush) are Mn oxides in thin, brown-to-black veneers and Fe oxides in orange veneers up to 0.1 mm thick that cover rock surfaces in arid and semiarid regions (Dorn 1991). The varnish is about 60% clay minerals and 20% to 30% oxides of Mn and Fe. The amount of Mn determines the color of the varnish. It is black and high in Mn when pH is low or neutral during its formation. It is orange and high in Fe when the pH is high or alkaline.

Rock varnish begins in fungal microcolonies or other microbial nucleation centers that occur in tiny pits in the rock surface. During a wet period, microorganisms proliferate on trace amounts of C and secrete enzymes that oxidize Mn and cement clay. High pH inhibits Mn-oxidizing bacteria, so the Mn/Fe ratio is low and the varnish yellows. Iron, rather than Mn, is also precipitated in anaerobic crevices. Two specific microorganisms involved in forming rock varnish are *Metallogenium* and *Arthrobacter*.

Acid Mine Drainage

Abandoned mines and mine spoils represent a major source of environmental contamination through acidic drainage. Coal is often associated with pyrite. When mining exposes the coal to air, a combination of autooxidation and microbial Fe and S oxidation produces acid. The subsequent acidic drainage is environmentally harmful to life and physical structures. Stream pH may decrease from near neutral to pH 2 to 4.5. Sulfate concentrations in streams may range from 1,000 to 20,000 mg L^{-1}. Ferrous iron may become nondetectable in the water, but the streams become bloodred with precipitated Fe^{3+}. Perhaps 10,000 miles of waterways are affected by acid mine drainage in the United States.

Acid mine drainage is the result of chemical and microbial reactions. At neutral pH, pyrite (FeS_2) oxidation is rapid and spontaneous, but below pH 4.5 autooxidation slows. In the range of pH 3.5 to 4.5, the iron bacterium *Metallogenium* catalyzes pyrite oxidation, while at pH < 3.5, *Thiobacillus* species and *Leptospirillum* species become important. At this stage, microbially catalyzed oxidation is several hundred times higher than spontaneous oxidation.

The first step in the production of acid mine drainage is both chemical and microbial.

Step 1

Thiobacillus thiooxidans, Metallogenium (pH > 4.5)
Thiobacillus ferrooxidans (acid pH)

$$2FeS_2 + 7O_2 + 2H_2O \xrightarrow{} 2Fe^{2+} + 4SO_4^{2-} + 4H^+$$

Spontaneous

Step 2

Thiobacillus ferrooxidans (pH < 3.5)
Leptospirillum (pH < 3.5)

$$2Fe^{2+} + 1/2O_2 + 2H^+ \xrightarrow{} 2Fe^{3+} + H_2O$$

Rate-limiting step

Step 3

$$8Fe^{3+} + S^{2-} + 4H_2O \xrightarrow{} 8Fe^{2+} + 8H^+ + SO_4^{2-}$$

Spontaneous

$$FeS_2 + Fe_2(SO_4)_3 \xrightarrow{} FeSO_4 + 2S^0$$

Spontaneous

Step 4

$$2Fe^{3+} + 6H_2O \xrightarrow{} Fe(OH)_3 + 6H^+$$

Spontaneous

The $Fe(OH)_3$ forms a red precipitate that is a classic indicator of acid mine drainage in streams.

How can acid mine drainage be prevented? Removal of the source is the most obvious answer. Sealing off the pyrite so that it can't oxidize and flooding abandoned mines are two solutions. Reclaiming land so that pyrite doesn't remain exposed also helps. However, it is very difficult to prevent pyrite from oxidizing once it has been disturbed and exposed to air.

You could coat the exposed pyrite with an impermeable coating of ferric phosphate to prevent exposure to air. You could also curb acid mine drainage by suppressing the activity of Fe- and S-oxidizing bacteria . In the laboratory, this has been done with low concentrations of organic acids and addition of antimicrobial compounds. However, this procedure appears to be impractical on a field scale.

An alternative mechanism is to place the acid mine drainage in a lagoon with large quantities of organic waste (sawdust, for example). Aerobic and anaerobic bacteria will lower the redox potential and produce degradation intermediates that can be used by SO_4^{2-} reducers. Hydrogen sulfide generated by the SO_4^{2-} reducers will reduce the Fe^{3+} to Fe^{2+} chemically and then precipitate it as FeS. The pH should rise and both Fe^{3+} and SO_4^{2-} should decline. This seems to work on a small scale but has not been tried on a large scale.

Bioleaching

It would be nice if metal-oxidizing microorganisms could be put to some use, so that the acid they generate isn't always a problem. That's the concept behind bioleaching. Bioleaching is an ancient process. The earliest example of metal bioleaching was reported by Liu-An (177–122 B.C.) in the West Han Dynasty of China. He referred to bitter-tasting blue waters in a spring, which yielded a blue material on evaporation. When mixed with iron, the blue water yielded copper. This is the first recorded observation of metal recovery from leachate. Recovering Cu sulfide metals by microbial bioleaching was also practiced in the Rio Tinto mines in Spain during the 17th century.

The rationale for bioleaching is simple. Ores with low metal content are not suitable for direct smelting, but metals can be dissolved from ore-bearing rocks using metal-oxidizing bacteria. Bioleaching is employed primarily in the extraction of copper and uranium, but it has also been used to extract antimony, arsenic, bismuth, cadmium, cobalt, molybdenum, nickel, lead, selenium, titanium, and zinc-containing ores. Bioleaching accounts for 20% of the copper mined worldwide.

There are many microorganisms with the potential for bioleaching, but only a few are widely used. This is because the most economic bioleaching has been found with sulfide-containing ores that can be partially degraded by S-oxidizing bacteria. Both *Thiobacillus thiooxidans* and *Thiobacillus ferrooxidans* are found in leaching dumps. Heterotrophs have been used, but only in the laboratory. These include *Pseudomonas fluorescens, Pseudomonas putida,* and *Bacillus licheniformis.* They chelate metals with organic acids. Thermophiles are favored for bioleaching because of their more rapid growth; *Thiobacillus thermophytica* is an example. *Sulfolobus acidocaldarius* is an obligate thermophile that is used in the bioleaching of Mo because it is more tolerant of Mo (a heavy metal) than is *Thiobacillus.*

Summary

Iron is the fourth most abundant element in the Earth's crust and the most abundant metal. Iron exists as metallic iron (Fe^0), ferrous iron (Fe^{2+}), and ferric iron (Fe^{3+}). Microbial Fe interactions involve Fe scavenging and uptake, Fe oxidation and precipitation, and Fe reduction and solubilization. In aerobic environments, microbial Fe oxidation dominates in acidic environments while chelation dominates in neutral environments. In anaerobic environments, the Fe cycle is dominated by Fe-reduction reactions and precipitation of Fe sulfides.

In aerated conditions, bacteria can get energy from oxidizing Fe^{2+}. Some examples of Fe-oxidizing bacteria are *Crenothrix, Gallionella, Leptothrix,* and *Thiobacillus.* Gleying refers to a condition in which the Fe in soil is reduced and has a grayish-green color. It reflects the microbial use of Fe^{3+} as an electron acceptor for oxidation of C substrates. Ferric iron is generally insoluble, but it can be solubilized by acidification and complexation with organic material (chelation).

Manganese is an essential trace element for plants but is toxic at high concentrations. Manganese is cycled between oxidized (Mn^{4+}) and reduced (Mn^{2+}) states by microorganisms much like Fe. There is little evidence that energy is obtained from Mn oxidation by most microorganisms. Manganese oxidation can lead to Mn deficiency in plants.

Microbial metal transformations are essential to soil formation, producing metallic ores, extracting metals from low-grade ores, acidifying mine wastewater, and polluting water supplies.

Sample Questions

1. Identify or name each of the microbial transformations:
   a. $SO_4^{2-} \Rightarrow S^{2-}$ d. $Mn^{2+} \Rightarrow Mn^{4+}$
   b. $Fe^{3+} \Rightarrow Fe^{2+}$ e. $Fe^{2+} \Rightarrow Fe^{3+}$
   c. $Mn^{4+} \Rightarrow Mn^{2+}$ f. $S^{2-} \Rightarrow S^0$

2. In approximately what order will the following electron acceptors be used in flooded soil?

$$CO_2, Fe^{3+}, Mn^{4+}, NO_3^-, O_2, SO_4^{2-}$$

3. If the concentration of H^+ is $10^{-4.5}$, what pH does this represent and what group of microorganisms will predominate in this environment?

4. Under what conditions will ferrous iron exist in soil? Mn^{2+}? Mn^{4+}?

5. How does the redox status of soil affect the type of electron acceptors microorganisms use?

6. Write a redox reaction showing coupled oxidation and reduction.

7. In the following chemical reactions, identify the oxidizing and reducing agents:
   a. $S^{2-} \Rightarrow SO_4^{2-} + 8e^-$ c. $Mn^{2+} \Rightarrow Mn^{4+}$
   b. $Fe^{3+} + e^- \Rightarrow Fe^{2+}$ d. $1/2 O_2 + 2H^+ + 2e^- \Rightarrow H_2O$

8. What are the environmental conditions and mechanisms by which iron and manganese clog drainage lines?

9. In the following illustration, which curve (A or B) best describes the changing concentration of Mn^{2+} in flooded soil?

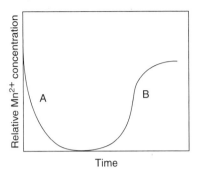

10. Briefly describe the microbiological phenomenon that is illustrated in the following figure (what is happening and why is it happening).

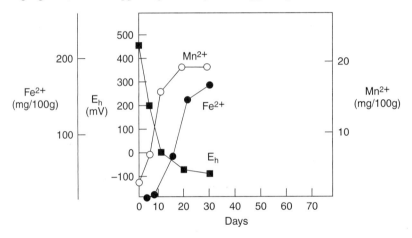

11. If the soil was drained on day 30 in the figure above, continue plotting the data to illustrate the subsequent fates of Mn^{2+}, Fe^{2+}, and E_h. Explain your reasoning.

Thought Question

You are draining a plot of land on your deserted island, which is iron rich. Inexplicably, the perforated bamboo pipes that you buried to carry off excess water seem to be doing a poor job. You also notice that the drainage water is a brownish-black color and smells bad. What do you think is happening to your drainage pipes and why?

Additional Reading

For more detailed reviews, read "Biology of Iron- and Manganese-Depositing Bacteria" by W. C. Ghiorse (1984, *Annual Review of Microbiology* 38:515–50) and "Dissimilatory Fe(III) and Mn (IV) Reduction by D. R. Lovely (1991, *Microbiological Reviews* 55:259–87). Also read the article by M. Silver et al. entitled, "Soil Mineral Transformations Mediated by Soil Microorganisms" (1986, In *Interactions of soil minerals with natural organics and microorganisms,* P. M. Huang and M. Schnitzer (eds.), 497–520. Soil Science Society of America, Madison, WI). R. Cullimore has a very readable introduction to biofouling in *Practical Manual of Ground Water Microbiology* (1993, Lewis Publishers, Chelsea, MI). To learn the techniques for isolating metal-transforming bacteria, read "Iron and Manganese Oxidation and Reduction" by W. C. Ghiorse (1994, In *Methods of soil analysis, part 2: Microbiological and biochemical properties,* R.W. Weaver et al. (ed.), 1079–96. Soil Science Society of America, Inc. Madison, WI).

References

Boogerd, F. C., and J. P. M. de Vrind. 1987. Manganese oxidation by *Leptothrix discophora* SS-1. *Journal of Bacteriology* 169:489–94.

Boopathy, R., and L. Daniels. 1991. Effect of pH on anaerobic mild steel corrosion by methanogenic bacteria. *Applied and Environmental Microbiology* 57:2104–08.

Doetsch, R. N. 1960. *Microbiology: Historical contributions from 1776 to 1908.* New Brunswick, NJ: Rutgers University Press.

Doetsch, R. N. and T. M. Cook. 1973. *Introduction to bacteria and their ecobiology.* Baltimore, MD: University Park Press.

Dorn, R. I. 1991. Rock varnish. *American Scientist* 79:542–53.

Lovely, D. R., and E. J. P. Phillips. 1989. Requirement for a microbial consortium to completely oxidize glucose in Fe (III)-reducing sediments. *Applied and Environmental Microbiology* 55:3234–36.

Marshall, K. C. 1979. Biogeochemistry of manganese minerals, In *Biogeochemical cycling of mineral-forming elements,* P. A. Trudinger and D. J. Swaine (eds.), 253–92. Amsterdam: Elsevier/North-Holland Publishing Co.

Nealson, K. H., and C. M. Meyers. 1992. Microbial reduction of manganese and iron: New approaches to carbon cycling. *Applied and Environmental Microbiology* 58:439–43.

Nealson, K. H., B. M. Tebo, and R. A. Rosson. 1988. Occurrence and mechanisms of microbial oxidation of manganese. *Advances in Applied Microbiology* 33:279–318.

Okereke, A., and S. E. Stevens. 1991. Kinetics of iron oxidation by *Thiobacillus ferrooxidans. Applied and Environmental Microbiology* 57:1052–6.

Sly, L. I., M. C. Hodgkinson, and V. Arunpairojana. 1990. Deposition of manganese in a drinking water distribution system. *Applied and Environmental Microbiology* 56:628–39.

Timonin, M. I. 1946. Microflora of the rhizosphere in relation to the manganese deficiency disease of oats. *Soil Science Society of America Proceedings* 11:284–92.

Tyler, P. A., and K. C. Marshall. 1967a. *Hyphomicrobium*—a significant factor in manganese problems. *Journal of the American Waterworks Association,* 59:1043–8.

Tyler, P. A., and K. C. Marshall. 1967b. Microbial oxidation of manganese in hydro-electric pipelines. *Antonie van Leeuwenhoek Journal of Microbiology and Serology* 33:171–83.

Uren, N. C., and G. W. Leeper. 1978. Microbial oxidation of divalent manganese. *Soil Biology and Biochemistry* 10:85–87.

Chapter 18

The Nitrogen Cycle— Nitrogen Mineralization

Overview

After you have studied this chapter, you should be able to:

- Define "N mineralization."
- Identify some of the principal mineralizable N-containing compounds in soil.
- Explain the pathways of N mineralization, and describe the mineralization process mathematically.
- Summarize the optimal conditions for N mineralization in soil.
- Describe how excessive N mineralization can sometimes have unwanted environmental consequences.

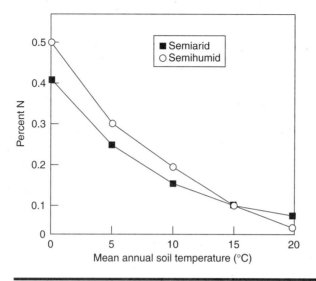

Figure 18-1 Relationship between soil N and mean average soil temperature in prairie soils. (Adapted from Jenny 1930; Stevenson 1982)

INTRODUCTION

Nitrogen is the mineral nutrient most often demanded by plants. This is not surprising since nitrogen is the most common constituent of plants after C and O. Hence, it is important to know about its microbial transformations in soil. Mineralization, as we define it, is the decomposition of organic N compounds to release inorganic N. Ammonification is another term for this process, because the immediate product is ammonia (NH_3, which quickly becomes NH_4^+ in soil solutions). Decomposition and rot are common terms that encompass mineralization.

In this chapter we examine N-containing compounds that are available for mineralization, the pathways by which they are mineralized, and a mathematical description for mineralization. We also look at how to study N-mineralization in soil and the effect that environmental factors have on mineralization rates.

THE EFFECT OF ENVIRONMENT ON NITROGEN IN SOIL

The total mineralizable N in soil depends on its original organic N content. The total N content, in turn, depends on the climate, vegetation, topography, age, and management of the soil (Jenny 1930). If you traveled from north to south (Canada to Arkansas) measuring soil N, you might plot data that look like those in Figure 18-1. As a rule, soil N decreases as the average soil temperature increases. For every 10°C rise in mean average soil temperature, the amount of soil N is two to three times lower (Stevenson 1982). The decrease is partly because microbial N mineralization rates increase about two-fold for every 10°C rise in temperature in the range of 20° to 60°C. Nitrogen mineralization

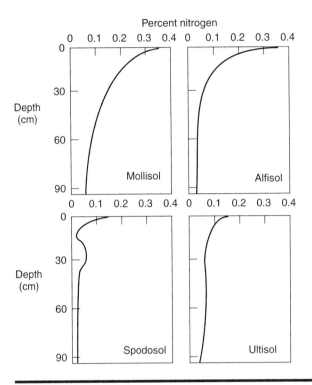

Figure 18-2 Distribution of soil N in the profile of several soil orders. (Adapted from Stevenson 1982)

increases as the temperature rises to a point at which microbial growth is affected. The optimum temperature for mineralization is between 40° and 60°C. This is substantially higher than most temperatures found in soil. As a consequence, NH_4^+ accumulates in compost heaps kept at 65°C because it is too hot for processes that remove the NH_4^+.

Soil N increases as the soil moisture increases. Notice in Figure 18-1 that there was consistently more soil N in the semihumid soils than in the semiarid soils until the mean annual soil temperature was greater than 15°C. Optimum water content for mineralization is between 50% to 75% water-holding capacity, or around −0.01 MPa (−0.1 atm).

As a rule, there is more soil N under permanent grass vegetation than there is under forest soils. The dense rhizosphere that develops in grassland soils promotes humus formation, which in turn promotes N immobilization (Stevenson 1982). Consequently, the profile of soil N might look something like Figure 18-2.

Soil management has a significant effect on the distribution of soil N. Cultivation invariably reduces the soil N content (Figure 18-3). No-tillage soil, for example, which maintains most of the plant residue from cropping at the soil surface, has higher soil N levels than does a tilled soil (Table 18-1). The level of N in soil doesn't go to zero, as extrapolating the curves in Figure 18-3 might suggest, because the soil eventually returns to a new equilibrium in which N inputs approximately balance N removal.

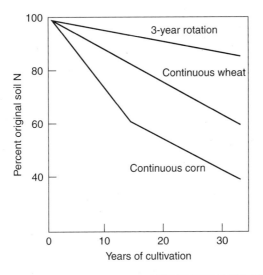

Figure 18-3   Decrease in the N content of unfertilized soil under different cropping systems. (Adapted from Salter and Green 1933)

Table 18-1   Total N, microbial biomass N, and mineralizable N in no-tillage and conventional tillage Alfisols after 24 years of continuous corn.

N Source	0–7.5 cm	
	No Tillage	**Conventional Tillage**
Total N (mg/kg)	1,900	1,300
Microbial biomass N (mg/kg soil)	32.0	15.0
Mineralizable N (mg/kg soil)	9.0	1.4

N Source	7.5–15 cm	
	No Tillage	**Conventional Tillage**
Total N (mg/kg soil)	1,300	1,200
Microbial biomass N (mg/kg soil)	8.0	11.0
Mineralizable N (mg/kg soil)	0.7	0.5

(From Handayani 1996)

ORGANIC NITROGEN IN SOIL

The predominant forms of organic N in soil are proteins, nucleic acids, chitin and peptidoglycan, and amino sugars. Bacteria, for example, are about 50% protein, 25% RNA, and 3% DNA on a dry weight basis (Ingraham et al. 1983). Proteins consist of chains of amino acids, each of which contains one or more amine groups. Nucleic acids (RNA and DNA) consist of the purines guanine and adenine, and the pyrimidines uracil, cytosine, and thymine. Each nucleotide consists of a N-containing base attached to ribose

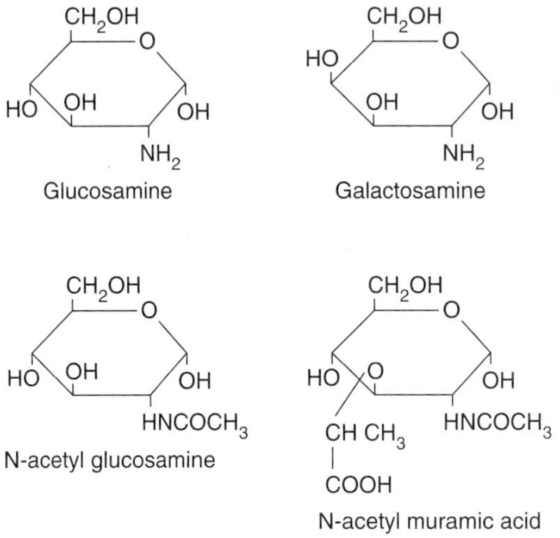

Figure 18-4   Some examples of common amino sugars in soil.

(a sugar with five carbons). Chitin is a polymer of N-acetyl glucosamine (glucose is a six-carbon sugar). Peptidoglycan, which is in bacterial cell walls, is a polymer consisting of alternating subunits of N-acetyl glucosamine and N-acetyl muramic acid (which is just a variation of glucosamine). Amino sugars are compounds such as glucosamine and galactosamine (typically hexose sugars with amine groups attached to them) (Figure 18-4). There is also organic N in soil whose identity is not known.

MINERALIZATION OF ORGANIC N

Mineralization of organic N-containing materials refers to the decomposition of these compounds to release NH_4^+, (Figure 18-5). Mineralization can never be eliminated in fertile land, since approximately 10^5 to 10^7 microorganisms per gram of soil are active mineralizers. Typically, the rate of NH_4^+ production ranges from 1 to 20 ppm N per day (mg NH_4^+ per kg of soil per day). This represents 1% to 4% of the total N released to plants during the growing season in temperate latitudes.

Let's look at some examples of mineralization. In chitin mineralization, a polymer, chitin, is broken down into its constituent subunits, N-acetyl glucosamine. These subunits are then further decomposed. Likewise, proteins are decomposed by enzymes called peptidases and proteases. These enzymes break the peptide bonds between amino acids in proteins and ultimately release free amino acids. Whether an amino acid is used as an energy source, a C source, an N source, or remains intact as a building block for new protein, depends on a complex series of metabolic controls. Carbohydrate C is preferentially used for energy and for building biomass, if it is available, so that microorganisms don't have to resynthesize new amino acids.

One important mineralization reaction in amino acid metabolism is catalyzed by the enzyme glutamate dehydrogenase, which can serve as a general model for how the decomposition of amino acids works.

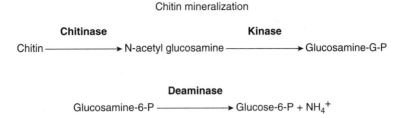

Figure 18-5   Mineralization of organic N.

$$\text{Glutamate} + NAD^+ + H_2O \xrightarrow{\text{Glutamate dehydrogenase}} \alpha\text{-Ketoglutarate} + NH_4^+ + NADH + H^+$$

Deamination, which is what this reaction represents, occurs when microorganisms need the C skeleton.

Proteins may sometimes persist in soils even though they are readily mineralizable. The proteins may be bound to clays that exhibit greater affinity for them than for microorganisms. Binding to clays may hide the specific peptide bonds that extracellular proteases require to decompose the proteins, or may change the shape of the proteins so much that they aren't recognized as substrates by proteases. Proteins may also be protected in micropores that are too small for microorganisms to enter.

Nucleic acids (RNA and DNA) are polymers of purine and pyrimidine nucleotides (guanine, adenine, uracil, cytosine, and thymine). Nucleic acids are second only to proteins in importance as N-substrates. Nucleases in soil are the enzymes that catalyze the decomposition of the nucleic acid polymer. Nucleases are both intracellular and extracellular. Nucleases in soil are also endo- or exonucleases, which means that they either degrade the nucleic acids from the ends (exonucleases) or break them up from the middle (endonucleases).

There's an obvious model here. In mineralization, macromolecules are broken down into subunits, and subunits are broken down into individual ions. These are multistep processes requiring different enzymes and, frequently, the combined activity of several different microbial species.

The decomposition product of nucleotides is urea. Urea is also present in animal excrement. Urea is one of the fastest-growing sources of solid fertilizer N used in agriculture worldwide. It is one of the most common N fertilizers used on lawns and golf courses. Urea is mineralized by an enzyme called urease, (Figure 18-6) which is commonly found in soil bacteria and as an extracellular enzyme in soils. Urea-degrading bacteria represent about 32% to 69% of the total bacterial population and urea-degrading fungi represent about 58% to 100% of the total fungal population; so, it's a common feature.

When urea enters the soil it is rapidly hydrolyzed. In the mineralization of urea by urease, H^+ is consumed, the pH increases, and the local microsite pH increases. This can have adverse consequences since ammonia (NH_3) volatilizes at high pH.

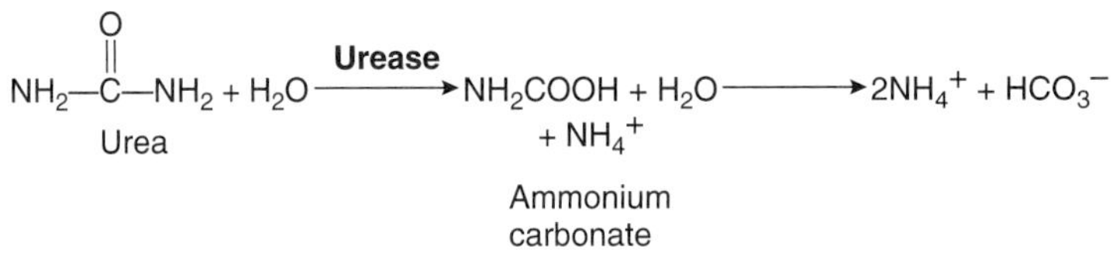

$$NH_2-\overset{\displaystyle O}{\overset{\displaystyle \|}{C}}-NH_2 + H_2O \xrightarrow{\text{Urease}} NH_2COOH + H_2O \longrightarrow 2NH_4^+ + HCO_3^-$$

Urea $+ NH_4^+$

Ammonium
carbonate

Figure 18-6   Mineralization of urea.

$$NH_4^+ \xrightarrow[\text{pH 9}]{OH^-} NH_3 + H_2O$$

FATES OF NH_4^+

Once NH_4^+ is produced, it has multiple fates. It can be taken up by plants. Ammonium can bind to the cation exchange complex where it is readily available. It can also be immobilized in the interlayers of clays, on the weathered edges of clays, and within the structure of collapsing clays such as vermiculite. This chemically immobilized N can slowly be made available to plants and is probably a major source of available N within the soil profile. Ammonium can react with organic matter to form quinone-NH_2 complexes. This is important in organic matter and humus formation. Ammonium can volatilize; up to 50% of the N in urea and manure can be lost as NH_3 gas at high pH.

Ammonium can be used as an energy source by autotrophic bacteria (nitrifiers). We discuss this process in Chapter 19. Ammonium is also assimilated by microorganisms. It is the preferred N source for most microorganisms, although some bacteria and fungi assimilate NO_3^- preferentially. Usually, NH_4^+ represses the enzymes necessary for NO_3^- reduction; we discuss that process in Chapter 20.

Since NH_4^+ can have several fates that prevent it from being observed, when we discuss N mineralization, we really mean "net mineralization," the difference between the organic N that is mineralized and the actual inorganic N that is recovered. "Gross mineralization" represents the total amount of organic N that is mineralized.

N MINERALIZATION KINETICS

The mineralization of N is well modeled by using first-order kinetics. The equation that describes the progress of N mineralization for most N-containing substrates is:

$$\delta S/\delta t = -kS$$

where S equals the amount of N-containing substrate and –k is a mineralization rate constant that is negative because the substrate is constantly disappearing.

You remember that in a first-order reaction, the rate of the reaction depends on the substrate concentration. So, if the concentration of the N-containing substrate is constantly declining, the mineralization rate will constantly change. A graph of mineralization will look something like Figure 18-7.

The equation that describes this line is $N_t = N_o e^{-kt}$, where N_t is the amount of N-containing substrate remaining at any given time, e is the Naperian base (with a value

How Do You Study N Mineralization in Soil?

Two things that soil microbiologists are interested in when they study N mineralization in soil are the amount of mineralizable N and the rate at which that mineralizable N gets transformed. There are several chemical methods of estimating mineralizable N that range from extraction in hot water to treatment with chemical reagents (Bundy and Meisinger 1994).

There are two common laboratory methods for estimating mineralizable N biologically. In the aerobic method, a mixture of sand and soil is incubated at an appropriate temperature and periodically leached with a 1 M KCl solution. The KCl removes any inorganic N that has been produced by mineralization and nitrification. This continues for several months or until the amount of inorganic N available (typically measured as NO_3^-) is insignificant. In the anaerobic method, a soil sample is saturated with water and incubated in a closed container for 1 to 2 weeks at 40°C. At the end of incubation, the resulting mixture is extracted with KCl and the NH_4^+ is determined. In both methods, the difference in inorganic N between the start and end of the incubation is taken as the measure of biologically mineralizable N.

Both laboratory methods disturb the soil samples, and soil microbiologists know from experience that when you disturb soil samples, you frequently overestimate almost all microbial processes. In-field methods of estimating N-mineralization rates employ a method like the buried bag technique. In the buried bag technique, a relatively undisturbed plug of soil is placed in an impermeable bag (a plastic sandwich bag works fine) and reburied in the location from which it came (Hart et al. 1994). Two to three weeks later, the bag is recoverd and its contents are extracted with KCl. The difference in inorganic N between the start and finish of the incubation divided by the time interval is the N-mineralization rate.

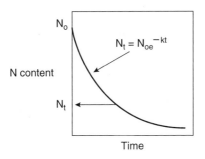

Figure 18-7   Plot of a first-order N mineralization reaction.

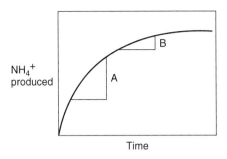

Figure 18-8   Schematic diagram of mineralization of multiple organic N compounds.

of approximately $2.7183 \ldots$) and N_0 is the amount of starting substrate. It's worth repeating that this equation may look complicated, but it's really not, especially when you've got a calculator that does natural logarithms for you.

In practice, when you study N mineralization, you're less likely to measure a substrate disappearing than you are to measure a product such as NH_4^+ forming. You're also likely to measure the mineralization of a mixture of N-containing compounds, rather than a pure substrate. Consequently, you're more likely to see something like the graph in Figure 18-8. In Figure 18-8, you will notice that for the same amount of time, less NH_4^+ is formed in region B than in region A. Why? It should be clear that all types of organic soil N are not equally available nor do they mineralize at the same rates. If you ranked them in order of their ease of mineralization, you might come up with a list something like this:

Urea > amino acids > proteins > nucleic acids > amino sugars > humified N.

What Figure 18-8 indicates is that with a mixture of organic N-compounds, the mineralization rate will gradually slow down because the most available (labile) N substrates are mineralized first and rapidly, and when they are gone, the less available N substrates are mineralized at a slower rate.

Summary

Mineralization is the decomposition of organic N compounds to release inorganic N. Ammonification is another term for this process. The total N content in soil is a function of the climate, vegetation, topography, age, and management of the soil. Soil N decreases as the average soil temperature increases. Conversely, soil N increases as the soil moisture increases. Cultivation reduces the soil N content.

The predominant forms of organic N in soil are proteins, nucleic acids, chitin and peptidoglycan, and amino sugars. Mineralization is never eliminated in fertile land, since approximately 10^5 to 10^7 microorganisms per gram of soil are active mineralizers. In N mineralization, macromolecules are broken down into subunits, and subunits are broken down to release NH_4^+. Urea is mineralized by an enzyme called urease to produce CO_2 and NH_4^+. Urease is commonly found in soil bacteria and as an extracellular enzyme in soils. Once NH_4^+ is produced, it has multiple fates: plant uptake, immobilization on clays, incorporation into humic materials, volatilization, and further oxidation. The mineralization of N is well described by using first-order kinetics. All types of organic soil N are not equally available nor do they mineralize at the same rates.

Sample Questions

1. What does it mean when N mineralization is described as a first-order reaction in soil?
2. Draw an illustration representing N mineralization in the environment.
3. What is the difference between net mineralization and gross mineralization?
4. What factors control deamination in soil?
5. Define "ammonification."
6. Why aren't proteins degraded immediately when they are bound to clays?
7. Draw a chart that illustrates the stages DNA passes through as it is mineralized to NH_4^+.
8. Based on the following figure, what is the relationship between urealytic microorganisms and urease activity in this soil?

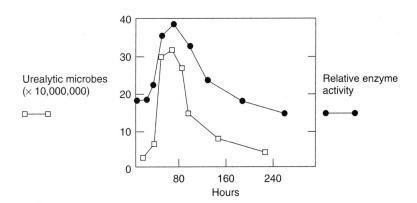

9. For the data above, plot your estimate of the subsequent change in NH_4^+ concentration in soil if urea is added again at 160 hours.
10. If the soil bulk density is 1.25 megagrams per cubic meter and the N content is 0.3%, how much total N is contained in this soil sample? If 4% is mineralized each year, how much inorganic N does this represent?

11. If the weight of soil in the upper 15 cm is 2×10^6 kg per ha, and it contains 0.3% N that mineralizes at a rate of 3% per year, how much inorganic N will this soil provide for a corn crop? What are some reasons why a farmer might still fertilize corn with N?

12. If $N_O = 20$ mg and $k = -.005$ h^{-1}, what is the approximate value of the N remaining after 15 hours? What does the reaction look like when you plot it?

Thought Question

Fresh substrates sometimes accelerate N mineralization in soil such that the amount of N mineralized is greater than the amount of N contained in the substrates added and the amount of N mineralized from the soil incubated alone. The enhancement is called priming. What are some reasons why this effect may occur?

Additional Reading

A good reference for anything to do with N in soil, including all of the aspects of the N cycle that are discussed in this textbook, is *Nitrogen in Agricultural Soils* (1982, edited by F. J. Stevenson, Soil Science Society of America, Inc. Madison, WI).

References

Bundy, L. G., and J. J. Meisinger. 1994. Nitrogen availability indices. In *Methods of soil analysis, part 2: Microbiological and biochemical properties,* R. W. Weaver et al. (eds.), 951–84. Madison, WI: Soil Science Society of America.

Handayani, I. 1996. Soil carbon and nitrogen pools and transformations after 23 years of no-tillage and conventional tillage. Ph.D. Diss. University of Kentucky, Lexington.

Hart, S. C., J. M. Stark, E. A. Davidson, and M. K. Firestone. 1994. Nitrogen mineralization, immobilization, and nitrification. In *Methods of soil analysis, part 2: Microbiological and biochemical properties,* R. W. Weaver et al. (eds.), 958–1018. Madison, WI: Soil Science Society of America.

Ingraham, J. L., O. Maaløe, and F. C. Neidhardt. 1983. *Growth of the bacterial cell.* Sunderland, MA: Sinauer Associates, Inc.

Jenny, H. 1930. A study of the influence of climate upon the nitrogen and organic matter content of the soil. *Missouri Agricultural Experiment Station Research Bulletin* 152:1–66.

Salter, R. M., and T. C. Green. 1933. Factors affecting the accumulation and loss of nitrogen and organic carbon in cropped soils. *Journal of the American Society of Agronomy* 25:622–30.

Stevenson, F. J. 1982. Origin and distribution of nitrogen in soil. In *Nitrogen in agricultural soil,* F. J. Stevenson (ed.), 1–42. Madison, WI: Agronomy Society of America.

Chapter 19

The Nitrogen Cycle—Nitrification

Overview

After you have studied this chapter, you should be able to:

- Define "nitrification," the process of NH_4^+ oxidation to NO_3^-.
- Explain several types of nitrification.
- Summarize the steps in NH_4^+ oxidation and NO_2^- oxidation.
- Identify the different bacteria involved in each step of nitrification.
- Describe the physiology of nitrification.
- List the environmental factors that regulate nitrification.
- Discuss the environmental consequences of nitrification.
- Name the inhibitors of nitrification and the rationale for their use.

INTRODUCTION

Nitrification is the microbial oxidation of NH_4^+ and organic N into NO_2^- and NO_3^-. The strictly biological nature of nitrification and the organisms involved were identified in 1889–1890 by Serge Winogradsky. Two types of nitrification are recognized: chemoautotrophic nitrification and heterotrophic nitrification. Although the end result is much the same, these are two distinct processes.

Chemoautotrophic nitrification is exclusively bacterial and is performed by a select group of lithotrophic bacteria. It is the dominant process in neutral to alkaline soils and is perhaps 100 to 1,000 times faster than heterotrophic nitrification. An important and useful characteristic of chemoautotrophic nitrification is that it is inhibited by low acetylene (C_2H_2) concentrations. Heterotrophic nitrification is performed by a diverse group of heterotrophic bacteria and fungi. It appears to dominate acidic soils (forest soils) but is a much slower process than is chemoautotrophic nitrification. Heterotrophic nitrification is not inhibited by low concentrations of acetylene.

We focus most of our attention on chemoautotrophic nitrification because it is the process that dominates most agricultural soils and is also the process about which most is known.

CHEMOAUTOTROPHIC NITRIFIERS

These nitrifying bacteria are gram-negative chemoautotrophs belonging to the family Nitrobacteriaceae. They exist by oxidizing NH_4^+ or NO_2^- to NO_2^- and NO_3^-, respectively.

$$NH_4^+ \Rightarrow NH_2OH \Rightarrow NO_2^- \Rightarrow NO_3^-$$

The energy source (NH_4^+ or NO_2^-) is the same as the electron donor. Nitrifiers gain carbon by the fixation of CO_2 or HCO_3^- using the Calvin cycle, just like a plant. Nitrifiers are difficult to isolate in pure culture because of their slow growth rate; most other bacteria grow faster and overrun the media.

Two nitrifier groups are recognized:

Ammonium oxidizers ($NH_4^+ \Rightarrow NO_2^-$)(prefix—nitroso);
Nitrosomonas, Nitrosolobus, Nitrosospira, Nitrosovibrio, Nitrosococcus
Nitrite oxidizers ($NO_2^- \Rightarrow NO_3^-$)(prefix—nitro);
Nitrobacter, Nitrococcus, Nitrospina

Some characteristics of the different genera are given in Table 19-1.

Three genera, *Nitrosomonas, Nitrosolobus,* and *Nitrobacter,* coexist and are important in soil. *Nitrosomonas europaea* is the easiest NH_4^+ oxidizer to isolate but may not be the dominant nitrifier in soil. Nitrifier populations in soil range from 0 to 10^6 g^{-1} (10^7 cells g^{-1} if NH_4^+ fertilizer is used). A reasonable estimate is 10^2 to 10^5 g^{-1}. Populations in acidic soils are much smaller than populations in neutral or alkaline soils. However, you generally find nitrifiers in soils more acidic than pure cultures can tolerate.

Table 19-1   Genera and characteristics of chemoautotrophic nitrifying bacteria.

Genus	Morphology	Growth Range (pure culture)	Habitat
Ammonium Oxidizers			
Nitrosomonas	Straight rods motile or nonmotile; one or two subpolar flagella	5°–40°C pH 5.8–9.5	Soil; marine; sewage; freshwater
Nitrosospira	Spiral shaped; motile or nonmotile; peritrichous flagella	25°–30°C pH 7.5–8.0	Soil
Nitrosococcus	Cocci in pairs or tetrads; motile or nonmotile tuft of peritrichous flagella	2°–30°C pH 6.0–8.0	Soil; marine; freshwater
Nitrosolobus	Lobular, pleomorphic; motile; peritrichous flagella	15°–30°C pH 6.0–8.2	Soil
Nitrite Oxidizers			
Nitrobacter	Short rods; motile or nonmotile; single polar flagellum	5°–40°C pH 5.7–10.2	Soil; marine; freshwater
Nitrospina	Long, slender rod; nonmotile	20°–30°C pH 7.0–8.0	Marine only
Nitrococcus	Spherical cell	20°–30°C pH 7.0–8.0	Marine only

PHYSIOLOGY

Nitrification is a multi-step process. The most reduced form of inorganic N serves as the starting point for nitrification, and the first product is hydroxylamine (NH_2OH):

$$NH_4^+ + H^+ + O_2 \Rightarrow NH_2OH + H_2O$$
$$\;\;(-3) \qquad\qquad\qquad (-1) \longleftarrow \text{Valence state}$$

The first step in NO_2^- formation is catalyzed by the enzyme ammonia monooxygenase (AMO). The oxygen in NO_2^- comes from molecular oxygen, so oxygen is not only required by these bacteria for respiration, but it is also a substrate. Nitrifiers deprotonate ammonium (NH_4^+) to ammonia (NH_3) before oxidizing it to hydroxylamine (NH_2OH).

The next step in NO_2^- formation is oxidation of NH_2OH by hydroxylamine oxidoreductase:

$$NH_2OH + H_2O \Rightarrow [HNO] \Rightarrow NO_2^- + 5H^+ + 4e^-$$
$$\;\;(-1) \qquad\qquad (+1) \qquad (+3) \longleftarrow \text{Valence state}$$

HNO, nitroxyl, is a reactive intermediate.

The last step in nitrification is the oxidation of NO_2^- to NO_3^-.

$$2NO_2^- + H_2O + O_2 \rightarrow 2NO_3^- + H_2O$$
$$(+3) \qquad\qquad\qquad (+5) \longleftarrow \text{Valence state}$$

The oxygen incorporated into NO_2^- actually comes from H_2O, not O_2.

A small amount of the NH_4^+ oxidized is lost as the gases nitrous oxide (N_2O) and nitric oxide (NO). It is not clear at which step this occurs or whether it is physiologically relevant. However, N_2O and NO are two gases that contribute to global warming and ozone destruction, so they have the potential for significant environmental effects. To be complete, our diagram of autotrophic nitrification should probably be revised as follows (Paul and Clark 1996):

$$
\begin{array}{ccc}
 & NO & NO \\
 & \Uparrow? & \Uparrow \\
NH_4^+ \Rightarrow NH_2OH \Rightarrow [HNO] \Rightarrow & NO_2^-\ NO_3^- \\
 & \Downarrow & \\
 & N_2O &
\end{array}
$$

Autotrophic nitrifiers are not very efficient at recovering the available energy from oxidizing the NH_4^+ and NO_2^- on which they grow. In *Nitrosomonas,* the ratio of NH_4^+ oxidized to C assimilated varies from 14–70:1. In *Nitrobacter,* the ratio of NO_2^- oxidized to C assimilated ranges from 76–135:1.

Notice that H^+ is formed in the oxidation of NH_4^+. Nitrification is an acidifying process in soil. As nitrification proceeds over the long term, the soil becomes more acidic, which in turn inhibits nitrification. Soils that are fertilized with NH_4^+ or organic N that eventually nitrifies periodically have to be treated to raise the soil pH.

ENVIRONMENTAL CONTROL

Because NO_2^- oxidizers obtain less energy from their substrate than do NH_4^+ oxidizers, they have to metabolize relatively more NO_2^- to maintain the same amount of growth. Consequently, NO_2^- oxidation in soil is quite rapid, and it is unusual for NO_2^- to accumulate in soil.

Nitrate is formed in soils, marine environments, manure piles, and during sewage processing. Control of nitrification can be investigated at several scales (Table 19-2) (Groffman 1991). The diagram in Figure 19-1 gives a schematic view of environmental controls (extrinsic factors) on nitrification.

Oxygen is important to nitrification because chemoautotrophic nitrifiers are obligate aerobes. Nitrification can occur in submerged soils only because enough O_2 diffusion occurs to keep the first few mm of sediment oxygenated. A few nitrifiers can also denitrify (Poth and Focht 1985), which means they can convert NO_3^- to nitrogen gas (N_2). In this process, N_2O and NO are also lost.

How Do You Enumerate Nitrifiers in Soil?

In 1873, Müller suggested that bacteria performed NH_4^+ oxidation in soil. In 1877, Schloesing demonstrated that NH_4^+ oxidation in sewage was biological since chloroform fumigation stopped it. Between 1879 and 1886, Warrington definitively showed what microorganisms required to nitrify and that there were two types of nitrifiers—NH_4^+ oxidizers and NO_2^- oxidizers (Doetsch 1960). But none of these early investigators could isolate nitrifiers from soil.

Serge Winogradsky succeeded in isolating nitrifiers in 1889 because he had one tremendous advantage—he recognized that these bacteria could be chemolithotrophs. So, an appropriate media for nitrifier isolation should be devoid of organic C. It should only contain minerals and bicarbonate as a C source.

It is still difficult to cultivate nitrifiers on solid medium because they grow so slowly. Consequently, when soil microbiologists attempt to count nitrifiers in soil, they usually use a Most Probable Number (MPN) procedure (Schmidt and Belser 1994). After serially diluting a soil sample, they distribute aliquots into mineral salts media that contain either NH_4^+ for NH_4^+ oxidizers or NO_2^- for NO_2^- oxidizers. Then they wait for several weeks.

One indication that nitrification is occurring is if the pH of the media changes from slightly neutral to acidic. So, a pH indicator (bromothymol blue) is usually added to the media that turns yellow if the media is becoming acidic.

Antibodies raised against nitrifiers isolated from soil have also been used to directly stain and visualize nitrifiers using fluorescent antibody techniques.

Table 19-2 Factors controlling nitrification activity at different scales of investigation.

Scale of Investigation	Controlling Factors
Organism	Ammonium, oxygen
Field	Ammonium supply, soil water
Landscape	Soil type, plant community type
Regional	Geomorphology, land use

Below 5°C and above 40°C the nitrification rate is very slow. Ammonium accumulates in nature because mineralization is less susceptible to environmental change, particularly temperature change, than is nitrification. Nitrifiers are also more susceptible to pesticides than are most other physiological groups.

Chief among the environmental influences on nitrification is pH. Nitrification is most rapid in neutral to alkaline soils, and nitrifying bacteria are rare or absent in acidic environments. Nitrification rates fall markedly below pH 6 and are negligible below pH 5 in pure culture. On the other hand, NO_2^- accumulation in soil is the result of high

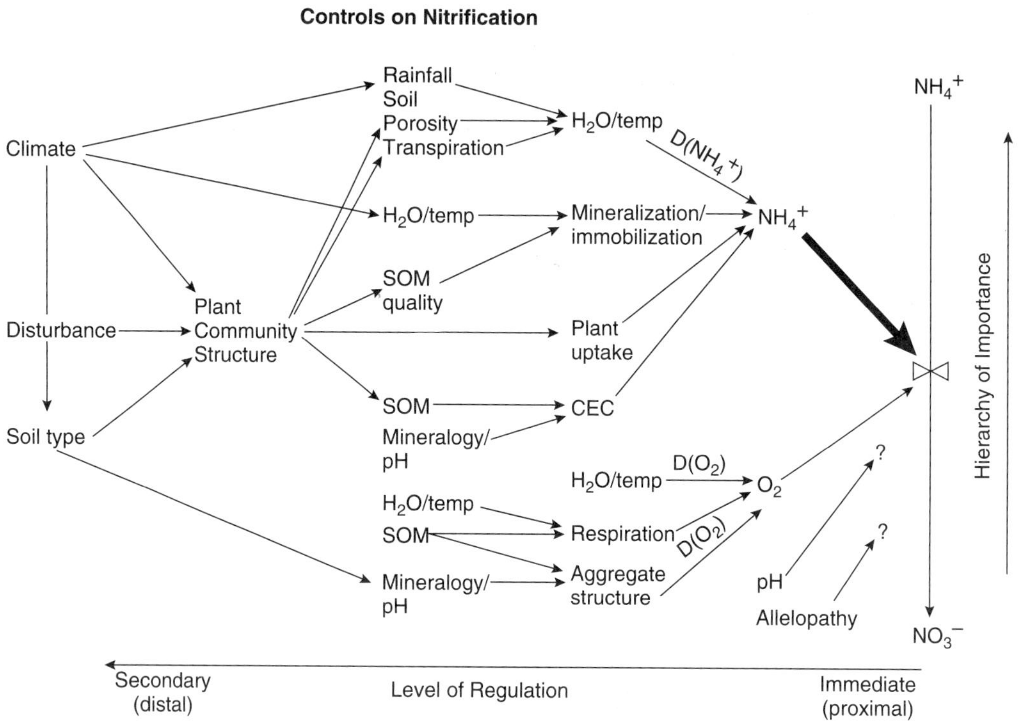

Controls on Nitrification

Figure 19-1   Immediate and secondary regulators of nitrification in the environment. (Adapted from Robertson, unpublished)

NH_4^+ concentrations and high pH. Since nitrification is a two-step process, anything that inhibits individual steps in nitrification causes intermediates to accumulate. *Nitrobacter,* for example is inhibited by NH_3. Chemical equilibrium between NH_4^+ and NH_3 leads to higher NH_3 concentrations if the NH_4^+ concentration rises. However, NO_2^- rarely accumulates in the environment; it is usually not the rate-limiting step in nitrification. As rapidly as NO_2^- forms, it is oxidized to NO_3^-.

In moist soils (soils with a high water potential) a decline in nitrification rates is mostly because either not enough NH_4^+ or not enough O_2 is available. In dry soils, nitrification rates are limited by the physiological effects of dehydration on the nitrifying bacteria. Diffusion of NH_4^+ and NO_2^- and movement of nitrifiers are also affected. At about -0.6 MPa, dehydration effects become critical. Figure 19-2 is a representation of how much overall nitrification is affected by either substrate limitation or by drying.

FERTILITY

Nitrate is the major form of N taken up by plants in soils because NH_4^+ is rapidly nitrified and NH_4^+ on the soil exchange complex is less readily available than is NO_3^-. Typical nitrification rates in soil are given in Table 19-3. Rapid nitrification rates are not

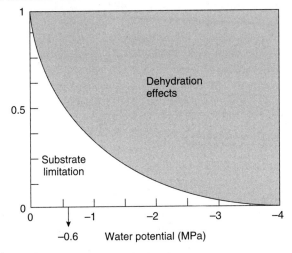

Fraction of decline
in nitrification rates

Dehydration
effects

Substrate
limitation

Water potential (MPa)

Figure 19-2 The effects of dehydration and substrate limitation on nitrification in soil. (Adapted from Stark and Firestone 1995)

Table 19-3 Nitrification in an intact core of Maury silt loam in two tillage methods as affected by soil moisture.

	Rate of NH_4^+ Consumption	
Moisture Content	No Till	Conventional Till
g/100 g Soil	mg of N Kg^{-1} of Soil d^{-1}	
34	3.6	4.2
31	4.4	3.7
26	4.0	2.7
22	3.2	2.6
18	2.9	2.7
14	0.8	1.7

(From Rice and Smith 1983)

always desirable because NO_3^- formation has been linked to eutrophication, nitrosamine formation, and human infant and animal methemoglobnemia. The pH of the infant gut favors production of NO_2^-. Nitrite binds and poisons hemoglobin, preventing it from accepting and transferring oxygen. Partly because of gut pH, and partly because adults take in less water per unit body mass than do infants NO_3^- toxicity is generally not a problem for adults. Consequently, federal regulations limit NO_3^- in potable water to 10 ppm NO_3^- N for the protection of infants.

NITRIFICATION INHIBITORS

Slow-release fertilizers such as sulfur-coated urea are used to limit nitrification in soil because they limit the rate at which NH_4^+ becomes available to nitrifiers. There are also many specific inhibitors of nitrification that differ in their mode of action. Chelating agents remove the Cu cofactor in ammonium monooxygenase (AMO). Some examples of chelating agents are:

Hydrazinecarbothiamide	NH_2CSNNH_2
Diethyldithiocarbamate	$(C_2H_5)_2NS_2Na$
Allylthiourea	$CH_2 = CHCH_2NHCSNH_2$
Thiourea	H_2NCSNH_2

Other chemicals also chelate Cu in ammonia monooxygenase and interfere with electron transport by cytochromes. Some examples are sodium azide (NaN_3) and potassium cyanide (KCN).

Pyridine compounds interact with the active site of ammonia monooxygenase. Examples include pyridine, nitrapyrin, and picolinic acid. The chemical nitrification inhibitor with the greatest commercial application in agriculture is nitrayprin (2-Chloro-6[trichloromethyl] pyridine), which is sold under the brand name "N-Serve."

Acetylene gas (C_2H_2) at very low concentrations (1–10 Pa) inhibits nitrification. It is a very useful inhibitor for laboratory studies of nitrification in soil because heterotrophic nitrification is unaffected by acetylene at similar concentrations.

Virtually all of these chemical inhibitors prevent nitrification at the level of NH_4^+ oxidation. This makes perfect sense because if they inhibit NO_2^- oxidation, several bad things will happen. First, NO_2^- is still an anion so it tends to leach whenever it rains. Second, NO_2^- is much more toxic to plant growth than is either NH_4^+ or NO_3^-, so plant toxicity occurs if NO_2^- accumulates. Third, NO_2^- in groundwater might rise to concentrations high enough so that it poses a problem to adults drinking the water. Fourth, NO_2^- is more reactive than is either NH_4^+ or NO_3^-, so it has a greater tendency to break down into gaseous products that affect the global atmosphere.

HETEROTROPHIC NITRIFICATION

Heterotrophic nitrification is the formation of NO_2^-, and NO_3^- from NH_4^+, NO_2^-, and organic N by heterotrophic organisms. Its importance is not clear. In some soils, heterotrophic nitrification accounts for half of nitrification. These are usually acidic soils in which autotrophic populations are low. Heterotrophic nitrifiers are both bacteria and fungi. Representative examples of heterotrophic nitrifying bacteria are *Arthrobacter, Aerobacter, Mycobacterium, Streptomyces,* and *Pseudomonas. Aspergillus* and *Penicillium* are two fungi linked to heterotrophic nitrification.

Heterotrophic nitrification is much less than autotrophic nitrification (Focht and Verstraete 1977), but since heterotrophic populations are higher, the total amount of nitrification from both processes may actually be quite similar. Nitrite appears only after growth has ceased, if NH_4^+ is in excess of cellular needs, and if the C:N ratio is low. Compared to autotrophic nitrification, heterotrophic nitrification is unaffected by C_2H_2.

So, measuring NO_2^- and NO_3^- formation in soil in the presence of low concentrations of C_2H_2 is one method of figuring out how much of the nitrification is due to heterotrophic nitrification.

Summary

Nitrification is the microbial oxidation of NH_4^+ to NO_2^- and NO_3^-. Two types of nitrification are recognized: chemoautotrophic nitrification and heterotrophic nitrification. Chemoautotrophic nitrification is exclusively bacterial and is performed by a select group of lithotrophic bacteria. It is the dominant process in neutral to alkaline soils. Heterotrophic nitrification is performed by a diverse group of heterotrophic bacteria and fungi. It appears to dominate acidic soils (forest soils).

Chemoautotrophic nitrification is a two-step process performed by ammonium oxidizers ($NH_4^+ \Rightarrow NO_2^-$) (prefix—nitroso; *Nitrosomonas, Nitrosolobus, Nitrosospira, Nitrosovibrio, Nitrosococcus*) and nitrite oxidizers ($NO_2^- \Rightarrow NO_3^-$) (prefix—nitro; *Nitrobacter, Nitrococcus, Nitrospina*). Nitrifier populations in soil range from 0 to 10^6 g^{-1}. A reasonable estimate is 10^2 to 10^5 g^{-1}. Populations in acidic soils are much smaller than populations in neutral or alkaline soils.

Nitrification is an acidifying process in soil. Some of the NH_4^+ is lost as the gases nitrous oxide (N_2O) and nitric oxide (NO). Rapid NO_3^- formation is not always desirable because it has been linked to eutrophication, nitrosamine formation, and human infant and animal methemoglobinemia.

Virtually all chemical inhibitors prevent nitrification at the level of NH_4^+ oxidation. The chemical nitrification inhibitor with the greatest commercial application in agriculture is nitrayprin. Heterotrophic nitrification is the formation of NO_2^- and NO_3^- from NH_4^+, NO_2^-, and organic N by heterotrophic organisms. It is of uncertain importance.

Sample Questions

1. What is nitrification? What are the steps involved? Illustrate/diagram this process and label all the appropriate steps. Give a specific genus of microorganism for each step.
2. Give hypotheses that explain why NO_3^- appears in groundwater after clear-cutting forests.
3. What are the major environmental controls on nitrification in soil?
4. What is heterotrophic nitrification?
5. What are the steps in nitrification that are most sensitive to either high or low pH?
6. What steps in the nitrification pathway are controlled by nitrification inhibitors?
7. How does urease activity benefit an autotrophic nitrifier growing in an acidic soil?
8. Define nitrification.

9. Draw your own diagram that shows the immediate (proximal) and secondary (distal) factors that control the rate or extent of nitrification in soil, and that illustrate the potential fate of nitrification products formed.

10. In a dry, acidic soil at 25°C, what is the effect on NO_3^- formation if the following environmental changes occur simultaneously?

   a. It rains.

   b. The soil is limed.

   c. The temperature of the soil rises to 35°C.

11. Based on the data in the following table, what is the effect of moisture content on nitrification in the two soils?

Nitrification as affected by soil moisture, in sieved and undisturbed Maury silt loam soils collected from no-till and conventional till plots.

		Nitrification Rate	
Soil Treatment	Soil Moisture Content	No Till	Conventional Till
	g/100 g Soil	mg of N kg^{-1} Soil d^{-1}	
Sieved	31	15.7	11.5
	26	14.1	15.3
	20	10.1	13.4
	15	8.1	11.5
	10	3.0	5.4
Undisturbed	34	3.6	4.2
	31	4.4	3.7
	26	4.0	2.7
	22	3.2	2.6
	18	2.9	2.7
	14	0.8	1.7

(From Rice and Smith 1983)

12. Why do you think sieving in the data above increases the amount of nitrification in both soils?

Thought Question

Saltpeter, or niter (KNO_3), is used to make gunpowder in combination with sulfur and charcoal. During the Napoleonic Wars, the importation of niter into France was cut off by a British blockade. To circumvent the blockade, saltpeter was made by mixing together soil, manure, and lime. This mixture was aerated and watered with urine and water. Nitrate was later extracted from the mixture with hot water. What process was occurring, what was the purpose of each step, and what was the purpose of each ingredient?

Additional Reading

Nitrifiers are a classic example of chemolithotrophs, so you might as well crack open Chapter 8 of John Postgate's book, *The Outer Reaches of Life* (1994, Cambridge University Press, Cambridge, England) and see what he has to say about them.

References

Doetsch, R. N. 1960. *Microbiology: Historical contributions from 1776–1948.* New Brunswick, NJ: Rutgers University Press.

Focht, D. D., and W. Verstraete. 1977. Biochemical ecology of nitrification and denitrification. *Advances in Microbial Ecology* 1:135–214.

Groffman, P. M. 1991. Ecology of nitrification and denitrification in soil evaluated at scales relevant to atmospheric chemistry. In *Microbial production and consumption of greenhouse gases,* J. E. Rogers, and W. B. Whitman (eds.), 201–17. Washington, DC: American Society for Microbiology.

Paul, E. A., and F. E. Clark. 1996. *Soil Biology and Biochemistry.* San Diego, CA: Academic Press.

Poth, M. and D. D. Focht. 1985. ^{15}N Kinetic analysis of N_2O production by *Nitrosomonas europaea:* An examination of nitrifier denitrification. *Applied and Environmental Microbiology* 49:1134–41.

Rice, C. W., and M. S. Smith. 1983. Nitrification of fertilizer and mineralized ammonium in no-till and plowed soil. *Soil Science Society of America Journal* 47:1125–29.

Schmidt, E. L., and L. W. Belser. 1994. Autotrophic nitrifying bacteria. In *Methods of soil analysis, part 2: Microbiological and biochemical properties.* R. W. Weaver et al. (eds.), 159–77. Madison, WI: Soil Science Society of America.

Stark, J. M., and M. K. Firestone. 1995. Mechanisms for soil moisture effects on activity of nitrifying bacteria. *Applied and Environmental Microbiology* 61:218–21.

Chapter 20

The Nitrogen Cycle—
Immobilization

Overview

After you have studied this chapter, you should be able to:

- Identify the most important mechanisms by which NH_4^+ is assimilated into the cell.
- Describe how NH_4^+ and other N-containing compounds control NO_3^- assimilation.
- Explain how the C:N ratio controls whether mineralization or immobilization occurs.
- Predict what will happen to the C:N ratio when an organic compound is added to soil.

INTRODUCTION

To be technically correct, when we discuss N immobilization in soil, we include fixation of inorganic N into mineral material and organic matter such as humus. For the purpose of this chapter, however, our definition of immobilization is much narrower. Immobilization (or assimilation—the two are synonymous terms) is the incorporation of inorganic N into organic N by soil microorganisms. The N is immobilized with respect to plants; if it is being turned into new microbial cells, it cannot be taken up by plants until those microbial cells decompose or mineralize. The consequences of inadequate N in soil can be devastating for plants, as Figure 20-1a,b illustrates.

NH_4^+ ASSIMILATION

There are two major pathways in which NH_4^+ ions are assimilated by microorganisms. The first NH_4^+ assimilation pathway is directly into glutamate by the enzyme glutamate dehydrogenase (GDH), (Figure 20-2).

This pathway operates at relatively high NH_4^+ concentrations (> 1 mM in solution or > 0.5 mg NH_4^+/kg soil).

At lower NH_4^+ concentrations (< 1 mM NH_4^+ or < 0.1 mg NH_4^+/kg soil) another pathway is used—GS/GOGAT, (Figure 20-3). (GS = glutamine synthetase; GOGAT = glutamine α-oxoglutarate amino transferase [Glutamate Synthase].) It is a two-step process in which the primary assimilation of NH_4^+ occurs in a reaction catalyzed by GS. The second step in the process forms glutamate from the transamination of α–ketoglutarate by glutamine, (Figure 20-4).

a   b

Figure 20-1   N-deficient corn (a) and tobacco (b). Note how the N-deficient corn is stunted with respect to its better nourished neighbor. One symptom of N deficiency is browning of lower leaves as the N in these older leaves is remobilized by the plant and transported to new leaves. Although this phenomenon is evident in the corn, it is much more pronounced in the tobacco (b), which has a high N demand. (Photographs courtesy of M. S. Coyne)

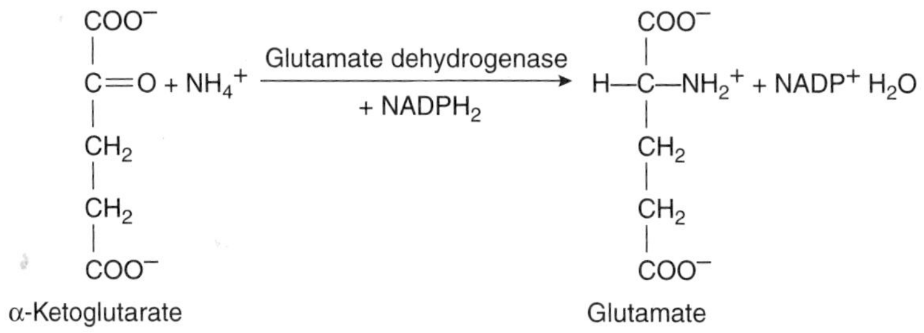

$$
\begin{array}{c}
\text{COO}^- \\
| \\
\text{C}{=}\text{O} + \text{NH}_4^+ \\
| \\
\text{CH}_2 \\
| \\
\text{CH}_2 \\
| \\
\text{COO}^-
\end{array}
\quad
\xrightarrow[\text{+ NADPH}_2]{\text{Glutamate dehydrogenase}}
\quad
\begin{array}{c}
\text{COO}^- \\
| \\
\text{H}{-}\text{C}{-}\text{NH}_2^+ + \text{NADP}^+ \text{H}_2\text{O} \\
| \\
\text{CH}_2 \\
| \\
\text{CH}_2 \\
| \\
\text{COO}^-
\end{array}
$$

α-Ketoglutarate   Glutamate

Figure 20-2   Reductive amination of α-ketoglutarate by glutamate dehydrogenase—a mechanism of NH_4^+ assimilation.

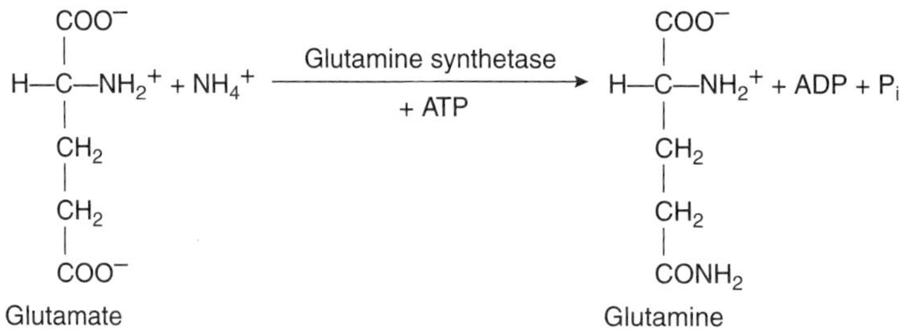

$$
\begin{array}{c}
\text{COO}^- \\
| \\
\text{H}{-}\text{C}{-}\text{NH}_2^+ + \text{NH}_4^+ \\
| \\
\text{CH}_2 \\
| \\
\text{CH}_2 \\
| \\
\text{COO}^-
\end{array}
\quad
\xrightarrow[\text{+ ATP}]{\text{Glutamine synthetase}}
\quad
\begin{array}{c}
\text{COO}^- \\
| \\
\text{H}{-}\text{C}{-}\text{NH}_2^+ + \text{ADP} + \text{P}_\text{i} \\
| \\
\text{CH}_2 \\
| \\
\text{CH}_2 \\
| \\
\text{CONH}_2
\end{array}
$$

Glutamate   Glutamine

Figure 20-3   Glutamine formation—the first step in NH_4^+ assimilation by GS/GOGAT.

$$
\begin{array}{c}
\text{COO}^- \\
| \\
\text{H}{-}\text{C}{-}\text{NH}_2 \\
| \\
\text{CH}_2 \\
| \\
\text{CH}_2 \\
| \\
\text{CONH}_2
\end{array}
+
\begin{array}{c}
\text{COO}^- \\
| \\
\text{C}{=}\text{O} \\
| \\
\text{CH}_2 \\
| \\
\text{CH}_2 \\
| \\
\text{COO}^-
\end{array}
\quad
\xrightarrow[\text{Amino transferase}]{\text{Glutamine α-oxoglutarate}}
\quad
2
\begin{array}{c}
\text{COO}^- \\
| \\
\text{H}{-}\text{C}{-}\text{NH}_2 \\
| \\
\text{CH}_2 \\
| \\
\text{CH}_2 \\
| \\
\text{COO}^-
\end{array}
$$

Glutamine   α-Ketoglutarate

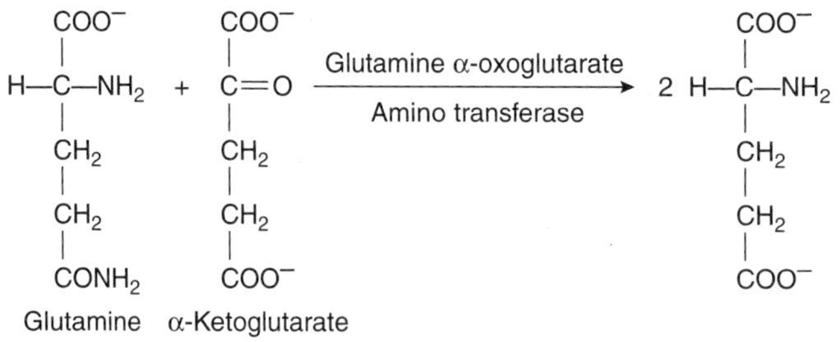

Figure 20-4   Glutamate formation—the second step in NH_4^+ assimilation by GS/GOGAT.

Glutamine synthetase-catalyzed reactions require ATP, but the affinity of GS for NH_4^+ is greater than that of GDH. The K_m is lower, which means that GS allows microorganisms to utilize lower concentrations of environmental NH_4^+. When both pathways are present, GS is repressed by high NH_4^+ concentrations.

Once N is incorporated as glutamate, it is transferred via transamination reactions to keto acids to form amino acids. Figure 20-5 shows an example.

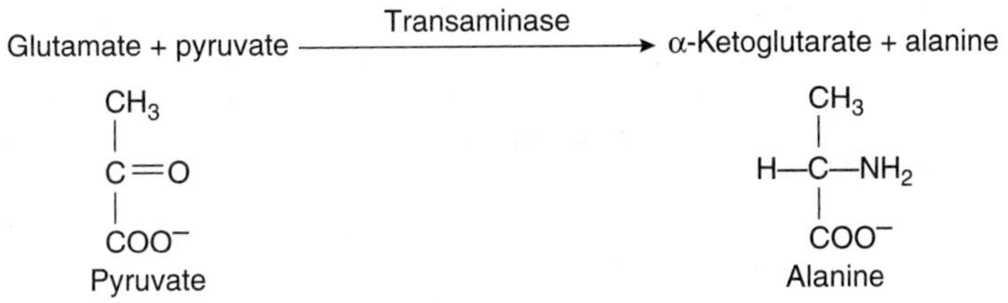

Glutamate + pyruvate $\xrightarrow{\text{Transaminase}}$ α-Ketoglutarate + alanine

$$
\begin{array}{c}
CH_3 \\
| \\
C{=}O \\
| \\
COO^- \\
\text{Pyruvate}
\end{array}
\qquad\qquad
\begin{array}{c}
CH_3 \\
| \\
H{-}C{-}NH_2 \\
| \\
COO^- \\
\text{Alanine}
\end{array}
$$

Figure 20-5 Immobilization of assimilated N into amino acids.

NO$_3^-$ ASSIMILATION

Many bacteria and fungi assimilate inorganic N as NO_3^-. To do this, they first reduce the NO_3^- to NO_2^- using an assimilatory NO_3^- reductase enzyme. Then they reduce the NO_2^- to NH_4^+ via an enzyme called assimilatory NO_2^- reductase (Cole 1987). There are many different types of NO_3^- reductases in the microbial and plant world, but they share a common feature; they are all molybdo-proteins. All molybdo-proteins have the metal molybdenum (Mo) as a cofactor. Tungsten (W) can replace Mo in molybdo-proteins, but the enzymes will lose their activity. So, W is a common inhibitor used to study NO_3^- assimilation reactions. Chlorate (ClO_3^-) also inhibits assimilatory NO_3^- reduction because it is an analog of the NO_3^- anion and is reduced in its place. The product, chlorite (ClO_2^-), is toxic to cells and kills them.

Assimilatory NO_3^- and NO_2^- reductases are synthesized only when soil N is limiting or when NO_3^- is present (Marzluf 1997). Compounds such as NH_4^+, glutamine, and glutamate are all preferentially used by soil microorganisms. Schimel and Firestone (1989) demonstrated this preferential use in an experiment that examined NH_4^+ and NO_3^- uptake by microbial biomass in the organic horizon of a coniferous forest soil.

The data in Table 20-1 show that regardless of the month (March, January, or April), the subhorizon (O2 or O3), or the concentration of added inorganic N (0.5 or 0.05 mM), more NH_4^+ than NO_3^- was incorporated into biomass. Schimel and Firestone suggested that this was biological assimilation because the sterile controls had < 20% of the N assimilation values of nonsterile treatments.

Even when microbial cells have synthesized assimilatory NO_3^- and NO_2^- reductase and are actively taking up NO_3^-, that uptake can be almost immediately inhibited by adding NH_4^+ to the environment. This inhibition could simply be because the NH_4^+ inhibited NO_3^- uptake, or because it suppressed either the synthesis or activity of NO_3^--assimilating enzymes. Rice and Tiedje (1989) did an interesting experiment to decipher this question by incubating soil slurries in conditions that caused NO_3^- uptake, monitoring the rate at which NO_3^- disappeared, and seeing if that rate changed when NH_4^+ and various other N-containing compounds were added.

Adding NH_4^+ had an almost immediate inhibitory effect (< 1 minute) on NO_3^- uptake. This result indicated that the mechanism of inhibition was by preventing NO_3^-

Table 20-1   N assimilation rates (μg N g^{-1} h^{-1}) from solution by organic horizon material.

Sample Month	O2 Subhorizon		O3 Subhorizon	
	0.5 mM N	0.05 mM N	0.5 mM N	0.05 mM N
NH_4^+-N Assimilation				
March 84	7.3	2.1	19.6	3.4
January 85	3.7	ND	7.8	ND
April 85	8.4	4.3	8.9	4.7
NO_3^--N Assimilation				
March 84	0.7	< 0.2	0.9	< 0.2
January 85	0.5	ND	0.4	ND
April 85	0.2	0.03	0.4	0.05

ND = not determined

(Adapted from Schimel and Firestone 1989)

from getting into the microbial cells, because an effect on enzyme synthesis wouldn't have been nearly as rapid. Compounds such as glutamine and asparagine had almost as great an inhibitory effect on NO_3^- uptake as NH_4^+, but took longer to cause the inhibition (> 4 minutes). This indicated to Rice and Tiedje that the mechanism by which these N-containing compounds inhibited NO_3^- uptake probably was an effect on the enzymes that reduced NO_3^- to the level of NH_4^+.

Once the NO_3^- is reduced to NH_4^+, it is probably assimilated via the GS/GOGAT pathway of N assimilation rather than by GDH (glutamate dehydrogenase). Betlach et al. (1981) demonstrated this by using radioactive ^{13}N to label NO_3^-, exposing cells to $^{13}NO_3^-$ under NO_3^--assimilating conditions, and then extracting the cells at various intervals to see where the radioactive N was appearing. They hypothesized that if the NO_3^- was incorporated into organic N by GDH, the first labeled organic compound they would see would be glutamate (see Figures 20-2 to 20-5 to refresh your memory about these pathways) and very little labeled glutamine would form. However, if the $^{13}NO_3^-$ was incorporated by GS/GOGAT, then the first labeled organic N compound they would see would be glutamine. What happened? Table 20-2 shows the result.

Betlach and his coworkers couldn't extract the cells fast enough to prevent $^{13}NO_3^-$ from being incorporated into glutamate, but since the relative amount of ^{13}N in glutamine kept declining, it indicated that their second hypothesis, that NO_3^- was assimilated via GS/GOGAT, was probably correct. Some of the additional compounds into which the labeled NO_3^- was assimilated are shown in Table 20-3. As you would predict, in the short term, most of these compounds are amino acids that form from the transamination of keto acids such as pyruvate.

Table 20-2 Incorporation of $^{13}N^-$ labeled NO_3^- into glutamate and glutamine in NO_3^--assimilating *Pseudomonas fluorescens*.

Time (seconds)	Relative Fraction of ^{13}N	
	Glutamate	Glutamine
10	1	0.75
75	1	0.35
300	1	0.11

(Adapted from Betlach et al. 1981)

Table 20-3 Distribution of ^{13}N in intracellular organic N pools after incubating *Pseudomonas fluorescens* 5 minutes with $^{13}NO_3^-$ in NO_3^--assimilating conditions.

Compound	Percent of ^{13}N Label
NO_3^-, NO_2^-	3.9
Aspartate	8.8
Glutamate	58.5
Asparagine	3.5
Glutamine	6.2
Alanine	17.6
Arginine	1.5

(Adapted from Betlach et al. 1981)

N AVAILABILITY

Mineralization and immobilization occur simultaneously. Organic N and NH_4^+ are continuously cycling back and forth. Whether mineralization or immobilization occurs during decomposition of plant residue depends on the C:N content of the material being decomposed by heterotrophic organisms.

The C:N ratio reflects the relative proportion of C and N in a given substance. More importantly, in microorganisms it indicates approximately how much N must be assimilated for every gram of C that is converted into biomass. Soil bacteria usually have a C:N ratio of 5:1 to 8:1. This is because microbial cell walls are about 45% C and 6% to 9% N. However, the C:N ratio of microorganisms is not constant. Fungi have a wide C:N ratio—4.5:1 to 15:1—because some fungi, especially phycomycetes, contain cellulose in their cell wall. Many bacteria produce polysaccharides under N-limitation that will raise their C:N ratio.

When a substance entering soil has a high C:N ratio (40:1 for example), net immobilization occurs because there is not enough N available in the substance to convert all of the C into biomass. To make up the deficit, microorganisms in the immediate environment assimilate all of the available inorganic N in their vicinity, which makes it unavailable for plants.

Table 20-4   Proportion of ^{15}N recovered in plant and microbial biomass.

Month	Treatment	Plant Biomass	Microbial Biomass	Total Recovery
		Fraction of Total ^{15}N Recovered		
February (early spring)				
	$^{15}NH_4^+$	8.7	45.8	76.4
	$^{15}NO_3^-$	19.6	38.1	84.0
April (late spring)				
	$^{15}NH_4^+$	10.7	60.7	125.9
	$^{15}NO_3^-$	25.9	49.8	120.0

(Adapted from Jackson et al. 1989)

Jackson et al. (1989) showed an example of plants and microorganisms competing for soil N when they compared the proportion of ^{15}N added to soil that was recovered by plant and microbial biomass in an annual grassland in northern California (Table 20-4). Nitrogen-15, unlike ^{13}N, is a nonradioactive isotope of N (i.e., it is a stable isotope), which makes it very useful for tracing the fate of various N-compounds in the environment.

The data in Table 20-4 show that the microbial biomass assimilated substantially more NH_4^+ and NO_3^- than plants did, regardless of the season. Although plants competed better for NO_3^- than for NH_4^+, microbial uptake was the major factor controlling NO_3^- availability to plants in this environment (Jackson et al. 1989).

At low C:N ratios (20:1 for example), net mineralization occurs. There is enough N in the substrate for the microorganisms to convert the C into biomass and the excess N accumulates in soil. The decomposition rate of added material is affected by the availability of N in plant and soil and fertilizer. The "priming effect" is a stimulated decomposition of existing soil organic material by the addition of fresh substrates. The more microorganisms that can grow on a substrate, the greater the microbial biomass that will be formed, and the greater the extent of decomposition that can occur. As plant decomposition proceeds, C:N ratios fall because C is lost through respiration while cellular N is retained. Eventually, the C:N ratios of the decomposing material will approximate the C:N ratio of the resident microbial population (Figure 20-6).

The C:N ratio is a good, but not absolute, indicator of whether N immobilization is likely to happen. It depends, in part, on the availability of C (Table 20-5). Trees can have a C:N ratio > 50:1 and still be mineralized because the effective C:N ratio of their tissue is lower by virtue of the lignin content. The lignin is virtually insoluble and indigestible by most microorganisms, so for all intents and purposes, it has no effect on metabolism of other substrates. As plants age, the lignin content goes up as well as the C:N ratio. Remember, mineralization rates are not constant because the availability of all mineralizable substrates is not constant. As a general rule, 1 kg of N is required to decompose every 100 kg of dry plant material with a low N content, such as straw.

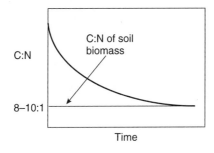

Figure 20-6   Decline in C:N of decomposing residue with time.

Mineralization, Immobilization, and Musical Chairs

You can use the analogy of musical chairs to get a better feeling for the role of the C:N ratio on N mineralization and immobilization. In mineralization, there are more players (N) than chairs (C) to hold them. So, some of the players can't sit, and no new players can join. The players with no chairs hang around waiting for something to do.

In contrast, when immobilization occurs, there are more chairs (C) than players (N). So, not only do all the chairs get occupied, but the open chairs attract more players from those in the room.

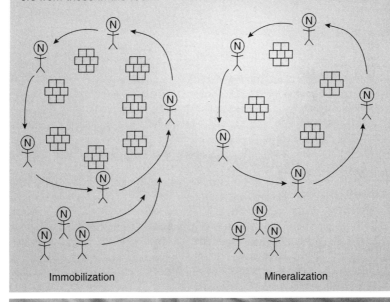

Immobilization   Mineralization

Table 20-5   C:N ratios of various organic materials.

Type of Residue	Percent C	Percent N	C:N

High C; high N
  Legume residue
  Legume green manure
  Young cereal green manure
	40	2.5	16:1

Low C; high N
  Composted strawy manure
  Feces
	25	3	8.3:1

High C; low N
  Straw
  Mature cereal
  Leaves
	40	0.5	80:1

Low C; low N Composted straw	25	0.8	31.3
Tree bark Tree wood			454:1 615:1

(From McCalla et al. 1977)

Immobilization Example

1. Start with plant material that has a C:N ratio of 40:1.
    a. Assume the efficiency of microbial conversion of C into biomass is 40% and assume the microbial C:N ratio is 10:1.
2. If you had 100 g C, then 2.5 g N would be present if that material had a C:N ratio of 40:1.
    a. If all C was immediately available, 40 g C (40%) would go to biomass and 60 g C (60%) would go to respiration.
3. However, only 2.5 g N is available, so if microbial C:N were 10:1, only 25 g C could be converted into biomass.
4. Two things could happen:
    a. Immobilization of soil NH_4^+ and NO_3^- could occur.
    b. The amount of mineralizable N will limit this round of decomposition and only 25 g of C will turn into biomass.
5. If 25 g of C represents 40% conversion into biomass, then 37.5 g C represents 60% loss of C through respiration.
    a. Only 62.5 g of C decomposes, leaving 37.5 g of C in soil.
6. Assume all of the microorganisms died.
    a. There would be 25 g of microbial C.

b. There would be 37.5 g of C left in undecomposed plant material.
c. This makes 62.5 g of C, which will still have only 2.5 g N (if none came from soil).
d. 2.5 g N will permit incorporation of 25 g C into biomass.
    1. 37.5 g C will be lost through respiration.
    2. The sum equals 62.5 g C, which is what you started with.
7. The C:N ratio of the starting material is 62.5 g C:2.5 g N, or a C:N ratio of 25:1.
8. As decomposition proceeds, C:N will decline until it eventually approximates the C:N ratio of the soil biomass—about 10:1. The break-even point is 20:1 to 30:1.

Summary

Immobilization, or assimilation, is the incorporation of inorganic N into organic N by soil microorganisms. There are two major pathways by which NH_4^+ is assimilated by microorganisms. The first pathway is directly into glutamate via catalysis the enzyme glutamate dehydrogenase (GDH). The second pathway is assimilation into glutamine and then glutamate by GS/GOGAT. Once N is incorporated as glutamate, it is transferred via transamination reactions to keto acids to form amino acids.

Many bacteria and fungi are able to assimilate inorganic N as NO_3^-. To do this, they first reduce the NO_3^- to NO_2^- by an enzyme called assimilatory NO_3^- reductase. Then they reduce the NO_2^- to NH_4^+ via an enzyme called assimilatory NO_2^- reductase. There are many different types of NO_3^- reductases, but they are all molybdo-proteins. Compounds such as NH_4^+, glutamine, and glutamate, rather than NO_3^-, are all preferentially used by soil microorganisms if they are available. They inhibit NO_3^- use, first by preventing NO_3^- from getting into the microbial cells, then by inhibiting NO_3^- reductase synthesis or activity.

Whether mineralization or immobilization occurs during decomposition of plant residue depends on the C:N content of the material being decomposed. When a substance entering soil has a high C:N ratio net immobilization occurs because there is not enough N available in the substance to convert all of the C into biomass. At low C:N ratios, net mineralization occurs. There is enough N in the substrate for the microorganisms to convert the C into biomass and the excess accumulates in soil. The break-even point for mineralization of organic N based on C:N ratios is approximately 20–30:1.

Sample Questions

1. Diagram the pathway of NH_4^+ assimilation into microbial biomass via GS/GOGAT.
2. Will incorporation of material with a C:N ratio of 100:1 cause N in soil to mineralize or immobilize? Why?

3. Will the C:N ratio of organic material with a C:N ratio of 100:1 remain constant in soil? Explain your reasoning.

4. What is the priming effect?

5. What controls whether GDH or GS/GOGAT is used to assimilate N?

6. When is the C:N ratio misleading about when organic matter will be cause immobilization?

7. What will be the consequences of adding organic N with a very low C:N ratio to soil?

8. Assume that the microbial efficiency is 35%, that the microbial C:N ratio is 15:1, and that plant material has a C:N ratio of 55:1, all of which can be readily decomposed. Assume there is no available N except in the plant material. If you begin with 130 g of C in the plant material, how much C will remain (if any) after one round of degradation?

9. If 100 g of soil containing 5 ppm NH_4^+ is amended with 10 g of organic C with a C:N ratio of 35:1, will there be a net increase or decrease of inorganic N in soil (assuming it remains well-aerated)? You may assume that the microbial C:N ratio in this soil is 26:1 and the microorganisms have an efficiency of 35%.

10. Given the following information, what is the "break point" for mineralization vs. immobilization in terms of the C:N ratio of the plant material?
    a. The C:N ratio of the microbial community is 10:1.
    b. The microbial efficiency is 25% (75% of the decomposed C goes to CO_2).
    c. All plant components are being utilized for the first time.

11. Based on the data in the following table, briefly describe what effect NH_4^+ and various amino acids have on NO_3^- assimilation in soil.

NO_3^- concentration ($\mu g\ NO_3^-$–N mL^{-1})	N source	Percent Inhibition	Time of Response (min)
2	NH_4^+	55	1
	Asparagine	50	1
	Glutamine	48	4
	Glutamate	22	ND
	Arginine	51	6
	Aspartate	21	ND
	Alanine	8	ND
	Glycine	1	ND
10	NH_4^+	79	5
	Glutamine	72	3

ND = not detectable

(Adapted from Rice and Tiedje 1989)

12. For the data in the following table, provide hypotheses to explain the result of each treatment.

Rate of NO$_3^-$ reduction in cells grown on different N sources.

Treatment	nmol NO$_3^-$ Reduced per Minute
10 mM NO$_3^-$	78.8 ± 30.4
10 mM NO$_2^-$	26.8 ± 3.2
10 mM NO$_2^-$ + 10 mM Na$_2$WO$_4$	7.2 ± 3.6
0.5 mM NH$_4^+$	4.9 ± 1.0

13. Based on the data in the following table, draw a graph that shows the effect of pith particle size on cumulative NO$_3^-$ immobilization. What can you conclude about the effect of pith particle size on net N immobilization? Why do you think this occurs?

Net NO$_3^-$ immobilized during 256 days of incubation at 25° C when 0.3 g of cornstalk pith was added to 100 g of sand.

Days of incubation	Pith Particle Size (mm)			
	0.3–2.3	4.8	7.9	19
	mg NO$_3^-$-N immobilized			
4	1.12	0.95	0.53	0.17
8	1.85	1.84	1.89	0.41
16	3.38	2.72	2.01	0.37
32	4.17	3.76	2.85	1.15
64	2.67	2.06	2.06	0.68
128	2.58	2.21	2.08	0.81
256	1.38	0.92	1.17	0.29

(Adapted from Sims and Frederick 1970)

Thought Question

You and your friend Bruce Smith (a fellow soil microbiologist) are having a coffee break when someone walks up to you and says, "My father decided to amend the soil in his peach orchard with some organic matter, so he mixed in several tons of sawdust. Was that a good idea?" Your friend Bruce says, "It's a great idea. By the way, my name's John Doe. Glad to meet you" and walks away. Why do you think Bruce is unwilling to have his real name known based on his advice?

Additional Reading

Once more I refer you to *Nitrogen in Agricultural Soils* (1982, F. J. Stevenson, ed., American Society of Agronomy, Madison, WI.). If you can't find your answer there, you've asked a really tough question.

References

Betlach, M. R., J. M. Tiedje, and R. B. Firestone. 1981. Assimilatory nitrate uptake in *Pseudomonas fluorescens* studied using nitrogen-13. *Archives of Microbiology* 129:135–40.

Cole, J. A. 1987. Assimilatory and dissimilatory reduction of nitrate to ammonia. In *The nitrogen and sulfur cycles.* J. A. Cole, and S. Ferguson (eds.), 281–329. Society for General Microbiology, Cambridge, England: Cambridge University Press.

Jackson, L. E., J. P. Schimel, and M. K. Firestone. 1989. Short-term partitioning of ammonium and nitrate between plants and microorganisms in an annual grassland. *Soil Biology and Biochemistry* 21:409–16.

Marzluf, G. A. 1997. Genetic regulation of nitrogen metabolism in the fungi. *Microbiology and Molecular Biology Review* 61:17–32.

McCalla, T. M., J. R. Peterson, and C. Lue-Hung. 1977. Properties of agricultural and municipal wastes. In *Soils for management of organic wastes and waste waters,* L. F. Elliot et al. (eds.), 12–43. Madison, WI: Soil Science Society of America.

Rice, C. W. and J. M. Tiedje. 1989. Regulation of nitrate assimilation by ammonium in soils and in isolated soil microorganisms. *Soil Biology and Biochemistry* 21:597–602.

Schimel, J. P. and M. K. Firestone. 1989. Inorganic N incorporation by coniferous forest floor material. *Soil Biology and Biochemistry* 21:41–46.

Sims, J. L. and L. R. Frederick. 1970. Nitrogen immobilization and decomposition of corn residue in soil and sand as affected by residue particle size. *Soil Science* 109:355–61.

Chapter 21

The Nitrogen Cycle— Denitrification and Dissimilatory Nitrate Reduction

Overview

After you have studied this chapter, you should be able to:

- Explain that NO_3^- can be biologically reduced by two processes—denitrification and dissimilatory NO_3^- reduction to NH_4^+ (DNRA).
- Describe the different environments and regulation of each process.
- Summarize what denitrification is, what organisms perform it, the physiological basis for denitrification, and the environmental factors regulating denitrification.
- Discuss some of the environmental consequences of denitrification.

INTRODUCTION

Here's an obvious puzzle. If all reduced N compounds are ultimately oxidized as a source of N for growth, metabolism, or energy, why hasn't all of Earth's N been converted to NO_3^- over geologic time? The answer is that reductive plant and microbial processes continuously remove NO_3^- from the environment. The topic of this chapter is the way that NO_3^- is converted back to reduced inorganic forms by microorganisms, not as a stage in assimilation, which we've already discussed, but as part of respiratory metabolism. The two processes we focus on are denitrification and dissimilatory nitrate reduction to ammonium (DNRA).

ENVIRONMENTAL FATES OF NO_3^-

What can happen to NO_3^- once it forms? There are several fates, not all of which are desirable. Nitrate can be taken up by plants or accumulate in their absence in fallow soil. Nitrate can be leached or lost via runoff. Nitrate can also be reduced by several processes that are either assimilatory (in which case N is taken up into microorganisms) or dissimilatory (in which case N is lost from the microorganism):

1. Assimilatory reduction to organic N
2. Dissimilatory reduction to NH_4^+ (DNRA)
3. Nitrate respiration to NO_2^-
4. Dissimilatory reduction to gas (denitrification)

These reductive fates are not only distinguished by their products, but also by their regulation and the environments in which they occur (Table 21-1).

Table 21-1   Reductive fates of nitrate.

Process	Products Assimilatory	Energy Yielding	Regulated By	Typical Environment
Assimilation	NH_4^+	No	NH_4^+ organic N	Very low NH_4^+; aerobic or anaerobic
	Dissimilatory			
Respiratory denitrification	$N_2 > N_2O > NO$	Yes	O_2; NO_3^-	Anaerobic
Dissimilatory NO_3^- reduction to $NH_4^{+\dagger}$	$NH_4^+ >> N_2O$	Some microorganisms	O_2	Anaerobic
NO_3^- respiration[†]	NO_2^-	Yes	O_2	Anaerobic
Nonrespiratory denitrification	N_2O	No	?	Aerobic
Chemodenitrification	$NO >> N_2, N_2O$	No	pH; NO_2^-	High NO_2^-; acidic

[†]All known organisms that dissimilate NO_3^- to NH_4^+ are also nitrate respirers, but most NO_3^- respirers accumulate NO_2^-.

REDUCTIVE FATES OF NO_3^-

As we have seen nitrate assimilation is a reductive process in which NO_3^- is reduced to NH_4^+, and NH_4^+ is rapidly assimilated into amino acids by either glutamate dehydrogenase (GDH) or glutamine synthetase/glutamine oxo-glutarate aminotransferase (GS/GOGAT).

$$NO_3^- \Rightarrow NO_2^- \Rightarrow NH_4^+ \Rightarrow R–NH_2$$

Nitrate assimilation is repressed by NH_4^+ or any other source of available organic N.

In agricultural systems, denitrification is the most significant factor in N loss. This is the process we discuss in this chapter. However, there are special ecosystems where dissimilatory nitrate reduction to ammonium (DNRA) occurs, and this process merits some attention. Generally speaking, DNRA occurs in continuously anaerobic environments such as sediments and estuaries. The key difference between denitrification and DNRA is that NO_3^- is reduced to gas in denitrification and to NH_4^+ in DNRA. If the concentration of available C in the environment is high relative to the concentration of NO_3^-, then NO_3^- is used as an electron dump and NH_4^+ formation is favored. If the concentration of C in the environment is low relative to the concentration of NO_3^-, then NO_3^- is used as a terminal electron acceptor in respiratory metabolism because more energy is available (Table 21-2).

Let's look at an example of how C and NO_3^- might affect NO_3^- dissimilation in nature. *Klebsiella* dissimilates NO_3^- to NH_4^+ while *Pseudomonas* is a respiratory denitrifier and dissimilates NO_3^- to N_2. What happens if the two microorganisms are put in the same environment and forced to compete for NO_3^-? When C is limiting (low C/high N), *Klebsiella* is outcompeted for NO_3^-, and denitrification predominates. Where N is limiting (high C/low N), *Pseudomonas* is outcompeted (Rehr and Klemme 1989).

Table 21-2 Comparative energy yield and electron transfer in denitrification and dissimilatory reduction to ammonium.

Reaction	Number of Electrons Accepted per N	Formal Valence Change	G_0' (kJ/mole NO_3^-)
Denitrification			
$2NO_3^- + 5H_2 + 2H^+ \Rightarrow N_2 + 6H_2O$	5	$+5 \Rightarrow 0$	−1121
Dissimilatory Nitrate Reduction to Ammonium			
$NO_3^- + 4H_2 + 2H^+ \Rightarrow NH_4^+ + 3H_2O$	8	$+5 \Rightarrow -3$	−600

(Adapted from Tiedje 1994)

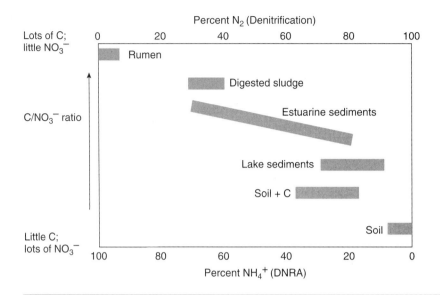

Figure 21-1   Partitioning of nitrate between denitrification and DNRA as a function of available carbon/nitrate ratio. (Adapted from Tiedje et al. 1982)

The same situation occurs in nature (Figure 21-1). Depending on the availability of C and NO_3^-, one process is favored over the other. In reducing environments, it's an advantage for microorganisms to lose as many electrons as they can; it allows them to regenerate oxidized pyridine nucleotides such as NAD and NADP for electron transport. In C-limited environments, reactions that generate the most energy are favored.

What are the consequences of NO_3^- reduction to NH_4^+? Ammonium is produced, which is useful for growth as an N source if the soil becomes aerobic again. Oxidized pyridine nucleotides (NAD, NADP) are generated. The pH of the environment increases. A toxic compound (NO_2^-) is removed from the intracellular and extracellular environments. Energy is generated. Anaerobic growth on a nonfermentable substrate such as glycerol is allowed (Cole 1987).

Nitrate respiration is the first step in both DNRA and denitrification. All microorganisms that dissimilate NO_3^- to NO_2^- can therefore be considered NO_3^- respirers.

Typically, though, this refers to microorganisms that convert NO_3^- to NO_2^- and no further. During NO_3^- respiration trace amounts of nitrous oxide (N_2O) are produced (Smith and Zimmerman 1981).

Chemodenitrification is an acid-catalyzed destruction of NO_2^- at pH < 5.

$$2H^+ + 3NO_2^- \Rightarrow 2NO + NO_3^- + H_2O + N_2O \text{ (trace)}$$
$$M_{red} + NO_2^- + 2H^+ \Rightarrow M_{ox} + NO + H_2O + N_2O \text{ (trace)}$$
$$M = Fe, Mn, Cu$$

Chemodenitrification usually affects only NO_2 and is important only in certain circumstances or environments such as acidic forest soils, when NO_2^- crystalizes in frozen soils, or when dry soils are wet up. One of the major differences between chemodenitrification and biological denitrification is that the major product of chemodenitrification is nitric oxide (NO). Another type of chemodenitrification is the oxidization of organic N by NO_2^-. This is the Van Slyke reaction, and it also occurs when the pH is 5 or lower:

$$HNO_2 + RNH_2 \Rightarrow N_2 + ROH + H_2O$$

DENITRIFICATION—A DEFINITION

Denitrification is the process by which nitrogenous oxides, principally NO_3^- and NO_2, are used as terminal electron acceptors in the absence of O_2 and are reduced to dinitrogen gases during respiratory metabolism. The extent of denitrification used to be estimated by N balances:

$$Denitrification = Inputs? - Losses?$$

Inputs were estimated by measuring N in precipitation, seeds, fertilizer, manure, and mineralization, while losses were estimated by measuring leaching, erosion, and crop removal. Much error is introduced by this method of estimation. Consequently, denitrification has been estimated to account for anywhere between 25% and 75% of the N lost from the soil. A more reasonable estimate is probably 20% to 30% (Tiedje 1988).

DENITRIFYING ORGANISMS

Denitrification is linked to bacteria. No actinomycetes or fungi denitrify if we think of denitrification as an energy-yielding version of respiratory metabolism. There have been recent reports of denitrifying fungi (Tiedje 1994), but these represent examples of nonrespiratory denitrification or nitrate respiration, which are distinctly different processes, as Table 21-1 shows. Denitrifying bacteria are primarily heterotrophs. While many heterotrophs reduce NO_3^- to NO_2^- (NO_3^- respiration), fewer are true denitrifiers ($NO_3^- \Rightarrow N_2$). Because denitrification occurs in the absence of O_2, denitrifiers are technically facultative anaerobes, although they grow better aerobically.

There are many genera that have denitrifiers, and denitrifiers represent many morphological and physiological types (Table 21-3). The principal denitrifiers isolated from soils are *Pseudomonas* species. They are the easiest denitrifiers to isolate because they grow on many different substrates. The most numerous denitrifiers vary from soil to soil. Because many heterotrophs are sensitive to acidity, but *Bacillus* are not, *Bacillus* are the denitrifiers frequently obtained from acidic soil. It is common to have $> 10^6$ denitrifiers g^{-1} soil—between 0.1% and 5% of the culturable soil population.

Table 21-3 Genera with denitrifying species.

Physiological Group	Example	Physiological Group	Example
Heterotrophs		**Phototrophs**	
Aerobes	Cytophaga		Rhodopseudomonas
	Pseudomonas		
	Alcaligenes	**Lithotrophs**	
	Flavobacterium		
	Paracoccus	H_2 oxidizers	Pseudomonas
			Bradyrhizobium
Oligophiles	Hyphomicrobium		Paracoccus
	Aquaspirillum		Alcaligenes
Fermenters	Wolinella	S oxidizers	Thiosphaera
	Azospirillum		Thiobacillus
	Bacillus		Thiomicrospira
Halophiles	Paracoccus	NH_4^+-oxidizers	Nitrosomonas
	Halobacterium		
Thermophiles	Bacillus		
Spore-formers	Bacillus		
Magnetotactic	Aquaspirillum		
N_2 fixing	Rhodopseudomonas		
	Agrobacterium		
	Rhizobium		
	Bradyrhizobium		
	Azospirillum		
	Pseudomonas		
Pathogens	Neisseria		
	Kigella		
	Wolinella		

(From Tiedje 1988)

DENITRIFICATION PATHWAY

The pathway of denitrification is:

$$NO_3^- \Rightarrow NO_2^- \Rightarrow NO \Rightarrow N_2O \Rightarrow N_2$$

We know that these are the steps in denitrification because the formation and consumption of each intermediate, starting with NO_3^-, has been observed in laboratory

A Short History of Denitrification

The first scientist to suggest that NO_3^- might be converted to gaseous products biologically, rather than chemically, was Schoenbein in 1868 (Focht and Verstraete 1977). Eighteen years later, in 1886, Gayon and Dupetit conclusively demonstrated that sand columns saturated with NO_3^- in oxygen-free conditions evolved gaseous products and that exposing the sand columns to chloroform inhibited biological activity and gaseous N loss. Such was the concern of some scientists that valuable NO_3^- would be lost, that Wagner, in 1895, suggested manure be treated with sulfuric acid to kill the microorganisms responsible for denitrification. However, calmer minds, like Dehrain in 1897, convincingly argued that denitrification was a normal part of N cycling in soil and that proper management of manure and NO_3^- addition to soil could diminish denitrification (Focht 1981).

By 1902, Weissenberg demonstrated that denitrification was essentially a respiratory process in which aerobic bacteria used NO_3^- in the absence of O_2. Giltnay and Alberson recognized the heterotrophic nature of denitrification by establishing the link between NO_3^- reduction and carbon oxidation. They also developed the first methods to isolate and culture denitrifying bacteria. Although Selman Waksman in 1927, suggested that denitrification had little significance in well-aerated soils, F. E. Allison, in 1955, concluded that roughly 30% of the N added to soil could not be accounted for and must be attributable to denitrification. It was difficult to measure denitrification in soil except by difference until Federova and colleagues (1973), Yoshinari and Knowles (1976), and Payne (Balderston et al. 1976) independently observed that acetylene preferentially inhibited the last step in denitrification and allowed N_2O to be measured with exquisite sensitivity by gas chromatography using the electron capture detector developed by Lovelock in 1957.

Recent advances in denitrification research have come at the cellular and molecular levels. The enzymes critical to denitrification were purified in the 1980s, which enabled scientists to use immunological techniques to analyze denitrifying enzymes. By the 1990s, sequences had been identified for several denitrification genes, and were being used to analyze the diversity of denitrifying bacteria in nature (Tiedje 1994).

experiments. Chemical inhibitors have been used to specifically block some of the steps and cause intermediates to accumulate. Mutants have been isolated or created for each step that also cause intermediates to accumulate.

Denitrification is a respiratory mechanism exactly like aerobic respiration. Nitrogen oxides take the place of O_2 as the terminal electron acceptor. The electron donor doesn't have to be organic C; *Thiobacillus denitrificans* is an example of an autotrophic denitrifier.

Each step in denitrification is catalyzed by a distinct enzyme:

1. $NO_3^- \Rightarrow NO_2^-$ Dissimilatory nitrate reductase
2. $NO_2^- \Rightarrow NO$ (maybe N_2O)   Dissimilatory nitrite reductase

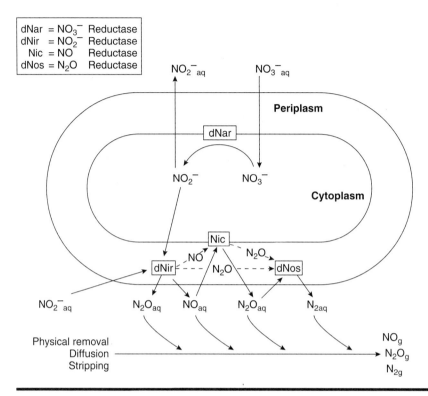

Figure 21-2 Cellular location of enzymes involved in denitrification. Denitrification enzymes are located either in the cell membrane or periplasm. A cytoplasmic assimilatory NO_3^- reductase can also produce NO_2^-, but is under different regulatory control. Aqueous (aq) denitrification products can be stripped from the cell or removed from vicinity of the cell by mass flow.

3. $NO \Rightarrow N_2O$ Nitric oxide reductase

4. $N_2O \Rightarrow N_2$ Dissimilatory nitrous oxide reductase

Bacteria may have the complete pathway or be truncated. That is, they can start with NO_2^- and accumulate N_2O. There are two types of NO_2^- reductase—one that contains Cu and another that contains a heme group (Coyne et al. 1989). Denitrifiers contain either one or the other. You can easily identify which type a denitrifier has, because the Cu-type NO_2^- reductase is strongly inhibited by the metal chelator diethyldithiocarbamate (DDC) and the heme type is not (Shapleigh and Payne 1985). The pathway after NO_2^- has been much disputed but the weight of evidence points to NO as being a true intermediate and not just a by-product.

Figure 21-2 shows how the various enzymatic processes are localized in the cell environment. Since each step in the denitrification pathway is associated with energy generation, the enzymes are all found at the cell membrane.

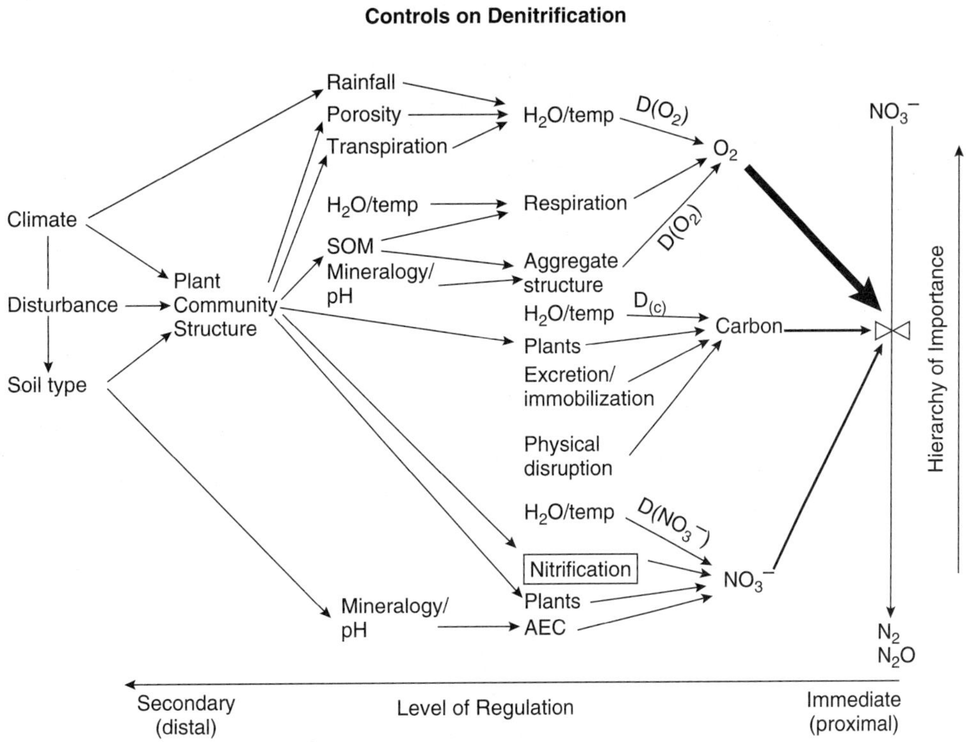

Controls on Denitrification

Figure 21-3   Environmental factors regulating denitrification. (Adapted from Teidje 1988)

ENVIRONMENTAL CONTROL

Denitrification varies enormously depending on O_2 status, moisture content, soil temperature, organic matter, C, and NO_3^-, all of which are interacting factors in the environment (Figure 21-3). A large denitrifier population is not conclusive evidence that denitrification occurs.

The most important control is the concentration of O_2. Oxygen inhibits denitrifying enzyme synthesis and electron flow to denitrifying enzymes. What controls the availability of O_2? Primarily water content, which is controlled by rainfall, texture, soil structure, and the plant community. Denitrifying populations are higher around plant roots because of the availability of C, but roots are not necessarily conducive to denitrification because more evapotranspiration occurs around plants, and also because there is greater NO_3^- demand by their roots. Christensen and Tiedje (1988) did an elegant study in which they showed that barley roots and a denitrifier didn't actually compete directly for NO_3^- because plant roots ceased NO_3^- uptake at 1.6% O_2 in solution while the denitrifier didn't start denitrifying until the O_2 saturation was less than 0.04%.

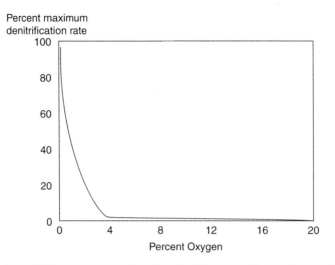

Figure 21-4 Effect of oxygen content on denitrification in soil. (Adapted from Tiedje 1989)

Denitrification becomes insignificant at moisture contents of −0.01 MPa, or less than 60% water-filled pore space. You typically find denitrification occurring at O_2 concentrations in soil higher than what pure culture studies indicate. This is because it is very difficult to measure the O_2 concentration that denitrifiers actually experience in microsites, which may be much lower than the O_2 concentration in the bulk soil (Figure 21-4). The optimum temperature for denitrification is 25°C or above, although the temperature range of denitrification is 5° to 75°C.

Nitrate or N-oxide availability also regulates denitrification. Nitrate is controlled by plant and microbial assimilation and by nitrification. The N-oxide concentration affects the denitrification products formed. At high NO_3^- concentrations, N_2 predominates. At low NO_3^- concentrations, N_2O predominates.

This actually represents an example of a general phenomenon that we should note. Control of the terminal product of denitrification (usually thought of as N_2O or N_2) is determined by any process that causes an intermediate to accumulate. If we looked at the ratio of these two gases, we might see something similar to what is outlined in Table 21-4.

One of the rate-limiting steps in denitrification is N_2O reduction. If NO_3^- and NO_2^- are reduced faster than N_2O, N_2O will accumulate. Anything that limits N_2O reduction also causes it to accumulate. Nitrous oxide reductase is more acid sensitive than are the other denitrification enzymes, so low pH affects its activity. Sulfide precipitates Cu and Ni in the environment; they are metal cofactors in N_2O reductase.

Insufficient C means that electrons are unavailable to further reduce N_2O. Carbon supply is determined by H_2O (solubility), plants, physical disruption, competition, and excretion by other organisms in soil. Denitrification is strongly dependent on C availability. Water-soluble C can account for 71% of the denitrification potential and most

Table 21-4   Factors affecting the terminal product of denitrification.

Factor	Effect on N_2O/N_2
NO_3^- concentration	Ratio increases if NO_3^- increases
NO_2^- concentration	Ratio increases if NO_2^- increases
O_2 concentration	Ratio increases if O_2 increases
pH	Ratio increases if pH declines
Sulfide	Ratio increases if sulfide increases
Carbon	Ratio decreases as C increases and increases as C decreases
E_h	Ratio decreases as E_h decreases. No effect below 0 mV

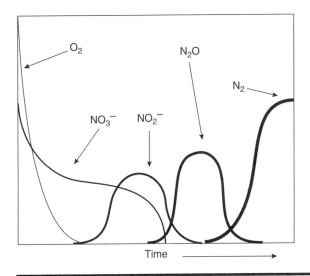

Figure 21-5   Sequential formation of denitrification products in a flooded environment. (Adapted from Cooper and Smith 1963)

denitrification in soil is associated with organic matter deposits. Parkin (1987) showed this nicely by subdividing soil into increasingly smaller fractions until he identified where most of the denitrification occurred. It turned out to be located around a decaying leaf. Since organic matter distribution in soil is extremely variable, it means that denitrification in soil is also extremely variable and can vary dramatically within short distances.

Redox potential (E_h) is really a reflection of O_2 availability in most soils. Each denitrification enzyme is regulated by a different O_2 concentration, either because its synthesis is inhibited, or because electron flow to it is restricted. Consequently, when a soil is flooded, you can see a progression of denitrification products formed as O_2 consumption, and diffusion limitations in soil gradually make it possible for each denitrification enzyme to be synthesized and expressed in turn (Figure 21-5).

How Do You Study Denitrification in the Environment?

The potential denitrification enzyme assay (Smith and Tiedje 1979) is a measure of how much denitrification can occur when conditions are optimized. Nitrate is added in excess, carbon (usually glucose) is added in excess, the soil sample is made into a slurry and continuously stirred so that there are no limts to NO_3^- diffusion, the sample is made anaerobic, and acetylene is added to inhibit N_2O reductase so that N_2O accumulating in the headspace of the sample container can be measured by gas chromatography.

Denitrifying bacteria grow and form more denitrifying enzymes under these conditions, so an antibiotic, chloramphenicol, is added to inhibit new protein synthesis. Consequently, in the assay, only the currently existing denitrification enzymes at the time of analysis are measured.

Denitrification can be measured in several ways in the field. Intact cores can be removed from the upper soil profile, placed in a chamber, and incubated anaerobically with acetylene to evaluate the denitrification rate with the existing NO_3^- and C in soil. Soil covers can be used to measure the flux of N_2O from saturated soils as a measure of denitrification. In this method, a soil cover, sometimes as simple as a PVC pipe with a lid on it, is pounded into the soil to a depth of 5 to 10 cm and then N_2O accumulation in the headspace is periodically sampled. The N_2O measured accounts for only a small portion of the denitrification that goes on because most of the N_2O is further reduced to N_2 before it escapes the soil. One approach has been to purge the soil immediately surrounding the soil covers with acetylene. This temporarily inhibits N_2O reductase so the N_2O that is measured better reflects the actual denitrification rate.

When greater sensitivity is required, ^{15}N-labeled NO_3^- can be added to a soil sample and the accumulation of [15]N_2 measured by a mass spectrometer.

Once induced, the denitrification enzymes persist, though they may not be active if the soil is aerobic (Smith and Parsons 1985). This means that when the soil is flooded again, denitrification can commence rapidly, because the denitrification enzymes are already present and merely need the absence of O_2 to function.

ENVIRONMENTAL CONCERNS

Nitric oxide and N_2O are always released during denitrification. If they reach the atmosphere, they can have significant effects. Although ozone (O_3) is considered an air pollutant at ground level, it's a useful gas at a distance because ozone protects us from ultraviolet (UV) light.

$$O_3 \xrightarrow[< 300 \text{ nm}]{hv} O_2 + O$$

However, O_3 is destroyed by the following reactions:

$$N_2O + O \Rightarrow 2NO$$
$$NO + O_3 \Rightarrow NO_2 + O_2$$
$$NO_2 + O \Rightarrow NO + O_2$$
$$NO + O_3 \Rightarrow NO_2 + O_2, \text{ etc.}$$

The effect of increased NO and N_2O in the atmosphere is catalytic destruction of ozone, increased exposure to ultraviolet light, and potentially greater risk of skin cancer.

Summary

Nitrate can be taken up by plants, accumulate in fallow soil, be leached or lost via runoff, or be reduced by assimilatory or dissimilatory processes. In agricultural systems, denitrification is the most significant factor in N loss. Dissimilatory nitrate reduction to ammonium (DNRA) occurs in long-term anaerobic environments, such as sediments and estuaries. The key difference between denitrification and DNRA is that NO_3^- is reduced to gas in denitrification and to NH_4^+ in DNRA. If the concentration of available C is high relative to the concentration of NO_3^-, then NH_4^+ formation is favored. If the concentration of C is low relative to the concentration of NO_3^-, then NO_3^- is used as a terminal electron acceptor in respiratory metabolism.

Denitrifying bacteria are primarily heterotrophs. Because denitrification occurs in the absence of O_2, denitrifiers are technically facultative anaerobes. Denitrification is a respiratory mechanism exactly like aerobic respiration. Nitrogen oxides take the place of O_2 as the terminal electron acceptor. Each step is catalyzed by a distinct enzyme: $NO_3^- \Rightarrow NO_2^-$ (dissimilatory nitrate reductase), $NO_2^- \Rightarrow NO$ (maybe N_2O) (dissimilatory nitrite reductase), $NO \Rightarrow N_2O$ (nitric oxide reductase), $N_2O \Rightarrow N_2$ (dissimilatory nitrous oxide reductase).

Denitrification varies enormously depending on soil temperature, moisture content, O_2 status, organic matter, and NO_3^-. Denitrification becomes insignificant at moisture contents of -0.01 MPa or less than 60% water-filled pore space. Control of the terminal product of denitrification is determined by any mechanism that causes an intermediate to accumulate. Once induced, the enzymes persist in soil. The effect of increased NO and N_2O in the atmosphere is catalytic destruction of ozone, increased exposure to ultraviolet light, and potentially greater risk of skin cancer.

Sample Questions

1. What is denitrification?
2. What is the pathway of denitrification?
3. How does denitrification differ from dissimilatory nitrate reduction to ammonium?
4. What conditions are necessary for denitrification to occur in soil?

5. What are the immediate and secondary regulators of denitrification in soil?
6. What are five potential fates of NO_3^- in soil?
7. What is chemodenitrification?
8. What kinds of enzymes are involved in denitrification?
9. What controls whether denitrification or dissimilation to ammonium occurs in an environment?
10. The data plotted below represent measurements taken from a flooded soil over time.

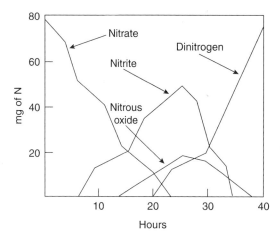

Answer the following questions based on the plot:
   a. What process is occurring?
   b. What are other potential sources of these compounds in the soil?
   c. Are these alternate sources credible?
   d. What accounts for the sequence in which each compound is formed?
11. What do you predict the fate of ^{15}N-labeled NO_3^- to be if it is added to anaerobic soil under the following conditions?
   a. High concentrations of NH_4^+ and organic C.
   b. Low concentrations of NH_4^+ and high concentrations of organic C.
   c. High concentrations of NH_4^+ and low concentrations of organic C.
   d. Low concentrations of NH_4^+ and organic C.
12. What are the intermediates of denitrification and what are their molecular weights?
13. Calculate the µg of N in the headspace of a soil cover that contains 100 ppm (µL L^{-1}) N_2O. Assume there are 750 mL of headspace and that 22.4 µL of gas contains 1 µmole of N_2O.
14. In a soil science seminar, the speaker displayed a graph that looked something like the following graph:

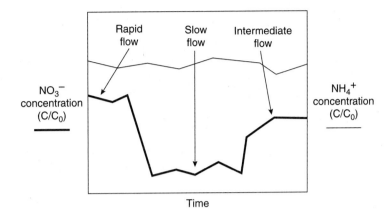

Flow refers to the flow rate of water seeping from a shallow spring. Based on what you know about N transformations in the environment, what is your hypothesis as to the process going on in this system?

Thought Question

What are the conditions in which you would like to see denitrification occur in soil?

Additional Reading

M. K. Firestone has a nice summary of denitrification in *Nitrogen in Agricultural Soils* (1982, F. J. Stevenson (ed.), American Society of Agronomy, Madison, WI.). The definitive book, though it is rapidly becoming dated, is *Denitrification* by W. J. Payne (1981, John Wiley and Sons, New York). A compendium of papers on denitrification from multiple perspectives is *Denitrification in Soil and Sediment,* edited by N. P. Revsbech and J. Sørensen (1990, Plenum Press, New York).

References

Balderston, W., L. B. Sherr, and W. J. Payne. 1976. Blockage by acetylene of nitrous oxide reduction in *Pseudomonas perfectomarinus. Applied and Environmental Microbiology* 31:504–508.

Christensen, S. and J. M. Tiedje. 1988. Oxygen control prevents denitrifiers and barley plant roots from directly competing for nitrate. *FEMS Microbiology and Ecology* 53:217–21.

Cole, J. A. 1987. Assimilatory and dissimilatory reduction of nitrate to ammonia, p. 281–329. In J. A. Cole and S. Ferguson (ed.). *The nitrogen and sulfur cycles.* Cambridge University Press, Cambridge, England.

Cooper, G. S., and R. L. Smith. 1963. Sequence of products formed during denitrification in some diverse western soils. *Soil Science Society of America Proceedings* 27:659–62.

Coyne, M. S., A. Arunakumari, B. A. Averill, and J. M. Tiedje. 1989. Immunological identification and distribution of dissimilatory heme cd_1 and nonheme copper nitrite reductases in denitrifying bacteria. *Applied and Environmental Microbiology* 55:2924–31.

Fedorova, R. I., E. I. Milekhina, and N. I. Il'yukhina. 1973. Evaluation of the method of "gas metabolism" for detecting extraterrestrial life. Identification of nitrogen-fixing microorganisms. *Isz. Akad. Nauk. SSR Ser. Biol* 6:797–806.

Focht, D. D. 1981. Soil denitrification. In *Genetic engineering of symbiotic nitrogen fixation and conservation of fixed nitrogen,* J. M. Lyons et al. (eds.), 499–515. New York: Plenum Press.

Focht, D. D., and W. Verstraete. 1977. Biochemical ecology of nitrification and denitrification. *Advanced Microbial Ecology* 1:135–213.

Parkin, T. B. 1987. Soil microsites as a source of denitrification variability. *Soil Science Society of America Journal* 51:1194–99.

Rehr, B., and J.-H. Klemme. 1989. Competition for nitrate between denitrifying *Pseudomonas stutzeri* and nitrate ammonifying enterobacteria. *FEMS Microbiology and Ecology* 62:51–58.

Shapleigh, J. P., and W. J. Payne. 1985. Differentiation of c,d_1 cytochrome and copper nitrite reductase production in denitrifiers. *FEMS Microbiology Letters* 26:275–79.

Smith, M. S., and L. L. Parsons. 1985. Persistence of denitrifying enzyme activity in dried soils. *Applied and Environmental Microbiology* 49:316–20.

Smith, M. S., and J. M. Tiedje. 1979. Phases of denitrification following oxygen depletion in soil. *Soil Biology and Biochemistry* 11:261–67.

Smith, M. S., and K. Zimmerman. 1981. Nitrous oxide production by nondenitrifying soil nitrate reducers. *Soil Science Society of America Journal* 45:865–71.

Tiedje, J. M. 1988. Ecology of denitrification and dissimilatory nitrate reduction to ammonium. In *Biology of anaerobic microorganisms.* A. J. B. Zehnder (ed.), 179–244. New York: John Wiley & Sons.

Tiedje, J. M. 1994. Denitrifiers. In *Methods of soil analysis, part 2: Microbiological and biochemical properties,* R. W. Weaver et al. (eds.), 245–67. Madison, WI: Soil Science Society of America.

Tiedje, J. M., A. J. Sexstone, D. D. Myrold, and A. J. Robinson. 1982. Denitrification: Ecological niches, competition and survival. *Antonie van Leeuwenhoek* 48:569–83.

Yoshinari, T., and R. Knowles. 1976. Acetylene inhibition of nitrous oxide production by denitrifying bacteria. *Biochemical and Biophysical Research Communications* 69:705–10.

Chapter 22

The Nitrogen Cycle— Nitrogen Fixation

Overview

After you have studied this chapter, you should be able to:

- Compare the differences between industrial and biological N_2 fixation.
- Explain that N_2 fixation is performed exclusively by asymbiotic, symbiotic, and associative bacteria.
- Name some of the N_2-fixing species.
- Identify the characteristics of N_2 fixation.
- Describe the general physiology of nitrogenase, the N_2-fixing enzyme complex.
- Identify some ways of isolating N_2 fixers and show that they fix N.

INTRODUCTION

Nitrogen is the macronutrient that most limits terrestrial life, although in freshwater systems P frequently becomes a more limiting factor. Seventy-eight percent of Earth's atmosphere is N_2, but atmospheric N represents only about 1.2% of the N on Earth. If you include the Earth's mantle N in rocks and minerals represents 98% of Earth's total N (Stevenson 1982).

Even though, thermodynamically, N_2 oxidation to NO_3^- is theoretically favorable, it doesn't happen readily. This is because the triple bond between N atoms is stable and requires tremendous energy to break it apart. That energy is found in lightning bolts, and lightning contributes anywhere from 0.5 to 5.0 kg NO_3^- $ha^{-1}yr^{-1}$ to soil. Another source of atmospheric N deposition to soil is NH_3 from various sources. The unmistakable odor of swine lagoons and feedlots masks considerable amounts of NH_3. Deposition of reduced forms of N accounts for 0.5 to 10 kg ha^{-1} of N per year depending on how close one is to the source.

Obviously, atmospheric N deposition is insufficient to meet the needs of most crop production systems. Weathering of N in rocks is likewise too slow to meet the growth requirements of plants. Somehow, living organisms need to replace the N that is lost to the atmosphere through dissimilative processes such as denitrification. In other words, our N cycle needs closure.

The process that replenishes most available N to biological systems is asymbiotic and symbiotic N_2 fixation—fixing or synthesizing inorganic atmospheric N into organic N. Even though there are industrial sources of fixed N, they still account for less than half of the total N fixed by biological systems (65×10^6 metric tons yr^{-1} vs. 170×10^6 metric tons yr^{-1} based on a 1982 estimate). In this chapter, we focus on asymbiotic N_2 fixation and the general mechanisms common to the N_2-fixation process. We discuss symbiotic N_2 fixation in Chapter 27.

INDUSTRIAL N_2 FIXATION

The incredible nature of biological N_2 fixation becomes obvious if we compare it to its industrial counterpart. Industrial N_2 fixation is accomplished by the Haber-Bosch method:

$$CH_4 \xrightarrow[\text{Heat}]{\text{steam}} H_2$$

$$N_2 + 3H_2 \xrightarrow[\text{Fe}_3\text{O}_4 \text{ catalyst}]{450°C, 200 \text{ atm}} 2NH_3$$

This process was actually developed as the first step in supplying Germany with NO_3^- for explosives in World War I. Haber was also responsible for Germany's poison gas development in World War I.

The H_2 and energy required to generate such high temperatures come from fossil fuels. As the price of energy rises, so does the price of fertilizer N. In the mid-1970s, when the price of energy dramatically increased, it became very expensive to use fertilizer N. Consequently, interest in biological fixation renewed. Managing crop production using biological N_2 fixation wasn't as convenient as using inorganic fertilizer, but aside from the metabolic cost to the plant, it was free.

BIOLOGICAL N_2 FIXATION: ORGANISMS

Knowledge of biological N_2 fixation isn't exactly new. The beneficial effects of legumes were noted by the ancient Romans, and the beneficial effects of *Azolla* (water fern) in rice culture were noted by the Chinese over 2,000 years ago. However, it wasn't until the late 1800s that the first N_2-fixing organisms were isolated in pure culture. In 1888, *Rhizobium* (a symbiotic, aerobic bacteria) was isolated by Martinus Beijerinck. Around the same time, *Clostridium* (an asymbiotic, anaerobic bacteria) was isolated by Serge Winogradsky.

Table 22-1 list some of the asymbiotic and associative N_2-fixing organisms in nature. As you can see, N_2-fixing ability is found in many physiological groups. They have in common that they are all prokaryotic. There have been reports that eukaryotes can fix N, but these appear to reflect cases in which a bacterial endosymbiont is actually responsible (Knowles and Barraquio 1994).

Table 22-1 Representative genera of microorganisms involved in asymbiotic, associative, and symbiotic N_2 fixation.

Physiological Group	Type of Association	Host (if any)	Representative Genera
Heterotroph	Free-living; aerobic		*Azotobacter*
			Azotomonas
			Azotococcus
			Beijerinckia
			Derxia
			Pseudomonas
			Rhizobium
			Xanthobacter
	Facultative anaerobe		*Azospirillum*
			Bacillus
			Klebsiella
			Thiobacillus
	Anaerobe		*Clostridium*
			Desulfovibrio
			Desulfotomaculum
			Methanobacillus

(Continued)

Table 22-1   Continued

	Associative; aerobic	Rhizosphere	*Agrobacterium*
		Digitaria	*Azospirillum*
		Paspalum	*Azotobacter*
		Phyllosphere	*Bacillus*
		Phyllosphere	*Beijerinckia*
	Facultative anaerobe	Phyllosphere	*Enterobacter*
		Phyllosphere	*Klebsiella*
	Symbiotic; aerobic	*Glycine*	*Bradyrhizobium*
		Legume	*Rhizobium*
		Trema	*Rhizobium*
		Alnus	*Frankia*
		Myrica	*Frankia*
		Gunnera	*Nostoc*
Autotroph	Free-living; unicellular		*Gloeocapsa*
	Filamentous; nonheterocystous		*Trichodesmium*
	Filamentous; heterocystous		*Anabaena*
			Calothrix
			Nostoc
	Anaerobic; purple nonsulfur		*Rhodospirillum*
			Rhodopseudomonas
	Purple sulfur		*Chromatium*
	Green sulfur		*Chlorobium*
	Symbiotic; nonnodule		
		Lichen	
		Peltigera	*Nostoc*
		Collema	*Nostoc*
		Liverwort	
		Anthoceros	*Nostoc*
		Biasia	*Nostoc*
		Mosses	
		Sphagnum	*Halosiphon*
		Ferns	
		Azolla	*Anabaena*
		Intracellular	
		Oocystis	*Nostoc*
	Nodule		
		Gymnosperm	
		Cycad	*Nostoc*

(Adapted from Havelka et al. 1982)

The best-characterized organism in aerobic conditions is *Azotobacter.* In anaerobic conditions the best-characterized organism is *Clostridium.* Several cyanobacteria have also been well characterized. In addition, some work has been done on organisms associated with the rhizosphere of grasses; *Azospirillum* is an example of this type of associative N_2 fixer.

Azotobacter is a gram-negative, motile soil organism. Since it is heterotrophic, mannitol is used in the isolation method to culture it from soil. It is big compared to other prokaryotes (4 to 7 μm in diameter) and yeastlike in appearance. It forms cysts that serve as resting bodies. *Azotobacter* grows best in neutral to alkaline, mesophylic soils. It does not grow when the pH is below 6 and it is not present in acidic soils. Large populations are uncommon (0 to 10^3 g^{-1} is the typical range). Consequently, it does not contribute much fixed N to soil. *Beijerinckia* and *Derxia* take the place of *Azotobacter* in acidic, tropical soils. In the phyllosphere, one can find N_2-fixing organisms such as *Klebsiella,* which is a microaerophylic bacteria that has the best characterized genetics of any N_2 fixer. *Bacillus* is also found in the phyllosphere.

Anaerobic N_2 fixers include *Clostridium pasteurianum,* a heterotroph with a pH range intermediate between *Azotobacter* and *Beijerinckia, Desulfovibrio* and *Desulfotomaculum,* SO_4^{2-}-reducing organisms; *Methanobacillus,* a CH_4-producing organism— a methanogen; and *Rhodospirillum* and *Chlorobium,* two anaerobic and photosynthetic organisms.

Nitrogen fixation is associated with grasses such as *Paspalum notatum.* Other associations have been found in salt-marsh grass, sugar cane, rice, sorghum, maize, millet, and wheat. However, associative fixation is negligible to 20 kg ha^{-1} yr^{-1}, so grain crops still need to be fertilized to get decent yield. Inoculation of plant roots with asymbiotic N_2-fixing organisms has given ambiguous results. Positive results may be due to enhanced N_2 fixation, or may simply be because of the production of growth stimulants and exclusion of pathogens.

Cyanobacteria are usually free-living diazotrophs that fix only a few kg of N ha^{-1} yr^{-1}. They are not of great significance compared to symbiotic fixation in terrestrial systems, with one exception. In tropical or aquatic regions, cyanobacteria are hugely significant. Cyanobacteria are most common in aquatic and freshwater environments. They take three basic morphological forms (Figure 22-1):

1. Unicellular (*Chroococcus, Gloeocapsa,* and *Microcystis*)
2. Nonfilamentous (*Plectonema*)
3. Filamentous (*Anabaena, Nostoc*)

Cyanobacteria can be found in soils, sometimes just below the soil surface. Some can survive extreme desiccation and form dry crusts in grasslands and deserts that begin N_2 fixation when moistened. In rice paddies, *Anabaena* and *Nostoc* are critical components of the production system. Long-term sustainable rice production without N fertilization is possible because N_2 fixation by cyanobacteria in paddies produces 30 to 70 kg N ha^{-1} yr^{-1} and meets the needs of rice growth.

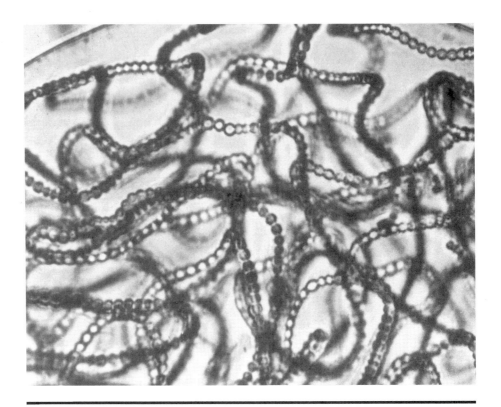

Figure 22-1   A N_2-fixing filamentous cyanobacteria—*Nostoc*. (Photograph courtesy of the Soil Science Society of America)

CHARACTERISTICS OF N_2 FIXATION

What are the characteristics of N_2 fixation? Three conditions must be met to prove that bacterial N_2 fixation occurs. First, there must be a large population of these organisms that is consistent with the observed N_2-fixation rates. Second, there must be rapid cell formation, which indicates that N_2 fixation is linked to growth. Third, the N must be atmospheric, not inorganic or organic; this is one of the most difficult conditions to meet.

Nitrogen fixation is energy intensive. Microorganisms must have a ready supply of electrons for significant N_2 fixation to occur. Electrons are necessary because N_2 fixation is a reductive process. Nitrogen gas (valence state of 0) gains electrons to become NH_3 (valence state -3). Reduced C is used by heterotrophs to supply the electrons while light energy is used by phototrophs to oxidize either water or a reduced sulfur compound to provide the necessary electrons in anaerobic conditions.

Nitrogen fixation requires an enzyme complex called nitrogenase, and nitrogenase must operate in a microenvironment protected from O_2. Nitrogen fixers must find or

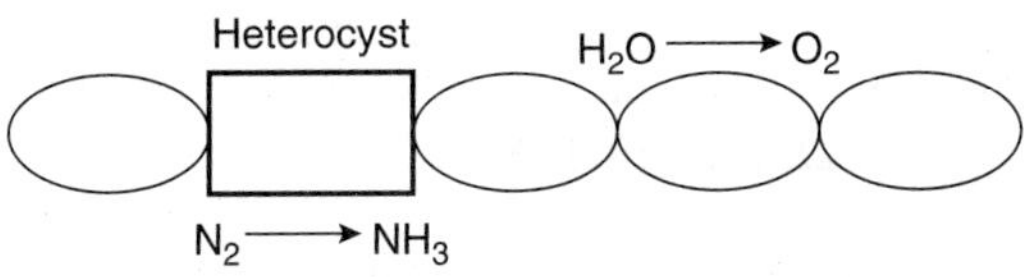

Figure 22-2   Spatial separation of N_2 fixation and O_2 formation in a heterocystous cyanobacteria.

produce an O_2-free or O_2-limited environment. This is a dilemma for aerobic N_2 fixers such as *Azotobacter* because O_2 respiration is required to produce electrons for fixation and for respiration. There are several mechanisms that are used to limit O_2. *Azotobacter* can limit O_2 diffusion by producing copious amounts of slime. It can limit O_2 by developing high respiration rates and consuming O_2 as rapidly as O_2 diffuses in. *Azotobacter* has the highest respiration rate of any microorganism known—500 to 1,000 cm^3 O_2 consumed per g^{-1} cells. *Azotobacter* also makes reversible changes in the arrangement of nitrogenase to convert it from an O_2-sensitive functional form to an O_2-tolerant, but nonfunctional, form.

Anaerobes such as *Clostridium* are protected from O_2 because they exist in an anaerobic environment. However, energy generation is a problem. Consequently, the amount of N fixed globally by anaerobes is small but significant where there are no other external sources of N.

Phototrophs, such as cyanobacteria, make O_2 much as plants do (Figure 22-2). The ability to fix N is most common in filamentous heterocyst-forming cyanobacteria. Nitrogen fixation occurs in the heterocysts, which are thick-walled, multilayered, vegetative cells that lack Photosystem II. Consequently, no O_2 is produced during photosynthesis in these cells. Photosystem II appears only in nonheterocystous, vegetative cells. Cyanobacteria that do not form heterocysts can fix N only when they're not photosynthesizing.

Nitrogen fixation does not occur when NH_4^+ or NO_3^- or organic N are readily available. A number of hypotheses have been offered to explain this. The presence of available N may direct electron flow away from nitrogenase—no electrons; no reduction. Another theory invoked for symbiotic systems is that NO_3^- reduction in nodules causes NO_2^- binding of compounds that protect nitrogenase from O_2.

PHYSIOLOGY

Biological N_2 fixation is catalyzed by an enzyme association called nitrogenase. This is strictly a prokaryotic enzyme complex. Nitrogenase is composed of two soluble proteins: the Fe protein—dinitrogenase reductase (also called Component II) and the MoFe protein—dinitrogenase (also called Component I). MoFe stands for molybdenum–iron, an essential cofactor in dinitrogenase. Some N_2 fixers have an alternative dinitrogenase that contains vanadium (V) instead of Mo. But this type of dinitrogenase is synthesized only in the absence of available Mo. The two enzymes in the complex work in tandem:

dinitrogenase reductase reduces dinitrogenase, and dinitrogenase reduces N_2. Both enzymes are required for fixation to occur.

An equation for N_2 fixation can be written as follows; Mg^{2+} is a required cofactor:

$$N_2 + 6e^- + 6H^+ + 12ATP + Mg^{2+} \longrightarrow 2NH_3 + 12ADP + 12Pi + Mg^{2+}$$

Electrons come from organic and inorganic sources, flow through several electron carriers, flow through nitrogenase, and ultimately reduce N_2:

$$e^- \Rightarrow \text{Ferredoxin} \Rightarrow \text{Flavodoxin} \Rightarrow \text{Fe protein} \Rightarrow \text{MoFe protein} \Rightarrow N_2$$

Each electron transferred requires two ATP. Unfortunately, nitrogenase is not perfectly efficient. It also reduces other compounds such as H^+, N_2O, N_3^-, and CN^-. Hydrogen evolution is a general by-product of N_2 fixation, particularly at low electron flux:

$$2H^+ + 2e^- + 4ATP \rightarrow H_2 + 4ADP + 4Pi$$

The overall equation for nitrogenase-catalyzed N_2 fixation is therefore:

$$N_2 + 16ATP + 8e^- + 8H^+ + Mg^{2+} \longrightarrow 2NH_3 + H_2 + 16ADP + 16Pi + Mg^{2+}$$

Some organisms have an uptake hydrogenase (Hup) that allows them to reoxidize H_2 and regain part of the lost energy. In studies with *Rhizobium* comparing Hup^+ and Hup^- strains, there was increased N_2 fixation and plant yield with Hup^+ strains that had the uptake hydrogenase.

Nitrogenase also catalyzes the transformation of substrates such as acetylene:

$$\text{Acetylene } (C_2H_2) + 2H^+ + 2e^- \longrightarrow \text{Ethylene } (C_2H_4)$$

This is useful because it can be used to estimate N_2 fixation rates:

$$2e^- + C_2H_2 \longrightarrow C_2H_4$$
$$6e^- + N_2 \longrightarrow 2NH_3$$

Ideally, the ratio of acetylene reduced to nitrogen fixed is 3:1. But this ratio is never ideal, as you've already guessed, because it doesn't take into account reduction of alternative substrates such as H^+. So, N_2 fixers can be compared in terms of rates of C_2H_4 evolution, but to actually figure out how much N is being fixed, the acetylene reduction rates have to be calibrated with ^{15}N for each organism.

What is the cost of N_2 fixation? First, there is the synthesis of nitrogenase and hydrogenase (if the organism has an uptake hydrogenase). Second, there is the cost of NH_3 assimilation, which typically begins with glutamine synthetase, an ATP-requiring process (see Chapter 20). Finally, there is the lost ATP to fuel nitrogenase that could otherwise be used for growth. Figure 22-3 summarizes the pathway of biological N_2 fixation.

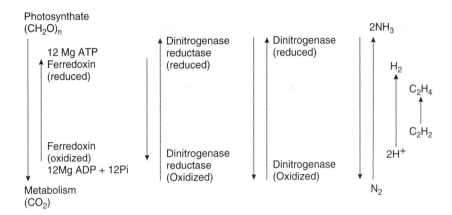

Figure 22-3   The pathway of electrons through nitrogenase during nitrogen fixation. The example uses a typical heterotroph that utilizes organic C (originally produced during photosynthesis) as its source of electrons to reduce N_2.

How Do You Perform an Acetylene Reduction Assay (ARA)?

One of the great developments in N_2 fixation research was when Hardy and colleagues (1968) introduced the acetylene reduction assay (ARA). Previously, to prove that a microorganism fixed N, one had to incubate it with $^{15}N_2$ and look for the accumulation of reduced ^{15}N-labeled products. This was prohibitively expensive for most soil microbiologists because of the cost of the ^{15}N and the mass spectrometer to measure it. Reduction of acetylene (C_2H_2) to ethylene (C_2H_4) was not only cheaper and faster to measure because it used an ordinary gas chromatograph, but it could be done with much greater sensitivity than measurement of ^{15}N.

The procedure for ARA is simple. The sample (soil, plant roots, or cell culture) is placed in a sealed container. Acetylene is introduced to the container at a concentration sufficient to saturate nitrogenase. Usually 10% of the headspace of the container is sufficient. The acetylene can be generated as required by mixing calcium carbide (CaC_2) with water and trapping the C_2H_2 that evolves in a plastic bag (CaC_2 is the same stuff that's used to fire cannons at football games and can be acquired at sporting goods stores—be careful, it's an explosive reaction that generates a lot of heat)(Weaver and Danso 1994). Periodically measure the C_2H_4 evolved by removing a gas sample from the sealed container and injecting it into a gas chromatograph equipped with an FID detector and a Porapak N column (Waters, Inc.).

Remember, the ARA is an indirect measure of N_2 fixation. Even though the conversion factor for moles of C_2H_4 produced (equal to moles of C_2H_2 reduced) to moles N_2 fixed is assumed to approximate 3–4:1, it actually varies anywhere from 1.5 to 25 (Weaver and Danso 1994). So to get a true value, you must ultimately resort to using ^{15}N or some other N isotope.

Summary

Nitrogen is the macronutrient that most limits terrestrial life. The process that replenishes the majority of available N to biological systems is asymbiotic and symbiotic N_2 fixation. In contrast to industrial N_2 fixation, which occurs at high temperature and pressure and consumes scarce fossil fuels, biological N_2 fixation occurs at room temperature and pressure and is free.

Nitrogen-fixing ability is distributed among many physiological groups, but it is exclusively prokaryotic. Nitrogen fixation is also energy intensive (for the bacteria) and requires O_2-free microenvironments. Biological N_2 fixation is catalyzed by an enzyme association called nitrogenase. Nitrogenase is composed of two soluble proteins: the Fe protein—dinitrogenase reductase and the MoFe protein—dinitrogenase.

The equation for N_2 fixation is $N_2 + 6e^- + 6H^+ + 12ATP + Mg^{2+} \Rightarrow 2NH_3 + 12ADP + 12Pi + Mg^{2+}$. Unfortunately, nitrogenase is not perfectly efficient. It also reduces other compounds such as H^+, N_2O, N_3^-, and CN^-. H_2 evolution is a general by-product of N_2 fixation, particularly at low electron flux. Nitrogenase-catalyzed reduction of acetylene (C_2H_2) to ethylene (C_2H_4) can be used to estimate N_2-fixation rates. Ideally, the ratio of acetylene reduced to nitrogen fixed is 3–4:1. But this ratio is never ideal, so N_2 fixers can be compared in terms of rates of C_2H_4 evolution, but to actually figure out how much N is being fixed, the acetylene reduction rates have to be calibrated with ^{15}N for each organism.

Sample Questions

1. Give an example of each of the following:
   a. An aerobic, free-living, N_2-fixing bacterium
   b. An anaerobic, free-living, N_2-fixing bacterium
   c. An anaerobic, photosynthetic, N_2-fixing bacterium
   d. An aerobic, free-living, N_2-fixing bacterium associated with plant roots
   e. An aerobic and photosynthetic N_2-fixing bacterium
2. How many types of nitrogenase are there?
3. What environmental factors regulate biological N_2 fixation?
4. How does biological N_2 fixation differ from industrial N_2 fixation?
5. What is nitrogenase, and what does it do?
6. How do asymbiotic organisms control O_2 concentrations to allow N_2 fixation to occur?
7. Why do N_2-fixing organisms commonly produce H_2? What is the role of hydrogenase in these organisms?
8. List one free-living, N_2-fixing bacteria and one plant–bacterial association that fixes atmospheric nitrogen. How do each of these meet the two primary requirements for N_2 fixation: significant energy availability and O_2 sensitivity of nitrogenase?

9. What is the role of dinitrogenase in the enzyme complex we call "nitrogenase"?
10. In the following table, what appears to be limiting cyanobacterial growth and why?

Effect of fertility on soil surface coverage by cyanobacteria.

Amendment	Percent of Soil Cover
Control (no amendment)	2.6
P, K, Na, Mg	8.6
48 kg N ha^{-1} + P, K, Na, Mg	7.0
192 kg N ha^{-1} + P, K, Na, Mg	0.2

11. You have been studying the effects of organic additions on nitrogen dynamics in soil. You obtain the following data. Based on your knowledge of control of N processes, explain trends in the data.

Treatment	N$_2$ fixation	Nitrification	Denitrification
		(Units m^{-2})	
Straw (C:N = 200) 20 Mg ha^{-1}	4	20	1
Manure (C:N = 30) 20 Mg ha^{-1}	0	200	40
NH$_4$SO$_4$ 20 Mg ha^{-1}	0	250	5

Mg = Megagrams = 1,000 kg

Thought Question

Now that you've gone through the nitrogen cycle, try to draw your own illustration of it including representative microbial groups for all pertinent steps.

Additional Reading

Many books have been written about N$_2$ fixation. I would start with *Nitrogen Fixing Organisms: Pure and Applied Aspects* by Sprent and Sprent (1990, Chapman and Hall, London). The Sprents have a very readable style and are able to clearly explain even complicated topics.

References

Hardy, R. W. F., R. D. Holsten, E. K. Jackson, and R. C. Burns. 1968. The acetylene-ethylene assay for N$_2$ fixation: Laboratory and field evaluation. *Plant Physiology* 43:1185–1207.

Havelka, U. D., M. G. Boyle, and R. W. F. Hardy. 1982. Biological nitrogen fixation. In *Nitrogen in agricultural soils,* F. J. Stevenson (ed.), 365–422. Madison, WI. American Society of Agronomy.

Knowles, R., and W. L. Barraquio. 1994. Free-living dinitrogen-fixing bacteria. In *Methods of soil analysis, part 2: Microbiological and biochemical properties,* R.W. Weaver et al. (eds.), Madison, WI: Soil Science Society of America.

Stevenson, F. J. 1982. Origin and distribution of nitrogen in soil. In *Nitrogen in agricultural soils,* F. J. Stevenson (ed.), 1–41. Madison, WI: American Society of Agronomy.

Weaver, R. W. and S. K. Danso. 1994. Dinitrogen fixation. In *Methods of soil analysis, part 2: Microbiological and biochemical properties,* R. W. Weaver et al. (eds.), Madison, WI: Soil Science Society of America.

Chapter 23

The Carbon Cycle— Organic Carbon Entering Soil

Overview

After you have studied this chapter, you should be able to:

- Describe the reservoirs of C on Earth and discuss how biogeochemists use changes in those reservoirs to assess C flow in the environment.
- Identify the structures of the most important organic C compounds that plants contribute to the soil environment.
- Explain how important structural polymers in nature are decomposed.
- List the components of lignin and explain its decomposition.
- Describe how lignin influences the decomposition of other C compounds.

INTRODUCTION

The simplest diagram or form of the C cycle might look something like that shown in Figure 23-1. This is a gross oversimplification in terms of geochemical cycling. However, it demonstrates the essential points. Carbon dioxide is fixed via photosynthesis into carbohydrate and other reduced compounds. These, in turn, are ultimately metabolized aerobically to CO_2 and anaerobically to CH_4 and CO_2. The purpose of this chapter is to describe, in more detail, the types of C entering soil, and the mechanisms or pathways by which they decompose.

TENETS OF MICROBIAL CARBON CYCLING

Microorganisms play the central role in converting reduced C compounds to their elemental forms. Microorganisms are nature's garbage disposal agents. Without them, decomposition would be unbelievably slow. In their search for energy, microorganisms feed on reduced organic compounds. As a result, they are the driving force behind C cycling, and C cycling is the driving force behind nearly all nutrient cycling reactions involving organic S, N, or P (Figure 23-2).

Any biologically synthesized organic compound can be decomposed by soil microorganisms. If it's naturally made, it's decomposed. This is the Theory of Microbial Infallibility, first articulated in this form by Martin Alexander, a soil microbiologist at Cornell University (Alexander 1973). We know this has to be true, otherwise in the course of geologic time there would be vast accumulations of undecomposed C compounds. Any organic compound that contains energy in reduced bonds is ultimately used as an energy source. In general, microbial life is limited by C.

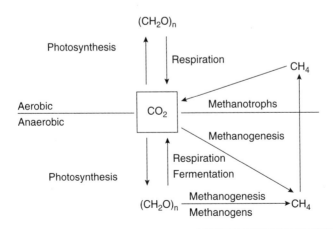

Figure 23-1   A simple diagram of the C cycle.

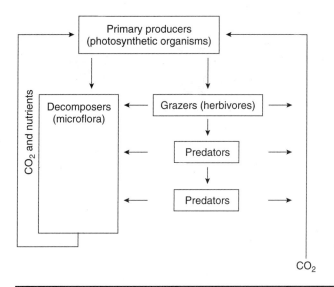

Figure 23-2   Interaction of autotrophs, heterotrophs, herbivores, and predators in the cycling of C.

ATMOSPHERIC CARBON

Where is the C on our planet? A rather small amount of C is actually available in the atmosphere, plants, and soil biomass, about 2.1×10^{15} kg (Schlesinger 1991). The remaining available C is in soil (3.5×10^{15} kg) or is dissolved in the oceans (38.0×10^{15} kg). Most of Earth's C is locked up in fossil fuels or carbonate (43.0×10^{18} kg). The CO_2 in the atmosphere has risen from about 250 ppm in 1850 to its present concentration of about 360 ppm (0.036% by volume). The increase in atmospheric CO_2 would have been more rapid except for the buffering capacity of the ocean (Schlesinger 1991). About 40% of the CO_2 released through industrial and human activity dissolves in the oceans as carbonate and bicarbonate. Some immobilization in terrestrial biomass may have also occurred.

The increase in atmospheric CO_2 seems to have come from burning fossil fuels, which fits into the framework of the industrial revolution, but this isn't the only cause. There have been other sources of CO_2 entering the atmosphere over the last 100 years. These sources are partly agricultural, due to the cultivation of forests and prairies, soil microbial respiration, forest clearing and burning, and the opening of new cropland in the Amazon, Americas, Siberia, Australia, and Africa. The increased cultivation of soil has resulted in a net loss of C from terrestrial ecosystems.

ORGANIC CARBON

The C cycle has two components that affect soil microbiology: a slow cycle, in which C turnover is measured in hundreds of thousands of years and involves rock weathering and dissolution of carbonates on land and in oceans; and a fast cycle in C turnover is measured in years or decades and is principally biological in nature. We are mostly interested in the fast cycle in this chapter. That's the cycle most directly affecting and most affected by soil microorganisms.

Most organic C in soil comes from plants. This C represents the residue of plants on the soil surface (Figure 23-3) and organic C coming from the decomposition of roots in soil. Plant C can be roughly characterized as follows (Killham 1994):

1. Carbohydrates (30% to 75% of dry weight)
     Cellulose (15% to 60%)
     Hemicellulose (10% to 30%)
     Sugars and starches (1% to 5%)
2. Lignin (10% to 30% of dry weight)
3. N-containing compounds (1% to 15% of dry weight)
     Proteins and amino acids
4. Waxes and pigments (1%)
5. Pectin (1%)
6. Others (5% to 20% of dry weight)
     Fats, oils, organic acids, hydrocarbons

The values vary because as plants age, the cellulose, hemicellulose, and lignin contents increase while the simple sugars, amino acids, proteins, fats, and oils decrease.

In upcoming chapters we focus on cellulose, hemicellulose, and lignin, the most important plant constituents, in terms of what they are, where they are, how they are

Figure 23-3   Plant residue on the soil surface of a no-tillage soil. (Photograph courtesy of M. S. Coyne)

decomposed, and what affects their decomposition. But first, we examine the common types of organic C entering soil and their chemistry.

Monosaccharides

Monosaccharides are simple sugars. They are the building blocks of many polymers and are nearly universal energy substrates for soil microorganisms. The only organisms that can't use them for energy are the lithotrophs and the phototrophs, and even some of them can, when conditions are right. Some groups of specialized bacteria grow exclusively on one or two C compounds (Crawford and Hanson 1984). Monosaccharides have two characteristic functional groups: an aldehyde and an alcohol:

$$
\begin{array}{cc}
\overset{\displaystyle O}{\underset{\displaystyle}{\|}} & \\
-C- & -\overset{|}{\underset{|}{C}}-OH \\
\text{Aldehyde} & \text{Alcohol}
\end{array}
$$

Xylose is a pentose, or 5-C, monosaccharide. Glucose is a hexose, or 6-C, monosaccharide, and is the most common sugar.

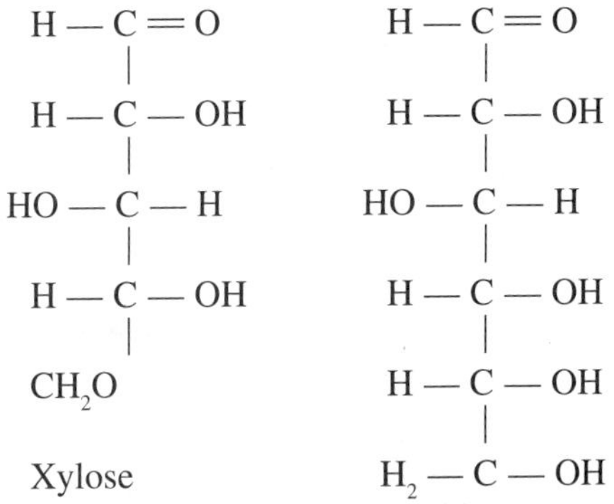

$$
\begin{array}{cc}
\text{H}-\text{C}=\text{O} & \text{H}-\text{C}=\text{O} \\
\text{H}-\text{C}-\text{OH} & \text{H}-\text{C}-\text{OH} \\
\text{HO}-\text{C}-\text{H} & \text{HO}-\text{C}-\text{H} \\
\text{H}-\text{C}-\text{OH} & \text{H}-\text{C}-\text{OH} \\
\text{CH}_2\text{O} & \text{H}-\text{C}-\text{OH} \\
\text{Xylose} & \text{H}_2-\text{C}-\text{OH} \\
 & \text{Glucose}
\end{array}
$$

The form of a sugar, such as glucose, is more commonly written as a ring structure because it is a better reflection of the actual shape in nature (Figure 23-4).

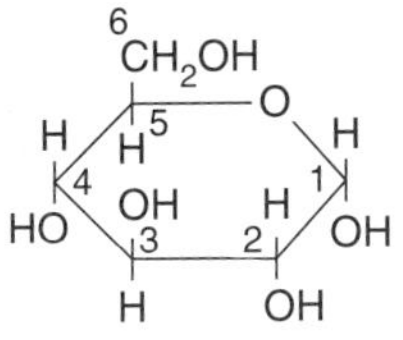

Figure 23-4   Haworth projection formula of glucose (α-D-glucopyranose) illustrating its ring structure. The C atoms in the hexose ring are numbered; the actual glucose ring is not planar.

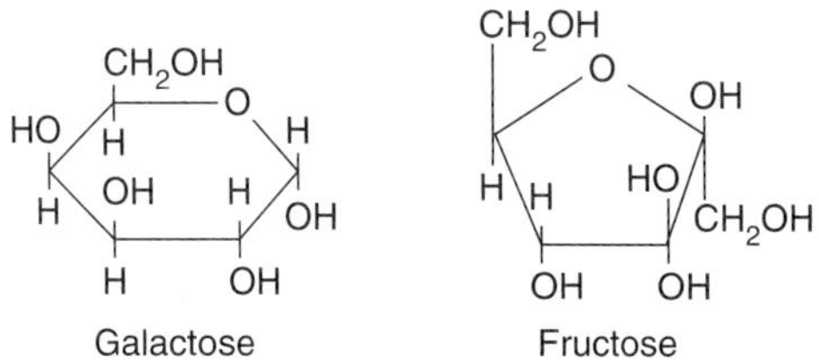

Figure 23-5 Haworth projection formula of galactose and fructose.

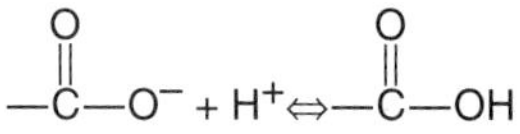

Figure 23-6 Carboxylic acid groups. The simplest carboxylic acid is formate (HCOOH). As the environment grows more acidic (H^+ concentrations increase), the carboxyl group becomes uncharged.

Galactose and fructose are two other common monosaccharides (Figure 23-5). Both are sugars and both have the same chemical formula ($C_6H_{12}O_6$) as glucose, so the three sugars are isomers of one another.

The chemical formulas of carbohydrates can be simplified to $(CH_2O)_n$. This is an easy way of indicating that the organic C is a carbohydrate without having to specify exactly what type of carbohydrate it is (since there can be many types).

The carboxyl is another important functional group in organic matter (Figure 23-6). It is a strong acid, so it gives the compounds containing it a charge that depends on the pH of the environment; the charge is negative in a neutral environment and neutral or positive in an acidic environment.

Glucuronic acid is an example of a monosaccharide that has a carboxyl functional group. Pectic acid is a polymer of galacturonic acid (Figure 23-7). Pectinic acid is a polymer of galacturonic acid in which some of the hydroxyl groups are replaced by methyl groups. Pectic compounds are major structural components in plant cell walls and in the space between plant cells (the middle lamellae). They are the compounds that help jams and jelly to solidify. You were introduced to amino sugars in Chapter 18. N-acetylglucosamine is one of the most common amino sugars encountered in soil (Figure 23-8).

Disaccharides

There are several noteworthy disaccharides with which you ought to be familiar because microorganisms in soil take up and metabolize disaccharides in the environment. Trehalose is a disaccharide that is characteristic of fungi. Trehalose is a nonreducing sugar, which means that carbon 1 of the ring structure does not have a hydroxyl attached to it; in contrast, a reducing sugar has a hydroxyl at the C-1 position (Figure 23-9).

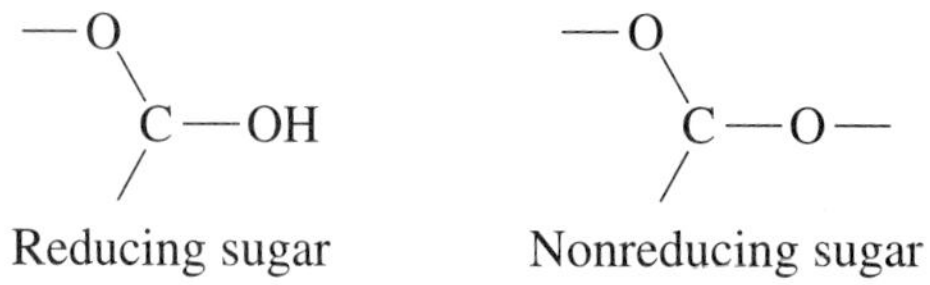

Reducing sugar   Nonreducing sugar

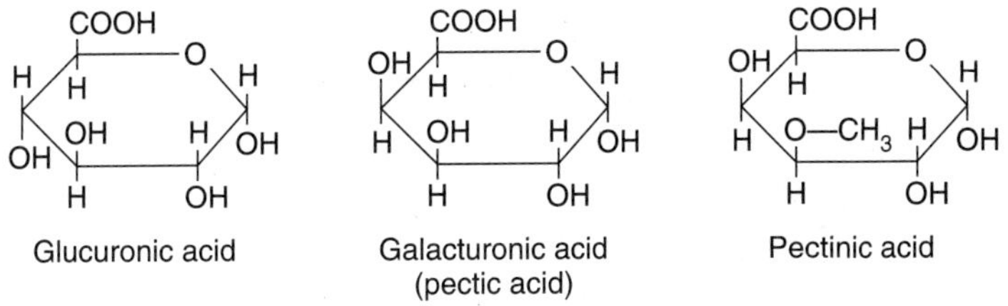

Figure 23-7   Glucuronic and galacturonic acids. Monosaccharides have a carboxyl functional group.

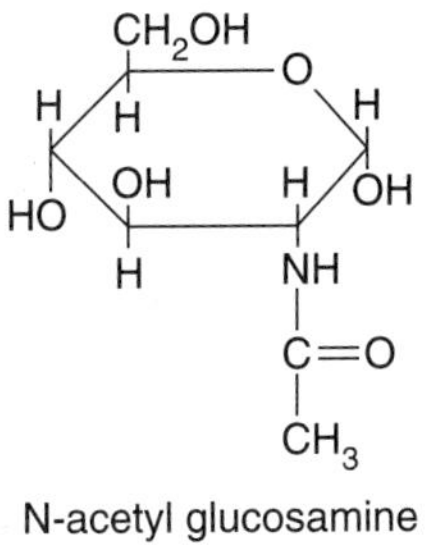

Figure 23-8   N-acetyl glucosamine, one of the building blocks of chitin and peptidoglycan.

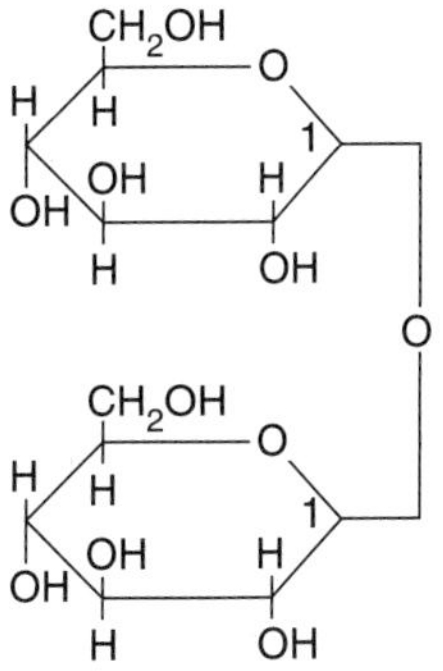

Figure 23-9   Trehalose—a nonreducing disaccharide composed of two glucose subunits.

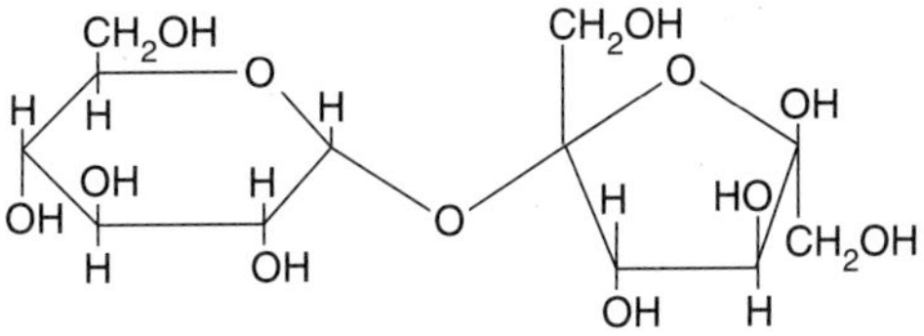

Figure 23-10   Sucrose—a disaccharide containing glucose and fructose.

The difference between reducing and nonreducing sugars is important in their biochemical analysis.

Sucrose is a disaccharide containing one glucose molecule and one fructose molecule (Figure 23-10). The enzyme invertase breaks sucrose apart to liberate these two monosaccharides.

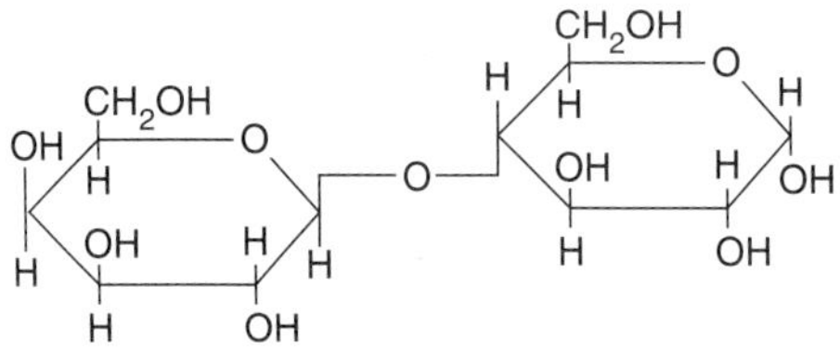

Figure 23-11   Lactose—a disaccharide containing galactose and glucose.

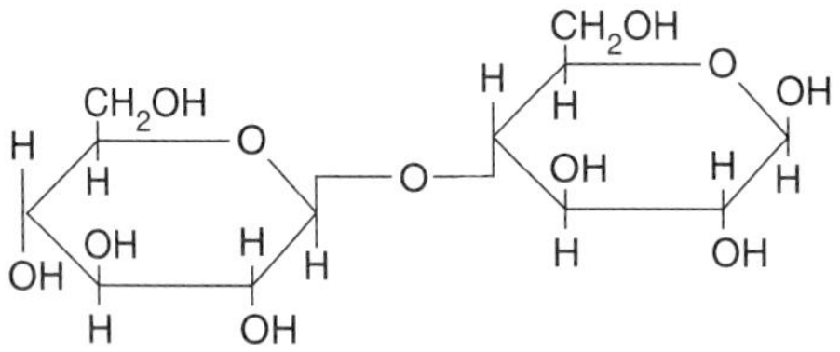

Figure 23-12   Cellobiose—a disaccharide containing two glucose molecules linked β (1→ 4).

Lactose is a disaccharide that contains one glucose molecule and one galactose molecule (Figure 23-11). The enzyme β-galactosidase breaks the bond between these two monosaccharides. β-galactosidase activity is an important enzymatic characteristic of indicator bacteria that is used to monitor water quality.

Cellobiose is a disaccharide containing two glucose molecules that are linked between carbons 1 and 4 of the ring structure. The orientation of this particular linkage is called β (1 → 4). The sugars in the lactose molecule in Figure 23-11 are also linked β (1 → 4). The type of linkage between individual sugars is important because it influences how easily they can be degraded. Cellobiose comes from the degradation of cellulose, a long polymer of > 1,000 glucose subunits all hooked together by β (1 → 4) linkages (Figure 23-12). Cellulose is probably the most abundant polymer on Earth because it forms a vital component of the structure of plants.

Maltose is a disaccharide that also contains two glucose molecules linked between carbons 1 and 4 of the ring structure (Figure 23-13). However, the orientation of this particular linkage is α (1→ 4). Maltose comes from the decomposition of starch, a long polymer of glucose subunits hooked together by α (1→ 4) linkages. Starch is a vital form of stored carbohydrates produced during photosynthesis.

Amylopectin is a highly branched version of starch in which the glucose molecules are not only hooked together by α (1→ 4) linkages, but are also connected by α (1→ 6) linkages (Figure 23-14).

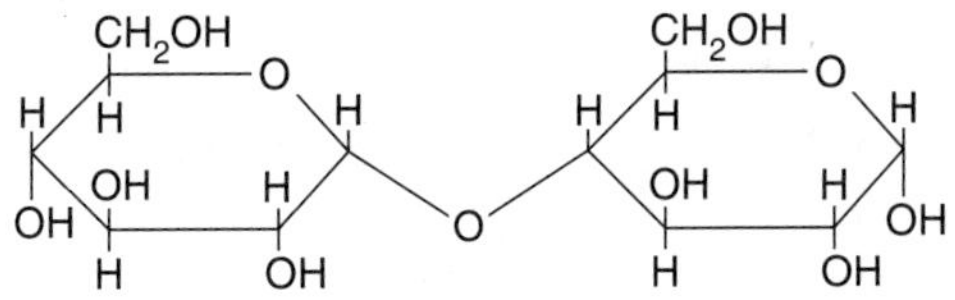

Figure 23-13   Maltose—a disaccharide containing two glucose molecules linked $\alpha\ (1 \rightarrow 4)$.

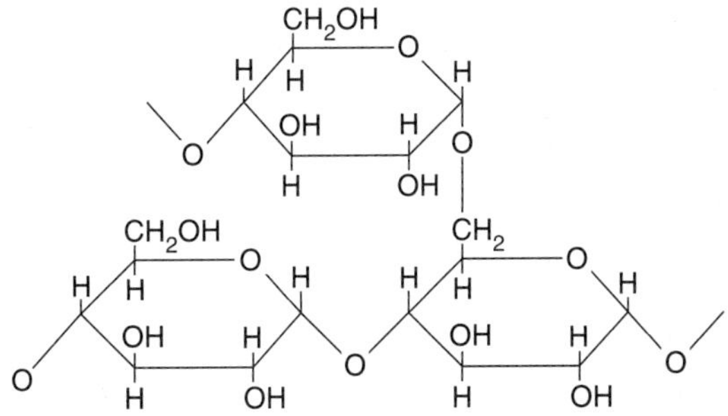

Figure 23-14   Amylopectin—a branched polymer containing glucose molecules linked $\alpha\ (1 \rightarrow 4)$ and $\alpha\ (1 \rightarrow 6)$.

LIGNIN AND LIGNIN STRUCTURE

Lignin is perhaps second to cellulose in terms of biomass. It protects cellulose and hemicellulose from enzymatic attack. Lignin is also an important structural component of plants. It provides structural rigidity; it provides resistance to compression and bending; it provides resistance to pathogens. Lignin is a cementing agent in wood. Wood that forms in spring is higher in lignin than wood that forms at the end of a tree's growth season.

Lignins are three-dimensional, amorphous, and highly branched molecules. Unlike cellulose, starch, or hemicellulose, they do not have identical linkages repeated at regular intervals. Lignins undergo more or less random reactions (condensations) that are both chemical and enzymatic, and yield polymers with no defined structure (Figure 23-15). They are the richest source of aromatic compounds in nature (aromatic compounds contain a benzene ring as part of their structure) and the bonds that lignins contain are difficult to cleave. This makes it difficult to extract lignin in an unmodified state.

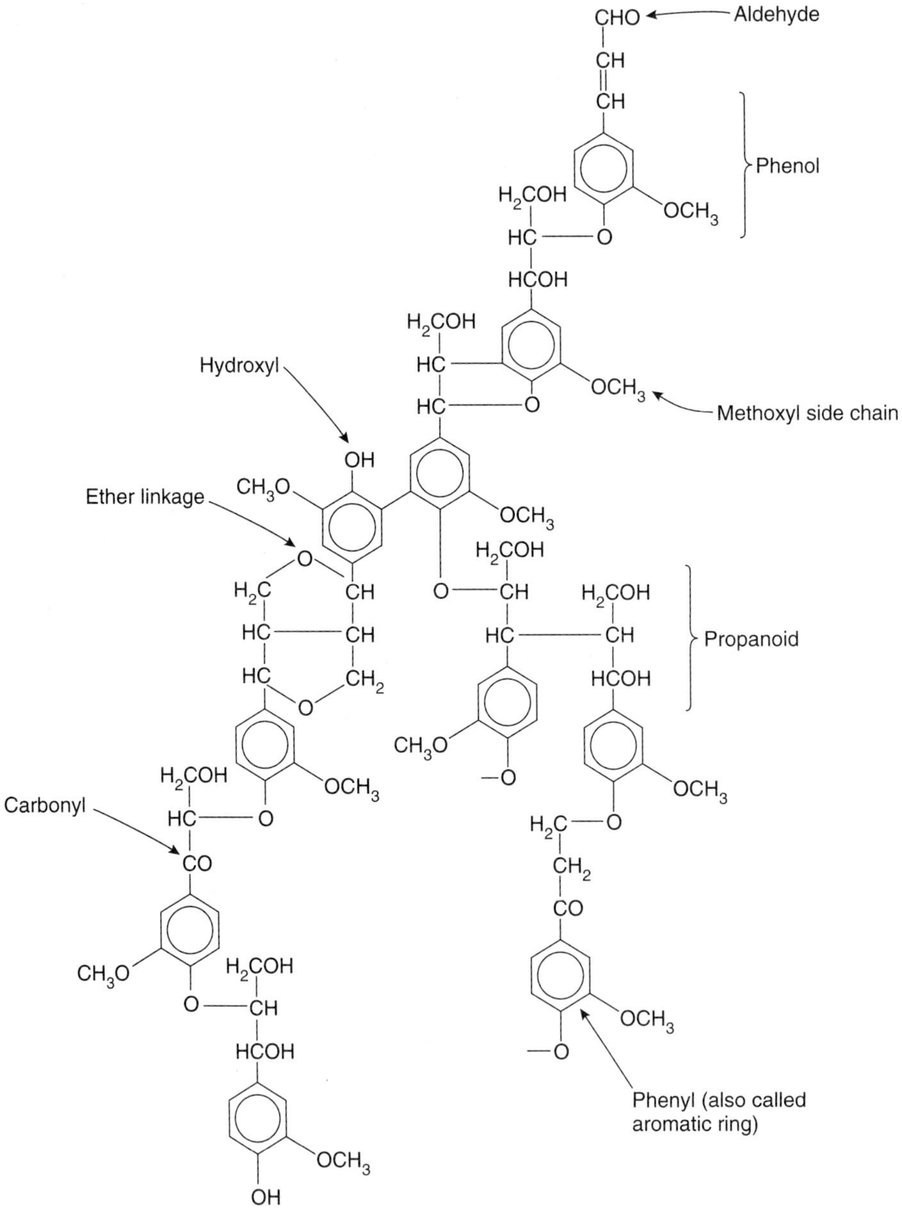

Figure 23-15   The complex, random structure of lignin. (Adapted from Paul and Clark 1996)

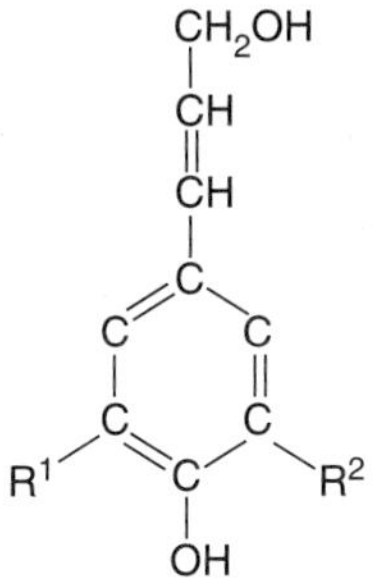

Figure 23-16 The phenyl propanoid, the basic building block of lignin.

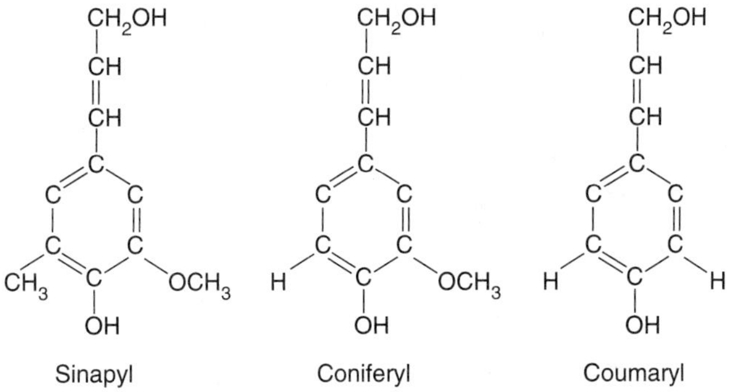

Figure 23-17 Sinapyl, coniferyl, and coumaryl, the phenyl propanoid precursors of lignin biosynthesis. (Adapted from Reddy and Forney 1978)

Lignin is a complex polymer. The basic subunit is an aromatic ring (phenyl group) with a three-carbon side chain (so, the basic subunit is called a phenyl propanoid) (Reddy and Forney 1978). Both the phenyl group and the propanoid group are modified in lignin itself (Figure 23-16). The R side chains differ among different plant species, and the relative proportion of units derived from each of these three basic building blocks is different in different plants, different in different plant tissues, and different at different plant ages (Figure 23-17). Hardwood lignin (beech, for example), is a 50/50 mixture of coniferyl and sinapyl building blocks. Softwood lignin (spruce, for example) is 85% coniferyl. Moss lignin is mainly coumaryl.

The lignin polymer is formed by polymerization of varying proportions of coumaryl, coniferyl, and sinapyl subunits and their derivatives. The most common linkage that arises from these reactions is shown in Figure 23-18.

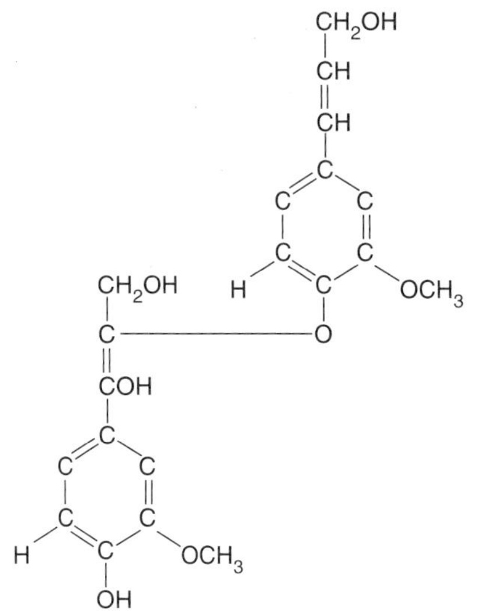

Figure 23-18   A common linkage formed during the polymerization of lignin precursors.

FATS, WAXES, RESINS, AND HYDROCARBONS

Fats, waxes, and resins represent a diverse group of C compounds that are soluble in ether or absolute ethanol. Hydrocarbons and long-chain fatty acids are probably the most important members of this group. Hydrocarbons have the general formula C_nH_{2n+2}. Methane (CH_4) and ethane (C_2H_6) are two of the simplest hydrocarbons. Fatty acids are important components of the cell membrane, and they may be saturated or unsaturated. A saturated fatty acid has no double bonds between carbons. An unsaturated fatty acid has one or more double bonds between carbons. An example of an unsaturated fatty acid is oleic acid:

$$CH_3 — (CH_2)_7 — CH = CH — (CH_2)_7 — COOH$$

MICROBIAL UTILIZATION OF PLANT CARBON

What are the problems associated with microbial degradation of C? Utilization of plant structural components by soil microorganisms presents significant problems. Physical barriers are imposed by meshworks of cell wall components such as polysaccharides, cutins, and suberins. Relatively inaccessible, stable crystalline regions in molecules such as cellulose and chitin impede degradation. Lignin slows the rate of decomposition mostly by physical protection. Filamentous fungi are important in decomposition partly because they penetrate these physical barriers. The greater the polymerization and rigidity of the C building blocks, the slower the decomposition. Water is required for hydrolysis and to increase the surface area available for enzymatic degradation. Some plant products have bacteriostatic and fungistatic properties.

Most plant material contains mixtures of different polymers such as cellulose, hemicellulose, pectins, and lignin. Their decomposition requires the combined action of many microorganisms, none of which has all of the enzymes required to completely decompose plant material. In plant tissue, decomposition is generally initiated by fungi and continued by bacteria and actinomycetes. So there is the combined action from a range of microorganisms and the combined action by the range of enzymes that these microorganisms produce.

Any environmental factor that affects the activities of the soil biota may influence the decomposition of organic residues. These factors include moisture, temperature, pH, O_2, inorganic nutrients, and clay. For example, decomposition of organic matter is more rapid in tropical than temperate regions partly because it is warmer and also because the microorganisms have a higher metabolic rate for longer periods.

Substrate quality also has an effect on decomposition. Two-thirds of the C added to soil is lost in the first year. If the C:N ratio is much greater than 20–30:1, immobilization of N will result, but another consequence is there could be inadequate N to sustain decomposition. Consequently, decomposition of C slows down. However, structural complexity is a more important determinant of substrate decomposition than is the C:N ratio. Lignin content slows decomposition because it has no N, but it is more important that lignins are structurally complex, large, and insoluble in water.

As a general rule, the ease with which C-compounds decompose is starch > hemicellulose > cellulose > lignin. Yet, surprisingly, soil carbohydrates are estimated to be 15% of the soil C even though in a free state they are readily degraded. Polysaccharides persist in soil for several reasons. They adsorb to clay and become less available. They interact with multivalent cations such as Fe, Al, and Cu, which render them somewhat toxic to microorganisms. Probably the most important reason is that polysaccharides aggregate soil particles, and in doing so, they form microaggregates that isolate them from microorganisms and become opened only when the soil is disturbed.

Summary

Soil microorganisms are the driving force behind C cycling, and C cycling is the driving force behind nearly all nutrient cycling reactions involving organic S, N, or P. Any biologically synthesized organic compound can be decomposed by soil microorganisms. If it is naturally made, it can be decomposed. This is the Theory of Microbial Infallibility. The C cycle has two components that affect soil microbiology: a slow cycle, involving rock weathering, and a fast cycle in which C turnover is measured in years or decades and is principally biological. Although fossil fuels are given as the source of increasing atmospheric CO_2 concentrations, there must have been a younger source of CO_2 entering the atmosphere over the last 100 years. That source may be decomposition of C in agricultural soils due to cultivation.

The major source of organic C entering soil is plants, and as plants age, their cellulose, hemicellulose, and lignin contents increase while the simple sugars, amino acids, proteins, fats, and oils decrease. Monosaccharides are simple sugars. They are the building blocks of many polymers and are nearly universal energy substrates for soil microorganisms. There are several noteworthy disaccharides: sucrose, lactose, trehalose, maltose, and cellobiose.

Organic C polymers include cellulose, probably the most abundant C compound on earth, hemicellulose, chitin, amylose, and lignin. Lignins are three-dimensional, amorphous, and highly branched molecules. Unlike cellulose, starch, or hemicellulose, they do not have identical linkages repeated at regular intervals. Any environmental factor that affects the activities of the soil biota may influence the decomposition of organic C.

Sample Questions

1. What do you think "malting" refers to?
2. Why do cellulose and amylose decompose at different rates in soil?
3. What is chitin, where is it found, and why is it important in the isolation of actinomycetes from soil?
4. What are the basic subunits of the following polymers?
   a. Starch   f. Hemicellulose
   b. Cellulose   g. Lignin
   c. DNA   h. Chitin
   d. RNA   i. Peptidoglycan
   e. Protein
5. What characteristics of plant structural components present significant problems to utilization by soil microorganisms?
6. What factors control the rate of plant litter decomposition?
7. Why might pure glucose inhibit cellulose decomposition?
8. Why is lignin difficult to degrade?
9. What are the major components of plant residue that enter soil?
10. In what ways does cellulose differ from starch?
11. In the following figure from Paul and Clark (1996), what can you say about the relative ease with which each of the three compounds are mineralized? Why is this so?

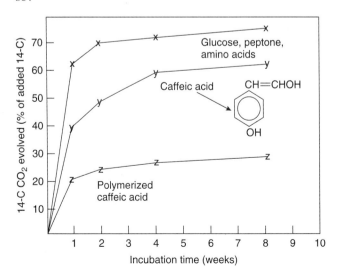

Thought Question

Draw a diagram of the C cycle as you perceive it. Indicate the name of each transformation process and give the genus of a representative microorganism that carries out each transformation.

Additional Reading

TAPPI is the research journal of the paper-producing industry in the United States. If you want to know what happens to much of the cellulose in trees, take a look at some of its articles. If you're interested in learning more about the biology of lignin decomposition, read this review article by Kirk and Farrell: "Enzymatic 'Combustion': The Microbial Degradation of Lignin" (1987, *Annual Review of Microbiology,* 41:465–505).

References

Alexander, M. 1973. Nonbiodegradable and other recalcitrant molecules. *Biotechnology and Bioengineering* 15:611–47.

Crawford, R. L. and R. S. Hanson. 1984. Microbial growth on C_1 compounds. Proceedings of the fourth international symposium, American Society for Microbiology. Washington, DC.

Killham, K. 1994. *Soil ecology.* Cambridge, England: Cambridge University Press.

Paul, E. A. and F. E. Clark. 1996. *Soil microbiology and biochemistry.* 2d ed. New York: Academic Press

Reddy, C. A., and L. Forney. 1978. Lignin chemistry and structure: A brief review. *Developments in Industrial Microbiology* 19:27–34.

Schlesinger, W. H. 1991. Biogeochemistry: An analysis of global change. New York: Academic Press.

Chapter

24

The Carbon Cycle— Mineralization and Residue Decomposition

Overview

After you have studied this chapter, you should be able to:

■ Explain the decomposition of cellulose, hemicellulose, starch, pectin, chitin, and lignin.

■ Discuss the environmental importance of hydrocarbons.

■ List the key intermediates in the mineralization of lignin and hydrocarbons and describe how they are metabolized by common pathways in nature.

INTRODUCTION

Carbon mineralization and residue decomposition are critical features of nutrient cycling. The organic C of plant residue is the principal source of energy for cell growth and metabolism in soil. Once plant residue, and other organic C are deposited in soil, how do they get into the microbial cell in useful forms? How are they metabolized? We can assume that in the case of plant residue, the initial stages of this process were performed by macrofauna and mesofauna that fenestrated and macerated the plant residue so that it was in a form physically available to microorganisms in soil. Carbon metabolism largely depends on whether microorganisms need C-containing compounds for growth and energy, whether they need the compounds as building blocks, or whether they need the compounds as a source of other nutrients such as N, P, or S. In this chapter, we examine some of the metabolic fates of C as it is mineralized by microbial cells.

C MINERALIZATION AND GROWTH

When organic C supports microbial growth, the microbial population increases. This process can be used to isolate specific microorganisms if C compounds are supplied that only they can use. The process is called enrichment culture, and it doesn't require organic C. Winogradsky isolated the first nitrifiers by enrichment culture in simple mineral salts media.

Metabolism can result in mineralization—conversion of an organic C compound to inorganic compounds such as CO_2 and the release of other inorganic nutrients such as NH_4^+, PO_4^{3-}, and SO_4^{2-} that the organic C compounds contain. The most obvious sign of mineralization is soil respiration (Figure 24-1). Metabolism can result in

Figure 24-1 Measuring CO_2 evolution using the soil cover method. The soil microbiologist periodically removes gas samples from the soil cover to measure the accumulation of CO_2 within. (Photograph courtesy of M. S. Coyne)

How Do You Measure Soil Respiration?

Respiration refers to the consumption of O_2 and the production of CO_2 by aerobic organisms. As O_2 is reduced, H_2O is produced, which is one reason why old-time physicians would hold a mirror over the mouths of a person on their deathbed. Condensation on the mirror indicated that the person was still breathing—and the undertaker was held at bay.

It's not very convenient to hold a mirror to soil as a measure of respiration. Instead, soil microbiologists most often measure the rate at which CO_2 is evolved from either intact soil or a soil sample. Figure 24-1 shows a soil microbiologist using the soil cover method to measure the CO_2 evolution rate. A ring is inserted several cm into the soil and a cover is placed on top of it. The cover has a septum in it so that the microbiologist can periodically remove gas samples, store them, and analyze them back in the laboratory with a gas chromatograph to detect changes in the CO_2 concentration with time. By knowing the volume of the soil cover and the soil area that is covered, the soil microbiologist can figure out the CO_2 evolution rate per unit soil area.

If you don't have access to a gas chromatograph, you can still measure the rate at which CO_2 is produced by trapping it in a strong base such as NaOH. This is a titrimetric method for CO_2 analysis. When the CO_2 is absorbed by the NaOH, each molecule of CO_2 neutralizes some of the OH^- ions that are present and forms carbonate (CO_3^{2-}). After you precipitate the CO_3^2 with excess $BaCl_2$ so that it doesn't react further, you titrate the remaining NaOH with HCl to determine how much OH^- remains. The difference between the total OH^- you started with and the OH^- you end up with, as determined by titration, is equivalent to the amount of CO_2 that was trapped (Zibilske 1994)

modification—production of biologically active or inactive compounds. Metabolism can result in incorporation into soil organic matter—polymerization reactions that cause a compound to accumulate in soil.

CELLULOSE AND CELLULOSE DECOMPOSITION

Cellulose is found in plants, trees, and the cell walls of some fungi. It is the most abundant constituent of plant residue, and it makes up about one-third of the biomass of annual plants and one-half of the biomass of perennials. Cellulose is probably the most abundant carbon compound on Earth. Cellulose is a linear polymer, is greater than 1,000 glucose subunits long, with the subunits linked by β ($1 \rightarrow 4$) bonds. It occurs in a semicrystalline state in which parallel chains of cellulose polymers are cross-linked by hydrogen bonds.

Cellulose decomposition is slow relative to the decomposition of some other carbon compounds. While many microorganisms decompose cellulose, few decompose the lignin found with it. This is one reason why wood persists in nature and makes a good building material. Both aerobic bacteria (*Pseudomonas, Chromobacteria*) and anaerobic bacteria (*Clostridium*) decompose cellulose. Cellulose decomposition is also found among the actinomycetes (*Streptomyces*) and myxobacteria. *Cytophaga* is an important bacterial cellulose decomposer in soils receiving straw or manure. Protozoa, especially the protozoa in the termite gut, also degrade cellulose. Without the protozoa, termites could not live on the wood they consume. The metabolic capacity for cellulose decomposition, however, is more common in fungi than in bacteria. Some examples of cellulose-decomposing fungi include *Trichoderma, Chaetomium,* and *Penicillium.*

Cellulose decomposition occurs via extracellular enzymes called cellulases. The microbial cell is impermeable to cellulose because the cellulose is such a large molecule, so decomposition must be started by extracellular enzymes. Cellulose decomposition has two distinct stages. In the first stage, the cross-links between cellulose polymers are broken. In the second stage, the depolymerization of cellulose occurs and the cellulose polymers are hydrolyzed by enzymes such as cellobiase to release cellobiose and glucose.

Cellulase, which is involved in the depolymerization of cellulose during this second stage, is actually an enzyme complex consisting of at least three enzymes:

1. Endo-β (1 → 4) glucanase, an endocellulase that breaks apart the cellulose polymer from the inside
2. Exo-β (1 → 4) glucanase, an exocellulase (cellobioydrolase) (probably more important in nature) that breaks cellulose apart by attacking the ends of the molecules and releasing cellobiose and other oligomers
3. β (1 → 4) glucosidase, also called cellobiase, a mostly intracellular enzyme that releases the glucose from cellobiose molecules that are finally small enough to pass through the cell membrane

HEMICELLULOSE AND STARCH

Hemicellulose is a major plant product that is actually a mixed group of compounds. It has no part in the biosynthesis of cellulose. Hemicelluloses are heteroglycans—polymers of hexoses, pentoses, and sometimes uronic acids—that contain two to four different types of subunits. Hemicelluloses are complex. They are composed of 50 to 200 sugar units that may be linked in a linear, branched, or multiple-branched configuration. The most common subunits in hemicelluloses are xylose and mannose. Xylans make up 30% of hardwoods and 12% of softwoods. Mannans are food reserves. Galactans are found in stress wood, the wood that forms, for example, in tree limbs with swings attached.

Starch is a food reserve of plants. It is made of the glucose polymers amylose and amylopectin. The glucose polymers are linked α (1 → 4) and α (1 → 6) which has important consequences in terms of enzymatic degradation because more microorganisms have enzymes that hydrolyze these α linkages than the β linkages in cellulose. Even though starch and cellulose composition is virtually identical chemically, starch decomposes more rapidly for this simple biochemical reason. Starch is processed by extracellular hydrolytic enzymes such as amylase.

PECTIN AND OTHER POLYMERS

Pectin (polygalacturonic acid) decomposition is studied because of its importance in the middle lamellae of plant cell walls. Pectinases used by Mycorrhizae and *Rhizobium* to initiate symbioses are also used by plant pathogens to infect plant tissue. Hemicelluloses in nature are frequently combined with other substances, which makes their degradation more difficult. In a pure state, hemicelluloses are easily decomposed by bacteria and fungi.

Chitin is the usual structural element in fungal cell walls and the exoskeleton of insects. It is a polymer of N-acetylglucosamine subunits that are linked in linear arrangement by β $(1 \rightarrow 4)$ linkages. Chitin is broken down by chitinases, which are a characteristic enzyme found in many actinomycetes. Peptidoglycan is also a linear polymer, but it consists of alternating N-acetyl glucosamine and N-acetyl muramic acid subunits that are joined by $\beta \rightarrow 4$ linkages. Peptidoglycan chains are linked by short (five amino acid units) peptide cross-linkages. Another name for peptidoglycan is murein. Gram-positive and gram-negative bacteria have peptidoglycan as the major constituent of their cell walls. Archaea do not have peptidoglycan in their cell walls, and this is one physiological characteristic that distinguishes them from the Bacteria.

LIGNIN DECOMPOSITION

The variety of intermonomer linkages in lignins differ greatly in their relative resistance to cleavage. Lignins commonly encrust cellulose and hemicellulose, which increases their resistance to decomposition. Lignin decomposition is primarily by fungi. White and brown rots are mainly basidiomycetes. Soft rots are mainly ascomycetes. White rot fungi are the most active lignin decomposers. *Coriolus versicolor* decomposes the aromatic ring, methoxyl groups, and longer side chains of lignin. *Phanaerochaete chrysosporium,* a white rot fungi, also completely decomposes lignin. These fungi are thought to decompose lignin while they use some other readily degradable C source as their primary energy source.

Examples of brown rot fungi are *Poria* and *Gloephyllum.* They degrade polysaccharides associated with lignin and remove the methyl (CH_3) and methoxyl (OCH_3) side groups. This oxidizes the phenol group, which turns brown; hence the name. Soft rot fungi such as *Chaetomium* and *Preussia* are important where it is wet—wet, but not anaerobic. Anaerobic lignin decomposition has not been confirmed.

Lignin decomposition is part of secondary metabolism. The enzymes that decompose lignin are called ligninases. They are extracellular enzymes that decompose lignin by oxidation. There are numerous ligninases; unfortunately, they are poorly characterized. Lignin metabolism is an event triggered by N, S, or C starvation. White rot fungi are actually inhibited by high N concentrations. Lignin decomposition is not inducible, so adding lignin doesn't initiate lignin decomposition, and very little lignin C is found in the cells of lignolytic fungi. Carbon 14-labeled lignin shows up in soil organic matter, but not in microbial biomass.

HYDROCARBONS

We discuss hydrocarbons from the simple to the complex because of their significant effect on environmental quality. We have to begin this discussion by defining what we mean by "hydrocarbons." The typical chemical formula of hydrocarbons can be generalized as C_nH_{2n+2}. Hydrocarbons with one to four carbons are gases: methane, ethane, propane, and butane. With 4 to 20 carbons, hydrocarbons are liquid at room temperature. At > 20 carbons, hydrocarbons are solid at room temperature; all are biodegradable.

Hydrocarbons are an environmental concern. For example, CH_4 contributes to global warming, and petroleum released into the environment kills fish, birds, and shellfish. Dissolved aromatic compounds in petroleum, even at low levels, disrupt the growth of some marine organisms such as coral. Cleaning up oil spills is estimated to cost $10 to $15 per gallon of spilled oil, which is a high price compared to the current price of gasoline.

METHANE OXIDATION—METHYLOTROPHY

At one time, most gas energy was supplied by coal gas:

$$\text{Steam + air} \xrightarrow{\text{Hot coals}} \text{Combustible gas, } H_2, CO, H_2O, \text{ impurities}$$

When cities switched from coal gas to natural gas, which consists of mostly CH_4, problems developed. The natural gas was drier than the coal gas, and gaskets in the pipes carrying the gas dried out, which created leaks. Plants near gas mains were killed even though CH_4 toxicity was low. This happened because the natural gas leaks stimulated hydrocarbon-degrading bacteria. These bacteria proliferated and used up the available O_2, killing plants by anoxia and by displacement of O_2 with CO_2.

Methane is unique in that it is the only hydrocarbon produced in large amounts by microorganisms. This is a process called methanogenesis, in which CO_2 is used as an electron acceptor during the anaerobic oxidation of H_2 (hydrogenotrophic methanogenesis) or in which acetate is split into CH_4 and CO_2 during fermentation (acetoclastic methanogenesis) (Schlesinger 1997). Methane is metabolized by microorganisms that are unable to use larger compounds. Paddy soils contain a surface film composed of bacteria that brings about CH_4 oxidation:

$$CH_4 + 2O_2 \Rightarrow CO_2 + 2H_2O$$

Methane oxidation is a sequential process:

$$CH_4 \Rightarrow CH_3OH \Rightarrow CHO \Rightarrow HCOOH \Rightarrow CO_2$$

One school of thought is that CH_4 utilization is restricted to specialized microorganisms that use CH_4 and CH_3OH but no other organic C as their carbon and energy source.

These bacteria and yeasts come under the general physiological heading of methylotrophs. A second school of thought is that many heterotrophic microorganisms oxidize CH_4. This includes fungi, bacteria, and actinomycetes.

Other gaseous hydrocarbons are metabolized by microorganisms. Ethylene (C_2H_4) is produced in soils largely by fungi. Ethylene is a plant growth hormone—a ripening agent and the basis for the expression "One bad apple spoils the rest," because damaged plant tissue produces ethylene. Ethylene affects root elongation, development of lateral roots, sprouting of bulbs, and enhances seed germination. Ethylene is also produced by combustion. Curiously, the concentration of ethylene is not rising in the atmosphere. Soils have a remarkable capacity to remove C_2H_4 from air. Ethylene metabolism is also an aerobic process.

PETROLEUM

Petroleum is a natural product resulting from the anaerobic conversion of organic matter under high temperature and pressure. Petroleum hydrocarbons can be categorized as saturates (straight chain and branched alkanes), cycloparaffins, and aromatics. Hydrocarbon metabolism by natural populations of microorganisms represents one of the primary mechanisms by which petroleum and other hydrocarbon pollutants are eliminated from the environment, although some photooxidation also occurs.

The principal factors limiting petroleum metabolism in the environment are: 1) the resistant and toxic components in the material itself; 2) low temperatures (particularly in marine environments); 3) few nutrients (P + K, N + S); 4) limited O_2 availability; and 5) the scarcity of hydrocarbon metabolizers (Atlas and Bartha 1993). Petroleum decomposition is controlled by its physical availability. Dispersing petroleum by adding an emulsifying agent increases the surface area and, consequently, increases the decomposition rate. Most petroleum decomposers produce emulsifying agents, and this is one reason why detergents are added to oil spills. Decomposition rates for many hydrocarbons are not concentration dependent—they are zero-order. The decomposition of higher molecular weight petroleum compounds such as naphthalene and phenanthrene are related to their water solubility rather than their concentration.

Hydrocarbon release into the environment causes excessively high C:N and C:P ratios that are unfavorable to microbial growth. Nitrogen and P limit microbial decomposition of hydrocarbons, particularly in soil, and fertilization improves decomposition rates. Crude oil decomposition is enhanced by adding urea phosphate, N, P, and K fertilizer, or NH_4 phosphate salts. Another important factor is pH. Decomposition is favored at a neutral pH, which favors the greatest population of microorganisms.

The temperature affects petroleum decomposition by affecting the physical nature and composition of oil, the rate of microbial metabolism, and the composition of the microbial community. At low temperature, petroleum viscosity increases and volatilization decreases. Volatile short-chain hydrocarbons that are toxic to microorganisms may accumulate. Water solubility of these toxic compounds also increases and impedes metabolism. In addition, microbial metabolism decreases as temperature decreases due to decreasing enzymatic activity. Metabolism increases as temperature increases to a maximum at about 30° to 40°C.

Decomposition of cyclic and aromatic hydrocarbons by microorganisms involves oxidation of the substrates by enzymes called oxygenases. The O_2 availability in soil depends on the rates of microbial consumption, soil type, waterlogging, and utilizable substrates. The concentration of O_2 is usually rate-limiting in petroleum metabolism in soil and gasoline metabolism in groundwater. Strictly anaerobic metabolism in the absence of O_2 and without alternate electron acceptors such as NO_3^- or Fe^{3+} is negligible and its ecological significance is minor.

ECOLOGY OF HYDROCARBON DECOMPOSITION

Why do hydrocarbons decompose in soil? About 0.02% of plant tissue may be considered as hydrocarbon or hydrocarbon-like. An important concept to remember is that of cross-acclimation. Exposure to one compound may cause an increase in the ability to metabolize similar compounds. This is because microorganisms conserve resources and metabolic abilities and because common metabolic pathways exist. For example, exposure to phenanthrene increases the subsequent metabolism of naphthalene. Enhanced decomposition may also be due to adaptation. Adaptation is induction or derepression of metabolic enzymes or genetic change resulting in new enzymatic capability.

An individual microorganism decomposes only a limited range of hydrocarbons. A mixed population with a broad range of enzymatic capacities is required to metabolize complex mixtures of crude hydrocarbons. This capacity is widespread among both bacteria and fungi. There has been no ecological significance to either algae or protozoa in hydrocarbon metabolism.

PATHWAYS OF HYDROCARBON DECOMPOSITION

Petroleum products differ in their susceptibility to decomposition (biodegradation). The ease of decomposition usually follows this pattern:

1. N-alkanes > Branched alkanes > Low molecular weight aromatics > Cyclic alkanes

N-alkanes of intermediate chain length—hydrocarbons with no double bonds and no branching—are metabolized most rapidly. This includes those hydrocarbons in the C_{10} to C_{24} range. Short-chain alkanes are toxic to many microorganisms, but are also volatile. Branching reduces the rate of decomposition. Long-chain alkanes become increasingly resistant to metabolism and, as the alkanes exceed a molecular weight of 500 to 600, they cease to be C sources. Aromatic compounds are metabolized more slowly than are alkanes.

An alkane is typically oxidized in a series of steps yielding a corresponding alcohol, ketone, aldehyde, and fatty acid (carboxyl) (Figure 24-2). The fatty acid is then subject to β-oxidation. β-oxidation is a stepwise process in which a fatty acid is sequentially reduced by removing two carbons at a time in the form of acetate. Remember that the general formula for a fatty acid is $CH_3 — (CH_2)_n — CH_3 — COOH$ and they may be saturated or unsaturated.

$$R-CH_2-CH_2-CH_3 \xrightarrow[-H_2O]{O_2 + 2H^+} R-CH_2-CH_2OH-CH_3$$

(Alkane)   (Secondary alcohol)

$$R-CH_2-CH_2OH-CH_3 \xrightarrow{-2H^+} R-CH_2-CO-CH_3$$

(Ketone)

$$R-CH_2-CO-CH_3 \xrightarrow[-H_2O]{O_2 + 2H^+} R-CH_2-O-CO-CH_3$$

(Acetylester)

$$R-CH_2-O-CO-CH_3 \xrightarrow{+H_2O} R-CH_2-OH + CH_3-COOH$$

(Primary alcohol) (Acetic acid)

$$R-CH_2-OH \xrightarrow{-2H^+} R-CHO \xrightarrow[-2H^+]{-H_2O} R-COOH$$

(Aldehyde)   (Carboxylic acid)

$$RCH_2CH_2COOH \xrightarrow{\beta-Oxidation} R-COOH + CH_3COOH$$

Figure 24-2   Steps in the sequential oxidation of an alkane.

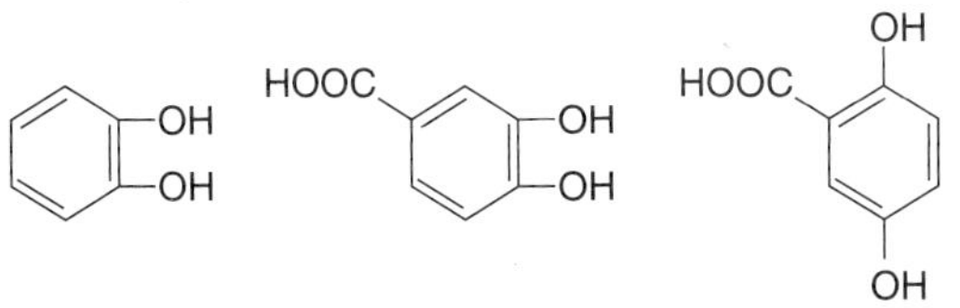

Catechol   Protocatechuic   Gentisic
                        acid   acid

Figure 24-3   Catechol, protocatechuic acid, and gentisic acid—three of the common intermediates of lignin and petroleum decomposition in soil.

The biochemical model that is followed during decomposition in nature is to change the chemical into a basic intermediate and thereafter metabolize it by common metabolic pathways. Metabolism of natural products in the environment frequently causes aromatic compounds to accumulate. Lignin and humus, for example, are composed of at least 50% aromatic components. There are also aromatic compounds in petroleum products. For aromatic compounds, these basic intermediates are catechol, protocatechuic acid, and gentisic acid (Figure 24-3) (Alexander 1977).

Metabolism of the common aromatic intermediates requires ring cleavage. Ring opening almost always requires O_2, and the ultimate products are succinic, fumaric, pyruvic, and acetic acids, and acetaldehyde, which can be utilized during aerobic and anaerobic metabolism. The two primary mechanisms of ring cleavage are ortho- and meta-cleavage (Figure 24-4). In ortho-cleavage, an oxygenase breaks the aromatic ring

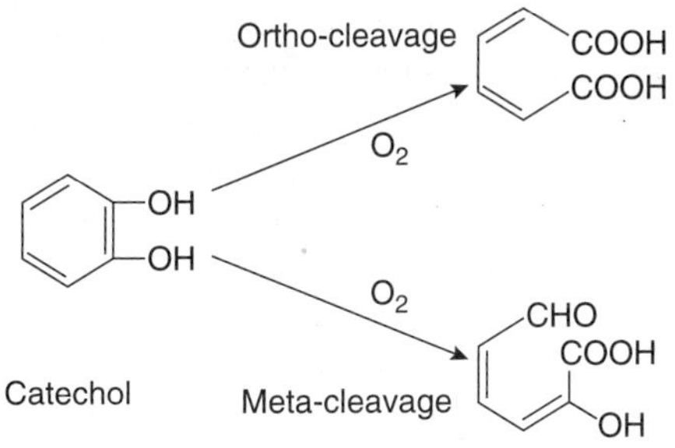

Figure 24-4   Ortho- and meta-cleavage of catechol—the first step in the metabolism of aromatic compounds.

between two adjacent hydroxyls. In meta-cleavage, oxygenase breaks the aromatic ring adjacent to one of the hydroxyl groups. Multicyclic compounds are broken one ring at a time.

Summary

Cellulose decomposition occurs via enzymes called cellulases that occur in many bacteria and fungi. Hemicelluloses are polymers of hexoses, pentoses, and sometimes uronic acids. The most common subunits in hemicelluloses are xylose and mannose. Lignins commonly encrust cellulose and hemicellulose, which increases their resistance to decomposition. The principal factors limiting petroleum metabolism in the environment are resistant and toxic components in the material itself; low temperatures; low P, K, N, and S; low O_2 availability; and scarcity of hydrocarbon metabolizers.

Lignin decomposition is part of secondary metabolism and is an event triggered by N, S, or C starvation. The enzymes that decompose lignin are called ligninases. Lignin and humus are composed of at least 50% aromatic components. There are also aromatic compounds in petroleum products. During decomposition, compounds decompose to basic intermediates and are thereafter metabolized by common metabolic pathways. For aromatic compounds, these basic intermediates are catechol, protocatechuic acid, and gentisic acid. Further metabolism of the common aromatic intermediates requires ring cleavage. The two primary mechanisms of ring cleavage are ortho-cleavage and meta-cleavage.

Sample Questions

1. What are the first likely decomposition products of amylose, cellulose, and hemicellulose decomposition?
2. What are the steps in the decomposition of cellulose?
3. What are the environmental factors limiting petroleum decomposition in nature?

4. What are some likely intermediates that form during the decomposition of lignin?

5. What is the difference between ortho- and meta-cleavage and why is this process important?

6. Why do soils have the ability to degrade hydrocarbons?

7. A soil cover has a diameter of 4 inches and a height of 25 inches above the soil surface. How many square meters are covered, and what is the internal volume of the soil cover in cm^3?

8. If the rate of CO_2 evolution is 25 ppm hr^{-1}, how much CO_2 will accumulate in the headspace of the soil cover in Question 7 in 2 hours? (Note that ppm = μL^{-1} and that there is 1 μmole of CO_2 for every 22.4 μL of CO_2.)

9. What is the CO_2 evolution rate in Question 8 in terms of ppm CO_2 m^{-2} hr^{-1}?

10. What are the molecular weights of CO_2, NaOH, and HCl? If the equivalent weight is equal to the molecular weight divided by the charge, what is the equivalent weight of NaOH and HCl?

11. You start with 25 mL of 1.0 N NaOH. After incubating it with a soil sample in a sealed container, you titrate it with 15 mL 1.0 N HCl to reach a pH of 7.0. How many equivalents of CO_2 are absorbed by the NaOH, assuming that the equivalent weight of CO_2 is 22 equivalents L^{-1}?

Thought Question

Given what you now know about enrichment culture and petroleum metabolism, how would you go about isolating an octane-degrading bacteria?

Additional Reading

A recent textbook that addresses biodegradation and bioremediation in detail, and at a reasonably elementary level, is *Biodegradation and Bioremediation* by Martin Alexander (1994, Academic Press, San Diego, CA).

References

Alexander, M. 1977. *Introduction to soil microbiology.* New York: John Wiley & Sons.

Atlas, R. M., and R. Bartha. 1993. *Microbial ecology: Fundamentals and applications.* Redwood City, CA: Benjamin/Cummings.

Schlesinger, W. H. 1997. *Biogeochemistry: An analysis of global change.* 2d ed. San Diego, CA: Academic Press.

W. B. Whitman (eds.) *Microbial production and consumption of greenhouse gases,* Washington, DC: American Society for Microbiology.

Zibilske, L. M. 1994. Carbon mineralization. In *Methods of soil analysis, part 2. Microbiological and biochemical properties,* R. W. Weaver et al. (eds.), 835–63. Madison, WI: Soil Science Society of America.

Chapter 25

The Carbon Cycle—Soil Organic Matter and Humus

Overview

After you have studied this chapter, you should be able to:

- Describe the major biosystems in which soil organic matter is stored and tell how that organic matter is affected by climate and cultivation.
- Explain why so much emphasis is placed on soil organic matter as a critical factor in soil productivity.
- Summarize the roles that organic matter plays in the soil ecosystem.
- Describe the major components of soil organic matter: humus and polysaccharides.
- Explain humus formation, fractionation of humus, and the role that polysaccharides play in improving soil properties.

INTRODUCTION

What type of organic matter enters soil? If we look at some of the basic constituents of plants, we can classify them into several identifiable fractions such as cellulose; hemicellulose; lignin; a water-soluble fraction composed of simple sugars, amino acids, and aliphatic acids (such as succinate and acetate); a protein fraction; and an ether- and alcohol-soluble fraction composed fats, oils, waxes, and resins. As plants age, the content of the first three fractions in plant material increases while content of the latter three decreases (Figure 25-1). This accounts for the changing proportion of each fraction in plant material. In this chapter, we look at how the mineralized components of plant material become incorporated into soil organic matter.

INCORPORATION OF ORGANIC C INTO SOIL

Let's look at what happens during oat straw decomposition (Figure 25-2). One thing you should notice is that the total C that is left is greater than the oat straw C that is left. This is because part of the C from the oat straw is used to create new microbial biomass.

After an extended incubation period, the proportions of different types of plant material retained in soil are usually similar, even though the initial plant composition may differ. If you add straw or glucose to soil, the glucose is used much faster than is the straw, but both substrates are ultimately incorporated into soil organic matter to about the same extent—about 20% (Figure 25-3). Some substrates are oxidized after a lag

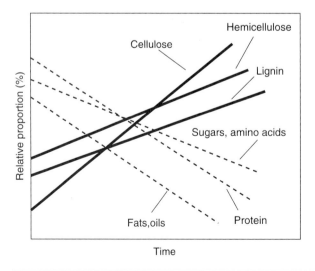

Figure 25-1   Changing composition of aging plant material. The proportion of soluble components in the plant material decreases while the proportion of insoluble and resistant components increases.

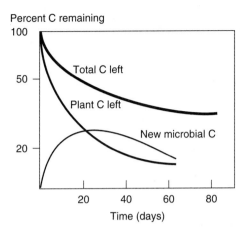

Figure 25-2   Schematic diagram of the changing concentrations of C and C pools during oat straw decomposition. (Adapted from Paul and Clark 1996)

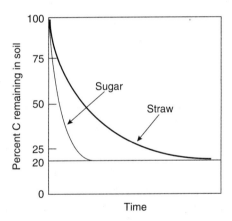

Figure 25-3   Decomposition and incorporation of sugar and straw into soil organic matter. After an extended incubation period the proportions of substrate retained in soil organic matter are usually similar even though the initial plant composition varies considerably.

period. This could reflect their solubility, difficulty in decomposing, or novel structure. Others, such as ethanol and acetate, are oxidized immediately, suggesting that soil microorganisms are exposed to these simpler compounds repeatedly and are always ready to decompose them.

The pH of the soil environment affects the rate of organic C incorporation, but not the ultimate amount that becomes incorporated. Table 25-1 shows that as the pH of the soil decreases from neutral to acidic, the C remaining in plant material in soil decreases more slowly, but after 5 years, there is relatively little difference.

Table 25-1   Retention of ryegrass carbon added to soil over time.

Soil Characteristics			Ryegrass Added (mg/100 g soil)	Percent Added Plant Carbon Remaining		
pH	Percent C	Percent Clay		1 year	2 years	5 years
8.1	1.0	18.0	125	31	23	17
7.8	2.4	18.0	125	32	26	18
7.8	2.4	18.0	252	29	25	–
6.9	4.6	20.0	125	31	25	19
4.8	3.9	21.0	125	31	26	20
3.7	4.0	21.0	125	42	30	20
6.2	1.6	8.0	125	27	21	15
3.7	2.9	5.0	125	36	26	17

(Adapted from Jenkinson 1971)

Complex materials that are added to soil undergo physical, biochemical, and biological changes during decomposition in which polymers are degraded to monomers, reduced compounds become increasingly oxidized, and rapidly growing microorganisms subsisting on easily degraded compounds are replaced by slower growing microorganisms subsisting on recalcitrant compounds. Figure 25-4 illustrates these changes.

SOIL ORGANIC MATTER (SOM)

Soil organic matter is composed of decomposing residues, by-products formed by decomposition, microorganisms, and resistant soil humic material. Different ecosystems have different amounts of soil organic matter, partly because of different temperature and decomposition rates. In Figure 25-5, the distribution of C in terrestrial sources is illustrated. In some ecosystems, such as tropical rain forests, about as much C is above ground as is below ground. In others, particularly the tundra, far more C is below ground than is above ground.

As Figure 25-5 shows, it is obvious that tropical ecosystems represent the greatest aboveground organic C pool while tundra represents the greatest below ground organic C pool. Decomposition of 1 petagram (10^{15} g or 10^{12} kg) is equivalent to enriching the CO_2 in the atmosphere by 0.47 ppm. So, rain forest destruction and global warming create a vicious cycle. As the tropical rain forests are cut down, CO_2 in the atmosphere increases. As CO_2 increases, the temperature warms. As the temperature warms, decomposition in tundra ecosystems increases, which leads to more CO_2 production, and so on.

The most important concept to learn from this chapter is that the greatest single factor controlling productivity in cultivated and uncultivated soil is the amount and depth

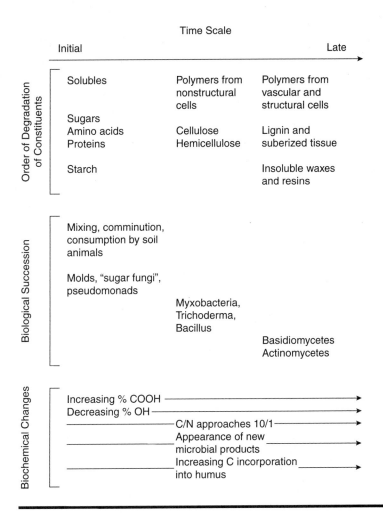

Figure 25-4 Physical, chemical, and biological changes that occur during the decomposition and incorporation of organic C into soil organic matter.

of soil organic matter. This concept has its limitations in the sense that a Histosol, or peat soil, even though it has abundant organic matter, is not necessarily the most productive soil for physical reasons. Likewise, although wetland soils have a lot of organic matter, they aren't necessarily productive for agriculture because they contain too much water. It's always possible to have too much of a good thing.

Soil organic matter is formed by the decomposition of plant components into low-molecular-weight compounds that then repolymerize. The C in soil organic matter comes from carbohydrates, which are primarily of microbial origin once decomposition

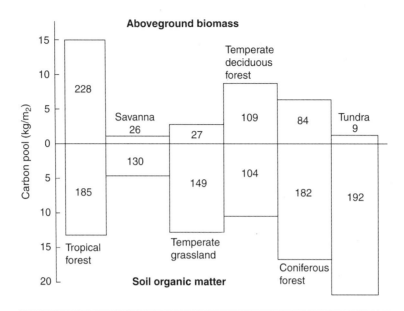

Figure 25-5 Different pools of organic C in the global environment. The *y*-axis represents an average content of C above and below the surface in each ecosystem. The depth of each bar is proportional to the area extent of each ecosystem. The value within the bar indicates the worldwide C storage in each ecosystem in petagrams (Pg) (1 Pg = 10^{15} g). (Adapted from Anderson 1991)

has been completed, or from nitrogen-containing compounds, aliphatic fatty acids and alkanes, oily compounds, and aromatic compounds that generally come from the decomposition of lignin (Figure 25-6). Thus, the exact nature of soil organic matter is not clear. It has no definite chemical formula, no defined structure, and no defined shape.

FUNCTIONS OF SOIL ORGANIC MATTER

What are the functions of soil organic matter? Soil organic matter contributes to the soil's cation and anion exchange capacity (CEC and AEC). Soil organic matter affects the retention, release, and availability of plant nutrients. It releases immobilized N, P, and S during decomposition. Soil organic matter is the storehouse for the major plant nutrients. Soil organic matter binds organic chemicals and pesticides, which affects their biological activity and toxicity. It lessens some of the harmful effects of chemicals dumped into the environment. Soil organic matter improves water percolation into, and retention by, soil. Soil organic matter is involved in the formation and maintenance of desirable soil structure. The more soil organic matter present, within limits, the better

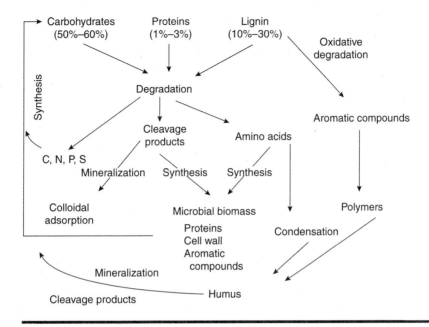

Figure 25-6 Steps in the cycling of soil C and the formation of soil organic matter and humus.

the tilth of soil. Soil organic matter is involved in the movement of tightly bound and normally insoluble metals with water-soluble components of soil organic matter. It is involved in the chelation of micronutrients and formation of spodic horizons (see chapter 12), so it contributes to forming taxonomically different soils. Soil organic matter absorbs solar radiation, which influences soil temperature. It provides color to soil and decreases its albedo (reflectance). In general, the more organic matter a soil has, the darker the soil will be and the more energy it will absorb. Soil organic matter provides C and energy for soil microorganisms.

SOIL ORGANIC MATTER AS A CARBON AND ENERGY SOURCE FOR MICROORGANISMS

If we consider soil organic matter as a carbon and energy source for microorganisms, particularly for heterotrophic organisms, then we are really talking about the ability of soil organic matter to serve two functions: to provide energy for growth and to provide carbon for the formation of new cell material. In this respect, when organic material is incorporated into soil, it causes several things to happen. First, there is an immediate and marked drop in the O_2 concentration as respiration increases. At the same time, CO_2 increases as a product of respiration. The combination of the two events causes the redox potential (E_h) to decline. Microorganisms cause the change in E_h through the consumption of O_2 and the liberation of reduced products.

Table 25-2 Levels of total C (g C kg^{-1} soil) in no-tillage and conventional tillage systems on an Alfisol (Maury silt loam) after 23 years of continuous corn cultivation.

Tillage System	Soil Depth (cm)	
	0–7.5 cm	**7.5–15 cm**
No tillage	19.53	10.58
Conventional tillage	11.47	11.47

(Handayani 1996)

Fungi are more efficient than bacteria in terms of the ratio of CO_2 respired to C assimilated. Fungi are 30% to 40% efficient in this regard while aerobic bacteria are only 5% to 10% efficient, and anaerobic bacteria are only 2% to 5% efficient. The rate at which CO_2 is released during soil organic matter mineralization varies greatly with soil type and is affected by such things as the organic matter concentration in soil, cultivation, and the sequence of degradation by bacteria and fungi.

Cultivation enhances soil organic matter destruction. A good illustration of this phenomenon is data that were collected from a long-term (25+ years) experiment with continuous corn that examined the effect of tillage on organic C contents (Handayani 1996). The data in Table 25-2 show that no-tillage soil, in which the soil is disturbed as little as possible, has almost twice as much total C as the conventionally tilled soil, in which the soil was moldboard plowed and disked to incorporate aboveground plant residue and provide a good seedbed. The C content decreases drastically with soil depth in the no-tillage system, but remains about the same in the conventional tillage system. This seems like an obvious result of the soil mixing that goes on during conventional tillage. Tillage has the effect of uniformly distributing C with soil depth.

The amount of soil organic matter doesn't change drastically during the year, particularly in soils that have been managed the same way for long periods. Figure 25-7 shows that the greatest change in total C levels occurs in a no-tillage soil, but only at the soil surface, where most biological activity takes place. Farther into the soil profile, the amount of soil C remains relatively constant.

Temperature and moisture also have an effect on the organic matter content of soil. At mesophilic temperatures (20° to 30°C), CO_2 production is 5 to 10 mg kg^{-1} soil day^{-1}. Wetting and drying enhance CO_2 evolution by breaking up soil aggregates and releasing encapsulated organic material. Tillage also has an effect on moisture content, which helps to explain why no-tillage soils tend to have more soil C than conventionally tilled soils. Figure 25-8 shows that in the surface 7.5 cm, no-tillage soil has consistently higher gravimetric water content than conventionally tilled soils have. Remember, the more water a soil has, the cooler it tends to be; the cooler it tends to be, the lower the metabolic activity of the microorganisms contained within it. The lower the microbial activity, the lower the overall C lost from soil during respiration. Total CO_2 evolution

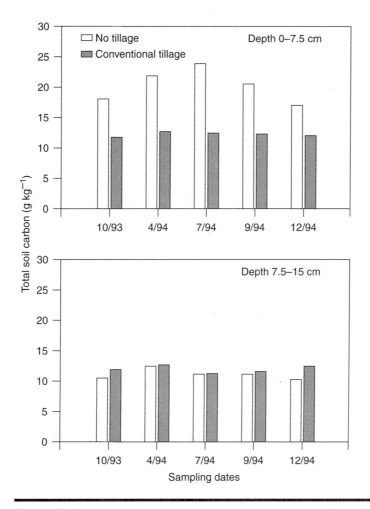

Figure 25-7   Temporal and spatial measurement of total C content in no tillage (NT) and coventional tillage (CT) systems. (Adapted from Handayani 1996)

from no-tillage soils is higher than in conventionally tilled soils simply because there is more C available to decompose. The pH, soil depth, and aeration also influence CO_2 evolution because they influence the microbial populations.

Overall, about 2% to 5% of the soil organic matter is mineralized each year. This varies dramatically depending on the environment. Fresh substrates accelerate and sometimes reduce the rate of soil organic matter decomposition. This is known as the priming effect (Jenkinson 1971). We discussed the priming effect when we examined N mineralization. Whether priming occurs, and what form it will take, depend on the type of soil and the type of organic C substrate added.

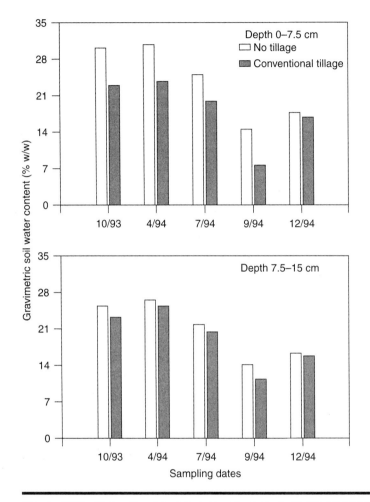

Figure 25-8 Temporal and spatial measurement of gravimetric soil water content in no tillage (NT) and conventional tillage (CT) systems. (Adapted from Handayani 1996)

HUMUS

The more ambiguous, less clearly defined fraction of soil organic matter is known as humus. Humus is the amorphous, colloidal, microbially altered, and relatively stable portion of soil organic matter. Humus forms from condensation of phenolic and amino compounds derived from organic matter breakdown and from condensation of amino-quinone intermediates (Figure 25-9). In the polyphenol theory of humus formation, lignin degrades to polyphenols such as catechol (this can be spontaneous in alkaline environments, but most likely is caused by enzymes), which are oxidized to quinones (Pauland Clark, 1996). The quinones, in turn, react with amino compounds and amino

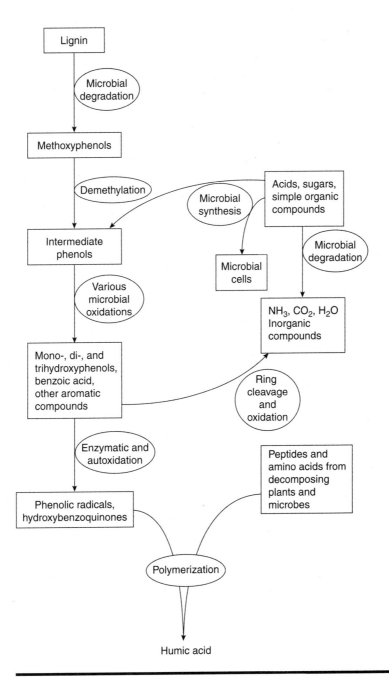

Figure 25-9   A schematic representation of humic acid synthesis in soil. (Adapted from Martin and Haider 1971)

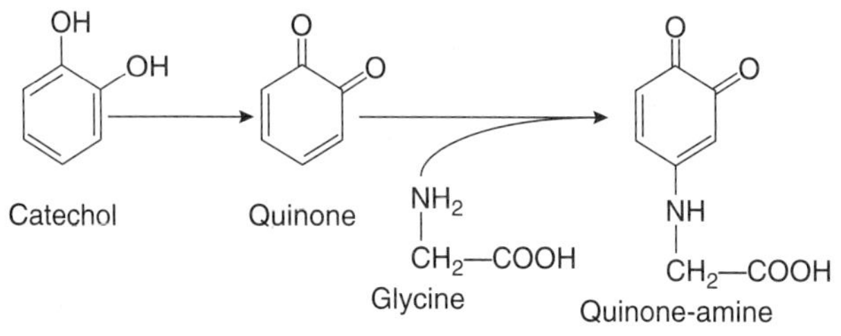

Figure 25-10 Schematic diagram of the polyphenol theory of humus formation.

acids to form nitrogenous polymers that combine to form larger nitrogenous polymers (Figure 25-10).

The primary characteristic of soil humus is its resistance to degradation. Humus removed from soil degrades more slowly than do most other organic compounds, but faster than humus in soil. Humus is resistant to degradation in soil because of its physical protection during the formation of microaggregates and micropores and its extensive interactions with soil minerals (clays, oxides, and amorphous material). The interactions directly reduce the availability of humus.

Humus also resists degradation because it can be very large. It forms immobile, sometimes hydrophobic, complex globular structures of low solubility. Because humic materials can be large and insoluble, they must first be attacked by extracellular enzymes. In addition, humus has a polyaromatic character with highly cross-linked, random, varied, and disordered structures that are due to chemical condensation reactions. This material can bind and inactivate attacking enzymes.

Fractionation of Humus

Humus is generally fractionated into three components following extraction in NaOH: humin, fulvic acid, and humic acid. Humin is the non-NaOH dispersible fraction. Humic acid is a NaOH soluble fraction of humus that is insoluble at pH 2. The molecular weight of humic acid varies from 10,000 to 100,000 and it is composed of aromatic rings, cyclic nitrogen compounds, and peptide chains of indeterminate structure. Humic acid has an overall composition of 57% C and 4% N. The functional groups of humic acid are COOH, phenolic OH, alcoholic OH, and ketones. Fulvic acid is soluble in NaOH and soluble at pH 2. It is smaller than humic acid, with molecular weights ranging from 1,000 to 30,000. Fulvic acid contains highly oxidized aromatic rings with numerous side chains. Fulvic acids in lower soil horizons are formed by the leaching of organic constituents from horizons above.

What is the contribution of functional groups in humus to cation exchange capacity? Fulvic acids are more acidic than humic acids. Humic materials are adsorbed to clay by polyvalent cations such as Ca^{2+} and Fe^{3+}. They also form associations with hydrous oxides. At low pH, the humus contributes to anion exchange capacity. We say that humic materials have pH-dependent charge because their overall charge alters in

How Do You Extract Humic Acid from Soil?

Extracting humic acid from soil is one of the easiest procedures a soil microbiologist can perform (Wolf et al. 1994). Weigh 40 g of soil into a 250-mL centrifuge tube and add 200 mL of 0.05 M HCl. Mix thoroughly. The acid removes carbonates and polyvalent cations that interfere with the extraction procedure. Centrifuge the mixture at 1,500 × g and discard the supernatant.

Add 200 mL of distilled water to the soil pellet, resuspend, centrifuge as before, and discard the supernatant. This step rinses residual acidity from the soil. Then add 200 mL of 0.5 M NaOH and flush out the air in the centrifuge tube with a stream of N_2. Seal the tube and shake it at room temperature for 12 to 24 hours. The N_2 helps prevent unwanted oxidation of the humic materials from occurring during extraction.

Centrifuge the tube as before. This time, decant and save the supernatant, which contains the dissolved humic materials. Filter the supernatant through glass wool to remove suspended organic material and collect it in a 1-L beaker. Repeat the extraction with NaOH two to three more times for maximum extraction efficiency. After the last extraction, add 200 mL of distilled water to the soil sample, shake for 10 minutes, centrifuge, and add this rinse water to the alkaline solutions you've been collecting in the beaker.

Add 2M HCl to the alkaline solution until the pH drops to 1.0. The dark precipitate that forms is humic acid. After settling for 24 hours, the humic acid fraction can be collected by centrifugation. The fulvic acid remains in solution in this procedure, but can be collected by drying the now-acidified solution. Lyophilization is the preferred method to dry both the humic acid and the fulvic acid fractions.

response to the change in the environment's pH—positive in acidic environments and negative in alkaline environments.

POLYSACCHARIDES IN SOIL

The main constituents of soil organic matter are humic and fulvic acids (about 80% of the soil organic matter). What we know about them principally comes from what we can say about them once they're extracted from soil. But extraction procedures recover only about 10% to 20% of the humic and fulvic acids in soil. During the extraction of fulvic acids, we can also recover a variety of carbohydrates from soil, such as monosaccharides (hexoses such as glucose and galactose and pentoses such as arabinose and xylose), disaccharides (sucrose and cellobiose), oligosaccharides (cellotriose), polysaccharides (cellulose and hemicellulose), amino sugars (glucosamine), sugar alcohols (such as mannitol), sugar acids (such as galacturonic and glucuronic acids), and methylated sugars.

The extraction of these compounds is something of a paradox because, away from soil, they degrade very quickly. Yet, 5% to 20% of the total organic matter in soil is composed of carbohydrates, particularly polysaccharides. How can polysaccharides and other relatively available organic matter persist in soil? There are several reasons:

1. They may be chemically protected by binding to humic acid and heavy metal cations.
2. They may be physically protected by binding to clay, by being mixed with lignin, or by being encapsulated in soil micropores.
3. There may be no extracellular enzyme present in their vicinity to metabolize them.
4. Microorganisms may preferentially grow on other C compounds.
5. There may be no microorganisms present in their vicinity.
6. There may not be enough N, P, or S to permit their complete decomposition.
7. They may be in an environment unfavorable for microbial growth.

The origin of most soil polysaccharides is microbial. They arise from the decomposition of organic material (mostly plant material) that enters soil. Polysaccharides have several attributes:

1. They contribute to aggregate stability.
2. They contribute to water relations (generally, but not always improving water availability).
3. They provide protection—a physical barrier for microorganisms against the environment (although phages can degrade some polysaccharide capsules).
4. They provide a food reserve for microorganisms; some organisms producing polysaccharides can also degrade them.
5. They are involved in plant/microbial interactions both in symbioses (the rhizobial capsule is composed of a variety of polysaccharides) and in pathogenicity (vascular wilt).

Polysaccharides surrounding microorganisms represent the outermost limit of the cell. They are the last barrier of excretion and the first barrier of uptake. They give the microbial cells some cation-exchange capacity. Carbohydrates can be acidic (uronic acids), negatively charged (pyruvate and succinate), or neutral (monosaccharide sugars). Polysaccharides make meshlike networks that act to sieve nutrients. Capsular polysaccharides are very hydrated and very hydroscopic, which is important to cells during water stress. Bacterial mutants that can't make a polysaccharide capsule are more sensitive to low water potential than are their encapsulated relatives. Polysaccharide production is not luxury production. Even under C-limited chemostat culture, *Xanthomonas* polymerizes 20% of the carbon it consumes to make extracellular polysaccharides. The normal case for bacteria in soil is production of a capsule.

Polysaccharides contribute to bacterial flocculation (clumping together). There are several possible reasons why microorganisms flocculate. Flocculation provides protection for the group, though not necessarily for the individual; it moderates environmental change; it promotes genetic exchange by increasing cell-to-cell contact. While flocculation may be good for microorganisms, it can have some undesirable effects as far as people are concerned. Flocculation and capsule formation can lead to clogging of filter beds in sewage treatment plants, clogging of water pipes in municipal water systems, and clogging of metal screens in groundwater wells.

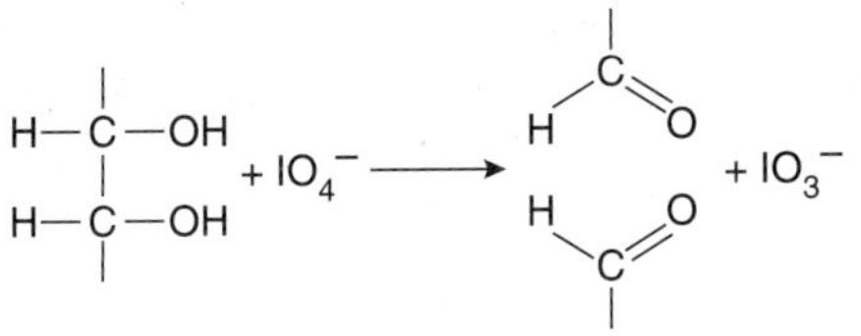

Figure 25-11 Periodate oxidation of a polysaccharide.

In soil, polysaccharides are stabilized by binding to humic acid, by binding to heavy metal cations, by binding to clays (steric hindrance), and by entering bonds between clay particles (intercalation). There are several lines of experimental evidence showing that polysaccharides and soil minerals interact. For example, one can demonstrate that there is a positive correlation between aggregate stability and soil polysaccharides and that aggregate stability can be destroyed by periodate. Periodate cleaves vicinyl glyols in polysaccharides, so if the polysaccharide is involved in aggregate formation, its removal by periodate should affect the aggregate (Figure 25-11). Polysaccharides added to unaggregated soil form soil aggregates. Higher molecular weight polysaccharides are more effective because the larger molecules have more contact points with soil and stretch farther. One can also use direct electron microscopy to observe polysaccharides binding with soil.

AGGREGATE STABILITY

Soil organic matter plays an important role in the formation and stabilization of soil aggregates. Aggregates are naturally occurring formations of soil particles in which the forces holding the particles together exceed the forces pulling them apart. The external forces pulling aggregates apart include freezing and thawing, and wetting and drying. Aggregates affect moisture, drainage, root growth, and reduce soil loss by wind erosion and runoff.

Most aggregation is in the top layer of soil. Macroaggregation is controlled by soil management (tillage). The integrity of large aggregates > 50 μm in diameter is not affected by periodate treatment, which shows that polysaccharide polymers are not what is holding them together. Microaggregates depend on persistent organic binding agents that are characteristic of soil and are independent of management. In other words, you can till a soil as much as you like without affecting the binding of microaggregates. Biological factors influencing aggregate formation include bacterial clay envelopes, earthworm activities, and bacterial and root polysaccharides.

The most important mechanism of interaction between the soil organic matter and the mineral fractions of soil is the formation of cation bridges between clay surfaces and organic polymers. Soil aggregates are stabilized by organic material through cross-linking by fibrils, cementing by humic substances, and binding by fungal hyphae and plant roots. Mixtures of live bacterial cells and clay particles appear to form aggregates 2 to 20 μm in diameter. Organic binding material can be classified as transient—mainly polysaccharides, temporary—roots and fungal hyphae, and persistent—resistant aromatic components associated with polyvalent metal cations and strongly adsorbed polymers.

SOIL CONDITIONERS

Soil conditioners are used to improve soil physical and chemical conditions by treatment with chemical or biological materials. These materials include adhesive polymers such as polyvinyl acetate, polyacrylamide, polyvinyl pyrolidone, polyethylene glycol, and polyvinyl alcohol. Well-rotted compost, to use the organic farmer's terminology, provides the same functions as a soil conditioner. The aim of adding soil conditioners to soil is to complement natural processes of soil aggregation by enhancing all factors that favor aggregation. The strategy is to enhance the strength of soil aggregates. All legitimate conditioners are subject to state and federal regulations.

INTERACTION WITH METALS

Soil organic matter reacts with metals in soil. Plants growing in soil with high organic matter show less aluminum (Al) toxicity than do plants in soil at the same pH with lower organic matter. It's thought that Al is exchanged from organic matter binding sites. The amount of Al adsorbed by peat increases with increasing pH. This is not surprising if you remember that the cation exchange capacity (CEC) of humic acids increases with increasing pH. Humic substances at a pH of 6.0 to 7.0 have an overall negative charge. Aluminum binding is probably due to carboxyl groups ($—COO^-$) on the humic substances. Humic substances bind divalent cations (M^{2+}) preferentially over monovalent cations (M^+). Their preference for binding divalent cations is something like this:

$$Cu^{2+} > Pb^{2+} >> Fe^{2+} > Ni^{2+} = Co^{2+} = Zn^{2+} > Mn^{2+} > Ca^{2+}$$

A quick method of immobilizing heavy metals such as Cu^{2+} and Pb^{2+} in soil is to mix the soil with a lot of organic material. This only works temporarily because the metals can be released again once the organic matter mineralizes.

Summary

The greatest single factor controlling productivity in cultivated and uncultivated soils is the amount and depth of soil organic matter. Complex materials that are added to soil undergo physical, biochemical, and biological changes during decomposition. Polymers are degraded to monomers, reduced compounds become increasingly oxidized, and rapidly growing microorganisms subsisting on easily degraded compounds are replaced by slower growing microorganisms subsisting on recalcitrant compounds. After an extended incubation period, the proportion of different types of plant material retained in soil is usually around 20% of the original amount added, even though the initial plant composition may differ.

The exact nature of soil organic matter is not clear. It has no definite chemical formula, no defined structure, and no defined shape. Cultivation enhances soil organic matter destruction. Two percent to 5% of the soil organic matter is mineralized each year. This varies dramatically depending on the environment.

Humus is the amorphous, colloidal, microbially altered, and relatively stable portion of soil organic matter. Humus forms from condensation of phenolic and amino compounds derived from organic matter breakdown and from condensation of amino-quinone intermediates. The primary characteristic of soil humus is its resistance to degradation. Five percent to 20% of the total organic matter in soil is composed of carbohydrates, particularly polysaccharides. Polysaccharides play an important role in the formation and stabilization of soil aggregates.

Sample Questions

1. "The driving force behind nearly all nutrient cycling reactions involving organic compounds is the search for energy tied up in reduced organic compounds." (E. A. Paul)
   a. Explain this quote.
   b. Discuss the relevance to N and P mineralization reactions.
2. What is soil organic matter?
3. What are two reasons why polysaccharides might not degrade in soil?
4. What is humus and how is it characterized?
5. What is the single most important factor that controls the productivity of cultivated and uncultivated soils? Give one reason why this is so.
6. What are five functions or roles of soil organic matter?
7. Soil humic materials commonly have mean residence times of 200 to 2,000 years. Why are soil humic materials so slowly degraded by soil microorganisms? Give at least four possible reasons or mechanisms.
8. Briefly explain the "Polyphenol Theory" of soil organic matter formation.
9. What types of evidence suggest that organic matter is involved in soil aggregation?
10. Does organic matter determine the stability of all-sized aggregates?
11. If CO_2 evolution from a typical soil is 5–10 mg CO_2 kg^{-1} soil day^{-1}, and if 1 milliequivalent of CO_2 is equal to 22 mg CO_2, how much 1N NaOH do you think you will need to trap all of the CO_2 from a 100-g sample of this soil over a 7-day period?

12. The following graph represents the relative survival of two bacteria in soil as the water potential declines. One bacteria (CAP$^+$) is able to form a capsule while the other (CAP$^+$) cannot form a capsule. Can you explain what the data imply and the mechanism behind the data?

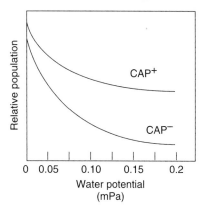

Additional Reading

There is no lack of additional reading material about soil organic matter. You can start by looking through some of the current and back issues of the journal *Soil Biology and Biochemistry* published by Elsevier and *Soil Science* published by the Williams and Wilkins Company. The Soil Science Society of America has two publications that are directly related to soil organic matter. The first is *Interactions of Soil Minerals with Natural Organics and Microorganisms* (1986, SSSA Special Publication No. 17, edited by P. M. Huang and M. Schnitzer, Soil Science Society of America, Inc. Madison, WI). The second publication is *Soil Fertility and Organic Matter as Critical Components of Production Systems* (1987, SSSA Special Publication No. 19, edited by R. F. Follett et al., Soil Science Society of America, Inc. Madison, WI).

References

Anderson, J. M. 1991. The effect of climate change on decomposition process in grassland and coniferous forests. *Ecological Applications* 1:326–47.

Handayani, I. P. 1996. Soil carbon and nitrogen pools and transformations after 23 years of no tillage and conventional tillage. Ph.D. Diss. University of Kentucky, Lexington.

Jenkinson, D. S. 1971. Studies on the decomposition of C^{14} labeled organic matter in soil. *Soil Science* 111:64–70.

Martin, J. P., and K. Haider. 1971. Microbial activity in relation to soil humus formation. *Soil Science* 111:54–63.

Paul, E. A., and F. E. Clark. 1996. *Soil microbiology and biochemistry.* 2d ed. San Diego, CA: Academic Press.

Wolf, D. C., J. O. Legg, and T. W. Boutton. 1994. Isotopic methods for the study of soil organic matter dynamics. In *Methods of soil analysis, part 2: Microbiological and biochemical methods*, R. W. Weaver et al. (eds.), 865–906. Madison, WI: Soil Science Society of America.

5

SOIL MICROBIAL INTERACTIONS

*S*oil microorganisms do not exist in isolation from one another. Like any other members of a community, they have a complex net of interactions. In this section, we briefly explore some of these interactions. Most of this section is devoted to symbiotic relationships between bacteria and plants—the legume–rhizobia symbiosis, and between fungi and plants—mycorrhizal symbioses. By the end of this section, you should have a good appreciation of how these microbial interactions form the basis of microbial ecology.

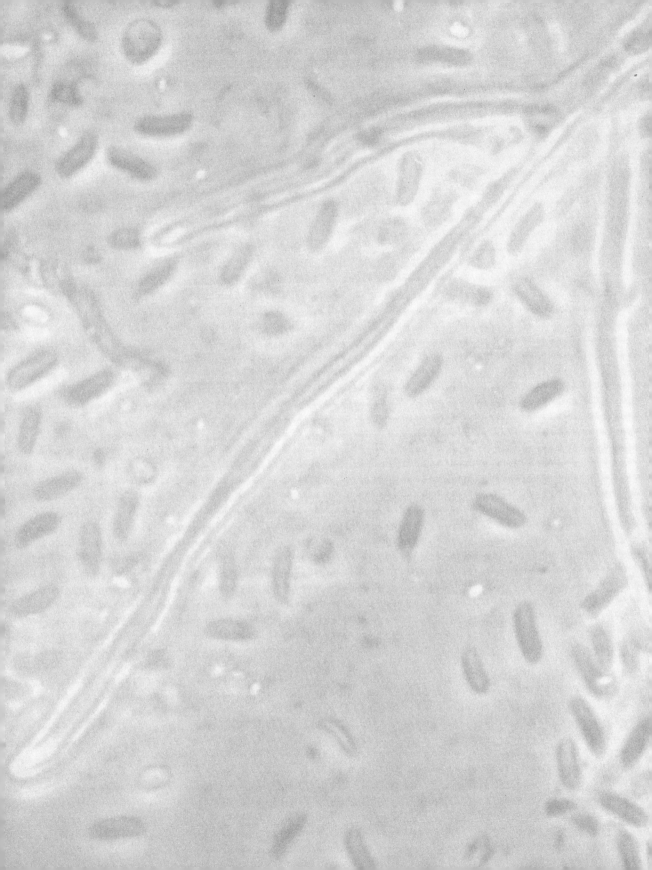

Chapter 26

Microbial Interactions— The Community Reflects the Habitat

Overview

After you have studied this chapter, you should be able to:

■ Describe seven critical microbial interactions: neutralism, amensalism, commensalism, mutualism, competition, parasitism, and predation.

■ Define synergistic, protocooperative, or symbiotic mutualistic interactions.

■ Discuss why parasitism and predation often involve soil microorganisms preying on larger organisms such as nematodes.

■ Explain why microbial succession is a common ecological feature in soil.

INTRODUCTION

Martinus Beijerinck once said, "Everything is everywhere; the environment selects." A good corollary to this is a saying attributed to Pasteur, "The community reflects the habitat." These are excellent rules of microbial ecology. Microorganism–microorganism interactions are at the heart of microbial ecology. Because of the interactions among microorganisms, the introduction of an alien organism into soil rarely leads to that alien's establishment. Microbial interactions and their effect on individual microorganisms are summarized in Table 26-1.

In this chapter we look at several microbial interactions and the mechanisms that underlie them.

NEUTRALISM

In neutralism, two microorganisms behave entirely independent of one another. There is no effect of one organism on the other—no direct interaction. Neutralism can be caused by physical, spatial, or temporal separation. Microorganisms may not interact because they are kept apart by physical barriers such as clay particles. They may not interact because they grow in different locations in the soil profile, or because one is an early colonizer of decaying vegetation while the other thrives at a later date. Neutralism is more likely at low density than at high density.

COMMENSALISM

In commensalism, only one microorganism actually benefits from the interaction; the other is unaffected. Commensal relationships are quite common, but are not obligatory. Some commensal relationships involve modifying the environment. Aerobes can completely consume O_2, which allows anaerobes to proliferate. Algae can photosynthetically produce O_2 for obligate aerobes. Yeast and other fungi growing on sugar can

Table 26-1 Types of microbial interactions in soil.

Interaction	Microorganism A	Microorganism B
Neutralism	No effect	No effect
Commensalism	+	No effect
Amensalism	No effect	−
Mutualism	+	+
Synergism		
Protocooperation		
Symbiosis		
Competition	−	−
Parasitism	+	−
Predation	+	−

+ = Positive effect
− = Negative effect

reduce the osmotic potential for more sensitive microorganisms. Microorganisms can dissolve minerals and release nutrients. Dissolving minerals in rocks make small pits that trap water, accumulate C, and act as sites for microbial growth. Acid production can help acidophiles grow better.

Microorganisms can excrete compounds that help other organisms grow. They can produce polysaccharides that increase adhesion, and can produce excess growth factors that are excreted into the environment. A microorganism that excretes excess amino acids and vitamins benefits auxotrophic microorganisms nearby. For example, fastidious microorganisms benefit from biotin release by neighboring microorganisms. Microorganisms can provide usable substrates by degrading polymers and by transforming insoluble compounds to soluble compounds. There can be chains of commensal relations during residue decomposition. Polymers can be decomposed to organic acids by heterotrophs. Organic acids can be decomposed to CH_4 by methanogens. Methane can be oxidized to methanol by methylotrophs. Methanol can be oxidized to CO_2 by microorganisms that grow on single C compounds.

Detoxification is a commensal activity. Hydrogen sulfide's toxicity vanishes when it is oxidized. When Mn is oxidized it becomes relatively insoluble, consequently, it becomes relatively less toxic. The toxicity of the heavy metal Cr (VI) can be lessened by reduction to Cr (III). Heavy metals can also be detoxified by transforming them into volatile forms that rapidly leave the environment. Antibiotics lose their biocidal activity when they are degraded by microorganisms.

One organism can supply a favorable habitat for another. Bacteria grow on algae, for example. One organism can also inadvertently help another organism disperse, for example, motile bacteria can carry viruses. Fungal hyphae can provide a path along which bacteria grow. Methane or O_2 gas bubbles produced by methanogens and plants, respectively, trap aquatic microorganisms at the gas–water interface and disperse the microorganisms as aerosols when the gas bubbles reach the atmosphere.

AMENSALISM

In amensalism, one species is suppressed while a second is not affected. Amensalism is the opposite of commensalism. Modifying an environment to inhibit another organism's growth, for example, is ammensal. A microorganism can have ammensal relationships with one group and commensal relationships with another. Antibiotic production is one of the best and most obvious examples of amensalism.

Despite the antagonistic relations we see between microorganisms growing on agar media (Figure 26-1), there are some compelling reasons why antibiotic production may not be an important factor in soil. First, there is no evidence that being able to produce antibiotics favors a microorganism's survival. Antibiotic producers are not more prevalent than nonantibiotic producers in soil. Second, there is no relationship between the predominant species in soil and their sensitivity to antibiotics. Third, even though alien organisms rapidly disappear from soil, their disappearance is not associated with antibiotic buildup. Fourth, antibiotic compounds are not routinely extracted from soil. This is not surprising since there are many mechanisms that inactivate antibiotics in soil; for example, degradation and adsorption to humus.

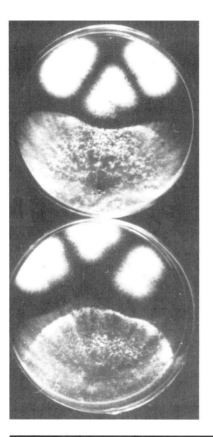

Figure 26-1   Antagonism of fungal growth by microbial colonies. (Photograph courtesy of the Soil Science Society of America)

MUTUALISM: SYMBIOSIS, SYNERGISM, AND PROTOCOOPERATION

Mutualism implies a relationship between microorganisms that is mutually beneficial. The relationship may or may not be obligatory. Symbiosis implies only a stable relationship between two organisms that is not necessarily beneficial, as when a legume is infected by an ineffective *Rhizobium*. However, symbioses are almost always used in the sense of beneficial relationships.

Some examples of symbioses that we discuss in detail are the legume–*Rhizobium* and actinorhizal symbioses and mycorrhizal symbioses. The former are examples of bacterial symbioses with higher plants and the latter is an example of fungal symbioses with higher plants. Both play crucial roles in crop productivity and soil fertility. Termites and protozoa also form symbioses, as do protozoa (ciliates) and green algae. Microorganisms convert cellulose to usable products in the termite gut. These microorganisms are not found free, so for them, the relationship is obligatory. *Paramecium* hosts *Chlorella* in its cytoplasm. Ciliates and bacteria also form symbioses, although it has not yet been shown to be beneficial to ciliates.

Symbioses—Lichens

The symbiosis between an algae and a fungi is called a lichen. Lichens are not particularly important in soil, but they are essential to soil formation. Lichens begin the biochemical weathering of rocks in inhospitable environments that eventually leads to soil formation. It is hard to escape noticing lichens since they grow on rocks, trees, and other surfaces (Figure 26-2a,b).

a

b

Figure 26-2   Lichens growing on a stone wall (a) and a tree (b). (Photographs courtesy of M. S. Coyne)

The fungi provide a habitat for the algae and, through respiration, slowly dissolve the underlying material on which they sit to provide inorganic nutrients. The algae provide carbohydrates via photosynthesis. However, most lichens don't appear green because the chlorophyll is masked by other pigments. Fungi in symbiotic unions with cyanobacteria have the added advantage that the cyanobacteria are also able to fix N_2.

There are three types of lichens. Crustose lichens are small and inconspicuous except for some coloration. They are frequently observed on rocks. Foliose lichens are quite conspicuous and form large, ruffled mats. Fruticose lichens are stalked or branching, so they seem to have a bushy appearance. Lichens grow slowly, perhaps a few millimeters per year. This is largely because they are one of the few organisms able to colonize such inhospitable surfaces as rock. Few other microorganisms can tolerate the extremes of hot and cold, desiccation, and exposure to ultraviolet light. One reason for the lichen's slow growth is that the carbohydrates that algae produce, instead of being used for growth, are used to maintain the lichen's water potential at a point low enough to survive its exposure to the atmosphere.

Synergism

Synergistic relationships are not obligatory. These are associations of mutual benefit to both species, but their cooperation is not obligatory for their existence or for their performance of some reaction. Herbicides are often degraded synergistically; one microorganism may degrade one part of a herbicide, while another microorganism may degrade another part. Metabolism and decomposition of lignin can also be considered a synergistic process.

Protocooperation

Microorganisms can cooperate with one another (protocooperation). The fact that microbial populations form colonies is probably evidence that they have made adaptations based on cooperative interactions. Why else would motile bacteria stay together? Two *Escherichia coli* colonies merging together eventually begin enzyme induction at the same time. Extracellular enzyme production also results in better substrate use in colonial populations compared to individual cells.

Cooperation occurs within a population as part of a growth cycle, as in the case of the slime mold *Dictylostelium,* and in the swarming behavior of the bacterium *Proteus mirabilis.* When food sources become limited, the amoeboid cells of *Dictylostelium* swarm together to make a single, multicelled mass. The cells unite to form a fruiting body, which is followed by sporulation and dispersal of the *Dictylostelium.* Some cells may find new food sources and growth can begin again. However, to form the fruiting structure, cooperation is necessary since some cells form the base, some the stalk, some the spores, and some form a food source until dispersion is complete. Most of the *Dictylostelium* cells that swarm never survive, but the multicelled organism that they form is able to reproduce itself, so the population as a whole persists.

COMPETITION

Competition is a condition in which two organisms suppress each other's growth as they struggle for limited nutrients. Inadequate C supplies are the likely basis for most competition in soil. Competition tends to bring about ecological separation of closely related populations. Darwin's finches are a good example; from one ancestral finch, many species of finches evolved to take advantage of different food niches on the Galápagos Islands. The trademark of competition is competitive exclusion—the principle of "first come, first served" (the microbial equivalent of musical chairs). Competitive exclusion precludes two populations from occupying the same ecological niche. If two organisms try to occupy the same niche, only one will win the competition.

PREDATION AND PARASITISM

Each major group in the soil community has parasites living on or in its cells. Bacteria have bacteriophages, for example, while fungi and bacteria both parasitize themselves. Predation is the attack of one organism on another. Bacteria are preyed upon by protozoa, viruses, slime molds, and other bacteria such as myxobacteria.

Nematodes are preyed upon by nematode-trapping fungi and other soil animals. Nematode-trapping fungi make specialized organelles to trap and infect nematodes. These structures are rings—constricting and nonconstricting (Figure 26-3a,b), sticky appendages, and spores that attach to the nematodes. Constricting rings are induced by touch or by raising temperatures to 50°C. Touching any one of the cells in the ring causes irreversible swelling. The touched cell grows three times its original size, followed by swelling of the other cells. The immobilized nematodes are then penetrated by hyphae and the fungus sporulates from the nematode.

The best known nematode-trapping fungi is *Arthrobotrys oligospora.* Other species of nematode-trapping fungi are *Arthrobotrys conoides, Dactylaria candida,* and *Meria coniospora.* Nematode-trapping fungi can be enriched by adding nematodes to soil. Saprophytic growth of the fungi on laboratory media does not induce growth of the trapping features, although nutrient deficiency induces trap formation. Nematodes also induce formation of traps.

SUCCESSION

Succession of microbial communities is a common event. It occurs in the degradation of complex polymers. It occurs in compost piles and in the fermentation of silage. Sauerkraut is the product of cabbage fermentation under controlled environmental conditions. It requires the exclusion of O_2, high salt concentration, and mesophilic temperatures. Table 26-2 illustrates the changing population of lactic acid bacteria in forage sorghum used to make silage. Good silage is acidic (pH 3.7), has a high lactic acid content (72 g kg^{-1}), and has a low butyric acid content (0.12 g kg^{-1}). At the start of fermentation, *Leuconostoc* species dominate, while at the end of incubation, *Lactobacillus* species predominate.

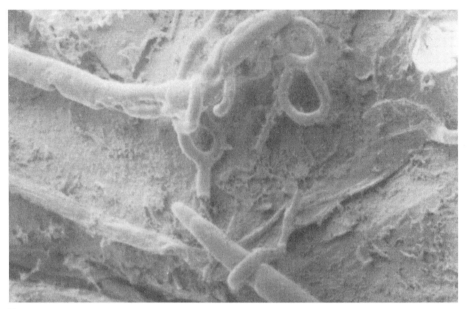

a

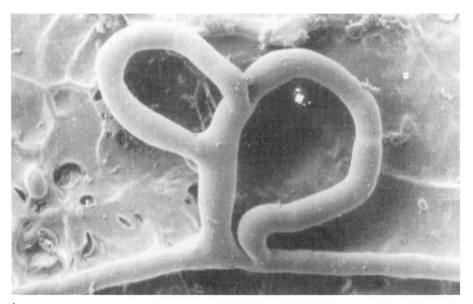

b

Figure 26-3 Nonconstricting rings formed by nematode-trapping fungi. (a) Several nematodes trapped by rings; (b) a close view of one of the ring structures. (Photographs courtesy of G. Koening)

Table 26-2   Change in the lactic acid bacteria populations with time in fermenting forage sorghum.

| | Incubation Time (days) | | | |
Bacteria	0	4	8	100
	Proportion of Lactic Acid Bacteria Sampled (%)			
Lactobacillus planetarum	35	84	87	44
Leuconostoc species	59	0	4	0
Lactobacillus fermentum	6	6	0	7
Lactobacillus brevis	0	10	9	49

(Adapted from Tjandraatmadja et al. 1991)

The Poor Man's Guide to Making Sauerkraut

To make sauerkraut, you need a big container, salt, a knife, a wooden dowel, plastic wrap, a rubber band, water, and one head of cabbage. Rinse the cabbage to wash off any visible debris, remove the outer leaves, and use your knife to section the cabbage and cut out the inner core. Next, chop the cabbage leaves finely and liberally mix them with salt. Put the cut cabbage into the large container in layers, vigorously tamping down each layer with the wooden dowel. Continue until the container is filled to within an inch or two of the rim.

Put a layer of plastic wrap over the top of the container and push it down so that it lies on top of the cabbage and lines the inside of the container. Secure the excess plastic at the top of the container with a rubber band. Pour water onto the plastic lining the container. The pressure of the water against the plastic makes a reasonably airtight seal. Put your container under the sink and let it ferment for several weeks. Periodically add more water to the plastic liner to maintain the airtight seal. When the sauerkraut is appropriately tart, eat it.

Each component of this process has a purpose. The cabbage is both the substrate and the inoculum for fermentation. All of the necessary microorganisms already exist on the leaf surfaces. Cutting the cabbage increases the surface area for bacteria to ferment and initiates the loss of cell constituents. Salt helps this process by increasing the permeability of the plant tissue. Tamping with the dowel packs the plant material together, further damaging tissue, and helps to minimize air pockets. The plastic wrap and water make an air barrier so that the fermenting cabbage remains in anaerobic conditions. Everything else is left up to the microorganisms.

During sauerkraut production, a succession of microbial communities occurs. Initially, the dominant bacterium is *Enterobacter cloacae,* which produces CO_2 and volatile acids. Next comes *Leuconostoc mesenteroides,* which outgrows other organisms and produces up to 1% lactic acid. The volatile acids it forms prevent yeast growth. *Lactobacillus plantarium* develops next and continues lactic acid production up to 2%. If the temperature is too hot, *Leuconostoc* is inhibited, and lactic acid production is inhibited, which allows undesirable fungi to grow. If it is too cold, *Flavobacterium* and *Enterobacter* dominate and produce poor-quality sauerkraut.

Summary

Microorganism–microorganism interactions are at the heart of microbial ecology. There are seven critical microbial interactions: neutralism, amensalism, commensalism, mutualism, competition, parasitism, and predation. In neutralism, two microorganisms behave independently of one another. Neutralism is caused by physical, spatial, or temporal separation. In commensalism, only one microorganism actually benefits from the interaction; the other is unaffected. Commensal relationships are quite common, but are not obligatory. In amensalism, one species is suppressed while a second is not affected.

Mutualism implies a relationship between microorganisms that is mutually beneficial. The relationship may or may not be obligatory. Mutualism may be synergistic, protocooperative, or symbiotic. Symbiosis implies a stable relationship between two organisms that is generally beneficial. There are symbioses between microorganisms, symbioses between microorganisms and plants, symbioses between microorganisms and protozoa, symbioses between microorganisms and insects, and symbioses between protozoa and insects. The symbiosis between an algae and a fungi is called a lichen. Lichens are essential to soil formation in inhospitable environments.

Competition is a condition in which two organisms suppress each other's growth. Inadequate C supplies are the likely basis for most competition. Competition tends to bring about ecological separation of closely related populations. Each major group in the soil community has parasites living on or in its cells. Predation is the attack of one organism on another.

Sample Questions

1. What is synergism? Give an example.
2. Why is microbial succession the norm, rather than the exception, in soil?
3. What are arguments against antibiotics playing an important role in soil antagonism?
4. In what ways is predation functionally important in soil?
5. Give an example of a potentially important predator in soil. What is its prey? In what ways will soil physical characteristics limit this interaction?
6. Give specific examples of the following ecological relationships: amensalism, commensalism, mutualism, antagonism, competition, symbiosis, neutralism, and parasitism.
7. What is predation?
8. What are the physiological characteristics of lichens that distinguish them from other microorganisms?
9. Why is a lichen commonly used as an example of a mutualistic association? In what kinds of environments do you expect lichens to be important in nutrient dynamics?
10. Draw a figure that illustrates changing populations in a classic predator–prey relationship.

Thought Question

Nathan Bedford Forrest, one of the Confederacy's most famous generals during the U. S. Civil War, was attributed with the saying, "Get there the firstest with the mostest." Why is this motto an apt description of microbial competition in the soil environment?

Additional Reading

There is a wealth of reading material to expand your knowledge of microbial ecology. Here are some textbooks that I have found useful: *Soil Ecology* by Ken Killham (1994, Cambridge University Press, Cambridge, England) examines ecology from a soil perspective. *Microbial Ecology: Habitat, Organisms, Habitats, Activities* by Heinz Stolp (1988, Cambridge University Press, Cambridge, England) looks at microbial ecology by focusing on microbial diversity in many environments. One environment is the microbial mat that forms between the interface of soil and water. For an in-depth look at the ecology of this environment, you should read the articles in *Microbial Mats: Physiological Ecology of Benthic Microbial Communities* (1989, Y. Cohen and E. Rosenberg, eds., American Society for Microbiology, Washington, DC). An excellent text in microbial ecology for beginning students is *Microbial Ecology: Fundamentals and Applications, 3d ed.* by R. M. Atlas and R. Bartha (1993, Benjamin/Cummings, Redwood City, CA).

References

Tjandraatmadja, M., B. W. Norton, and I. C. MacRae. 1991. Fermentation patterns of forage sorghum ensiled under different environmental conditions. *World Journal of Microbiology and Biotechnology* 7:206–18.

Chapter

27

Symbiotic Nitrogen Fixation

Overview

After you have studied this chapter, you should be able to:

- Describe several symbioses between bacteria and higher organisms that produce fixed N.
- Define what actinorhizal nodules are, and tell where they are found and what causes them.
- Identify the species of nodule-forming rhizobia and explain what a cross-inoculation group is.
- Describe how nodules form.
- Explain how N_2 fixation and plant growth are linked.

INTRODUCTION

In nitrogen fixation, atmospheric N (N_2) is converted into organic N by the enzyme complex nitrogenase. We discussed the mechanism in Chapter 22 when we examined free-living, N_2-fixing microorganisms. In this chapter, we examine N_2 fixation in associations of bacteria and higher plants. No topic in soil microbiology has been studied more than the symbiotic association between higher plants and bacteria that produces nodules—the specialized plant structures that specifically foster N_2 fixation.

Nitrogen fixation occurs in legume and nonlegume nodules. The symbiosis between *Rhizobium* and Parasponia (*Trema*), a nonleguminous shrub, causes nodules to form (Davey et al. 1973), but this seems to be an isolated case. *Nostoc* (a cyanobacteria) forms stem nodules on the herbaceous dicot *Gunnera*. However, the most important nodule-forming associations are the actinorhizal nodules that play an important role in forest fertility and the rhizobial nodules that form on legumes in cropland. These two associations are the focus of this chapter.

ACTINORHIZAL SYMBIOSES

Actinomycetes of the genus *Frankia* form N_2-fixing nodules on trees and shrubs. They are very important in the nitrogen enrichment of inhospitable environments such as mine spoils and reclaimed land and can contribute from 2 to 300 kg N ha^{-1} yr^{-1} to these environments (Myrold 1994). *Frankia* are promiscuous and form nodules on at least 25 genera of plants. These plants are typically woody shrubs and trees found in temperate climates, although actinorhizal associations occur from the Arctic to the tropics and from semidesert to rain forest ecosystems. Based on plant-trapping studies, *Frankia* populations can range from 0 to 4,600 infectious units per gram of soil. Their populations tend to be higher in slightly alkaline soils and in the rhizosphere of some plants such as birch, although these plants may never form nodules.

Typical plants that *Frankia* nodulate are pioneering species such as alder (*Alnus*) that grow in moist environments. The alder forms a large, branched actinorhizal nodule. In California, *Ceanothus* (deer brush) is a N_2-fixing shrub in chaparral ecosystems. *Ceanothus* also occurs in grasslands and forests, and is used as an ornamental plant. Actinorhizal nodules that form on *Ceanothus* are also branched, but are smaller (perhaps as large as your thumb) than alder nodules. *Myrica* grows in boglike conditions and also on landslides, eroded slopes, and mined areas. Like all actinorhizal associations, it forms perennial nodules (Figure 27-1). In wet areas, a normal root escapes from the apex of the *Myrica* nodule and grows upward (we call this phenomenon "negative geotropism"). These so-called "nodule roots" have aerenchyma (air-filled plant cells) that facilitate gas exchange under wet conditions and supply the nodules with air.

When an old nodule lobe stops growing, a new lobe develops from its apex, and the *Frankia* grow into the new lobe's cortical tissue. Figure 27-2 shows an example of *Frankia* growing in the cortical tissue of plant roots. This photograph shows infected root tissue of *Casaurina,* another plant that forms actinorhizal associations. The

Figure 27-1 Nodule macrophotograph of *Myrica cerifera* showing the branched (dichotomous) nature of the perennial nodule. A new nodule lobe branches from near the current lobe's apex when the lobe ceases growth. The *Frankia* grow into cortical tissue of the new lobe. Note the "nodule roots." (Photograph courtesy of R. Howard Berg)

Frankia grow toward the apex of actinorhizal nodules both inter- (within) cellularly and intra- (between) cellularly.

Nitrogen-fixing actinorhizal nodules aren't studied as much as other N_2-fixing associations because, until recently, it was impossible to grow the infecting *Frankia* apart from their host. Successful isolation of the first *Frankia* strain took more than 100 years (Myrold 1994). It has also proved difficult to get *Frankia* isolates to reinfect their hosts. In addition, growing the host is time-consuming, and it is more difficult to work with perennial trees and shrubs than with legumes. However, it appears that three distinct host groups can be identified based on infection patterns: strains that form nodules on *Alnus*, strains that form nodules on *Elaeagnus*, and strains that form nodules on *Casaurina*. These nodulation groups are not mutually exclusive.

Actinorhizal nodules are pink in color when cut open, but this is caused by anthocyanins, not by leghemoglobin as occurs in legumes. Nitrogen fixation occurs in symbiotic vesicles. The vesicles probably maintain the anaerobic conditions necessary for nitrogenase because the multilayered vesicle walls create an O_2 diffusion barrier (Silvester et al. 1990). Vesicles also form in N_2-fixing *Frankia* growing apart from the plant

Figure 27-2   Longitudinal section of the nodule cortex of an infected *Casaurina* species showing a typical gradient of infected cells. The *Frankia* grow inter- and intracellularly on their way to the nodule apex, invading parenchymal cells of the cortex in these modified lateral roots. Tannins are the dark objects in the uninfected cells. (Photograph courtesy of R. Howard Berg)

(Figure 27-3). The vesicle shape is host-plant-dependent. In *Myrica,* the symbiotic vesicles are club-shaped, but the same *Frankia* strain forms spherical vesicles in *Alnus.*

Virtually all actinorhizal nodules have hydrogenase activity, which means they recover some of the energy lost during H_2 formation by nitrogenase. Some *Frankia* strains are infective (they can form nodules) but are ineffective (they can't fix N). The nodules, in this case, are smaller than normal, have relatively few *Frankia,* and lack symbiotic vesicles. The capacity for N_2 fixation varies widely among Frankia–host combinations, but so far, no clearly superior *Frankia* strains have been

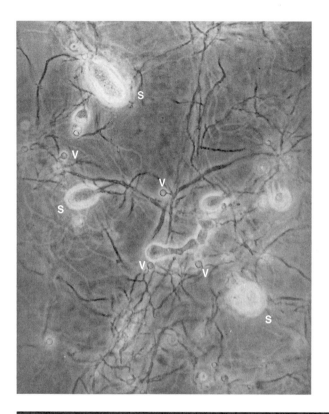

Figure 27-3 Phase-light micrograph of a N_2-fixing culture of *Frankia* (magnification 750×). There are vegetative hyphae with spherical swellings at the terminus of short branches; these are *Frankia* "vesicles" (V), the location of the O_2-sensitive nitrogenase complex. Oxygen protection is proposed to be a diffusion barrier in the vesicle wall; this also makes the vesicles phase bright. There are also large sporangia (S) formed in culture (and in senescent plant cells of the nodule). (Photograph courtesy of R. Howard Berg)

identified. *Frankia* nodules typically export citrulline to the plant xylem once fixation has taken place.

$$\underset{\text{Citrulline}}{\overset{\displaystyle \text{NH}_2 \qquad\qquad\qquad \text{NH}_2}{\text{CO—NH—(CH}_2)_3\text{—CH—COOH}}}$$

LEGUME-*RHIZOBIUM* SYMBIOSIS

The best-studied and most significant source of biological N_2 fixation in agricultural ecosystems comes from the symbiosis of bacteria, collectively known as rhizobia, and higher plants such as clover (Figure 27-4) or soybeans (Figure 27-5). Rhizobia are

Figure 27-4   Clover (*Trifolium*) a N$_2$-fixing forage legume. (Photograph courtesy of M. S. Coyne)

Figure 27-5   Soybean (*Glycine max*), a N$_2$-fixing grain legume (pulse). (Photograph courtesy of M. S. Coyne)

Table 27-1   The legume nodule symbionts.

Genus	Species	Biovar	Representative Host
Azorhizobium			
	A. caulinodans		Sesbania
Bradyrhizobium			
	B. elkanii		Soybean
	B. japonicum		Soybean
Rhizobium			
	R. leguminosarum		
		leguminosarum	Beans
		phaseoli	Beans
		trifolii	Clover
		viciae	Vetch
	R. etli		Beans
	R. fredii		Soybean
	R. galegae		
	R. haukuii		Astragalus
	R. loti		Trefoil
	R. meliloti		Alfalfa
	R. tropici		Beans
			Leucaena

(Weaver and Graham 1994)

free-living, gram-negative, aerobic and facultatively anaerobic, motile, chemo-heterotrophic bacteria. All rhizobia once were grouped in the genus *Rhizobium,* but new taxonomic methods were used to reclassify rhizobia in the 1980s. Taxonomists now recognize 3 genera and 11 species of rhizobia (Table 27-1).

Rhizobia persist in soil as saprophytic heterotrophs when they are not infecting their hosts. Depending on the season, crop history, and management practice, there may be 10 to 10^6 rhizobia per gram of soil. If legumes have ever been grown at a site, the higher value is more appropriate. If not, proper infection often requires inoculation (Figure 27-6).

There are 7,000 genera and 14,000 species of legumes, and perhaps 100 agriculturally important legumes are used, but not all legumes have been checked for their ability to fix N. This is particularly true of those legumes in tropical forests. Nitrogen fixation rates vary enormously. Up to 600 kg N fixed ha^{-1} yr^{-1} have been reported in forage legumes. Most rates are much less (Table 27-2). In general, grain legumes (pulses) fix less N than do forage legumes because they are grown for a shorter period and have fewer roots on which nodules can form. Nitrogen fixation supplies some-to-all of a plant's N needs.

Figure 27-6 The effect of inoculation on soybean (*Glycine max*). In the absence of inoculation by *Bradyrhizobium* strain 123, the soybeans are stunted and chlorotic. If a minimum of 14 *Bradyrhizobium* per seed are supplied, the plants are able to fix N, but improved growth doesn't occur until higher inoculation levels are used. (Photograph courtesy of the Soil Science Society of America)

How Do You Enumerate Rhizobia in Soil?

It's relatively easy to isolate rhizobia from nodules. After you dig up the plant and pick off some large nodules (the larger the better), you simply surface sterilize the nodule surface in a solution of commercial bleach (5.25% sodium hypochlorite), crush it, mix the contents with sterile water, and streak a sample on a suitable media such as yeast mannitol agar (Weaver and Graham 1994).

Isolating and enumerating rhizobia in soil is a problem. In soils where legumes have never grown, the native rhizobia population may be extremely low. Even where legumes have grown and plants are clearly nodulated, enumerating rhizobia by trying to grow them on agar media is complicated because they are outcompeted by other soil organisms—even when selective media are used. Most enumeration is done by the plant trap technique. The soil sample containing rhizobia is serially diluted, and aliquots of each dilution are dispensed into test tubes, pouches, or inert potting mixes (such as vermiculite) that contain germinated legumes or surface-sterilized seeds. After 2 to 3 weeks, nodules should begin to appear on the legume roots. It takes longer for nodules to appear on larger seeds.

The major drawback to this procedure is that the legume host you select will obviously determine the species of rhizobia you can enumerate. Consequently, you will underestimate total rhizobia with this technique.

Table 27-2 Estimates of N fixed by legumes.

Plant	Estimated N Fixed (kg ha^{-1} y^{-1})
Alfalfa (*Medicago*)	125–335
Clovers (*Trifolium*)	85–190
Lentil (*Lens*)	80–150
Pea (*Pisum*)	
Vetch (*Vicia*)	
Beans (*Phaseolus*)	
Lupines (*Lupinus*)	
Soybean (*Glycine*)	65–115
Cowpeas (*Vigna*)	65–130
Peanuts (*Arachis*)	
Lima beans (*Phaeolus*)	

(Data from Havelka et al. 1982)

CROSS-INOCULATION GROUPS

The classical means of characterizing rhizobia–legume associations has been the cross-inoculation group (i.e., defining the species of rhizobia by the type of legume it infects). This is a functional, rather than an ideal, system. It works because rhizobia show some specificity for the legume they infect. Table 27-3 presents the traditional organization of the cross-inoculation groups. Other legumes that are infected by rhizobia are birdsfoot trefoil (*Lotus corniculatus*), black locust (*Robina pseudoacacia*), and sesbania (*Sesbania rostrata*).

INFECTION AND NODULE FORMATION

Most nodules form through the infection of root hairs, although peanut nodules do not and *Sesbania rostrata* forms nodules on its stem. The N$_2$-fixing bacteria associated with stem nodulation are in the genus *Azorhizobium*. These exceptions to the general rule are important, but the mechanisms behind these infection processes are less well-studied than the root nodules that form via root hairs.

The first step in nodule formation is for the rhizobia in soil to recognize that a suitable host is present. The plant releases specific compounds—flavones—that attract, stimulate, or signal rhizobia. These flavones are specific for specific rhizobia strains. The next step is rhizobia attachment to the root hairs (Figure 27-7). The initial binding to root hairs is random and reversible (Dazzo et al. 1984). Subsequent binding is polar and irreversible. Rhizobia then multiply in the rhizosphere. During this period, the root hair curls and forms a structure called the shepherd's crook. As the shepherd's crook forms, the rhizobia begin the process of invading the plant root cell, which is apparent because an infection thread forms. The infection thread is a hollow, cellulose-lined tube in which rhizobia multiply. Usually only one type or strain of rhizobia is in an infection thread.

Table 27-3 Cross-inoculation groups.

Rhizobium–Bradyrhizobium Species	Plant Host
Fast Growing (Acid Producing)	
Rhizobium leguminosarum b.v. *leguminosarum*	*Lens* (lentil)
	Pisum (pea)
	Vicia (vetch)
Rhizobium leguminosarum b.v. *phaseoli*	*Phaseolus* (bean)
Rhizobium leguminosarum b.v. *trifolii*	*Trifolium* (clover)
Rhizobium meliloti	*Medicago* (alfalfa)
	Melilotus (sweetclover)
	Trigonella (fenugreek)
Slow Growing (Base Producing)	
Bradyrhizobium species	*Lupinus* (lupines)
	Cowpea
	Others
Bradyrhizobium japonicum	*Glycine* (soybean)
	Some cowpea
Rhizobia species ("Cowpea Miscellany")	*Vigna* (cowpea)
	Phaseolus (lima bean)
	Arachis (peanut)
	Pueraria (kudzu)
	Prosopis (mesquite)

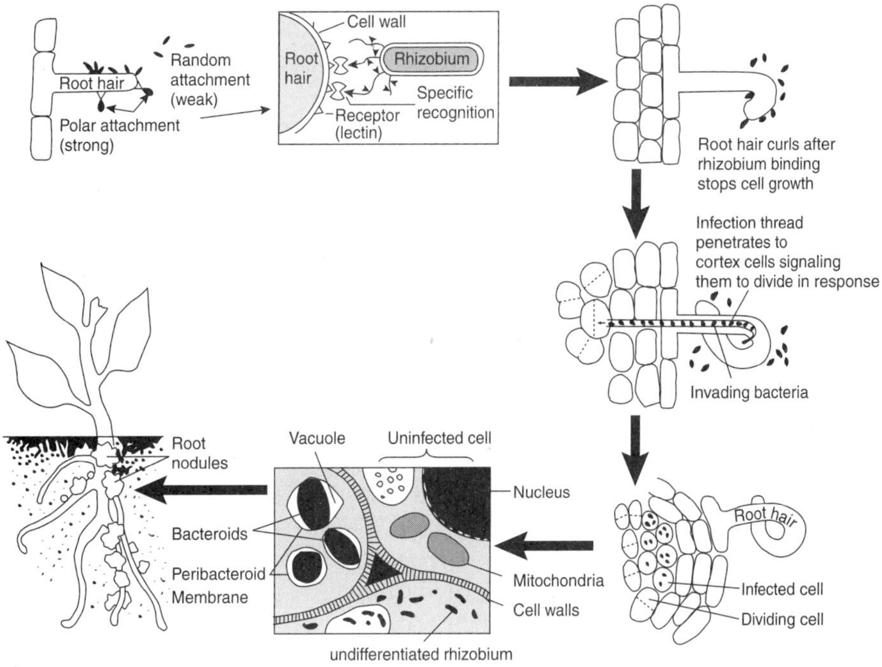

Figure 27-7 Steps in the formation of root nodules by rhizobia. The sequence reflects the lectin theory of recognition. (Adapted from Ahmadjian and Paracer, 1986)

A Short Biography of Fritz Haber and Carl Bosch

The Haber-Bosch process is the fundamental industrial procedure for making inorganic N fertilizer. It is named after Fritz Haber (1868–1934) and Carl Bosch (1874–1940). Haber's contribution was discovering that the reaction

$$N_2 + 3H_2 \Leftrightarrow 2NH_3$$

was thermodynamically favorable and could produce measurable NH_3 at a temperature of 450°C and a pressure of 200 atmospheres. Bosch's contribution was to translate the high temperature and pressure laboratory process to an industrial scale.

Fritz Haber was born in Breslau, Germany. His father was a chemical and dye merchant. After Haber received his university training in chemical technology, he briefly worked in his father's business before he and his father mutually agreed that his talents lay in academic research. First at the University of Jena, then at the Technical School of Karlsruhe (1894), and finally in Berlin at the Kaiser Wilhelm Institute (1911), Haber began a long and distinguished career in physical chemistry. His work was highlighted by receipt of the Nobel prize in 1918 for his research on the synthesis of NH_3. This achievement was marred by Haber's role in developing chlorine gas for chemical warfare during World War I.

Carl Bosch had wide-ranging interests including astronomy and astrophysics, and he was an amateur naturalist. He was born in Cologne, Germany and studied metallurgy and mechanical engineering at the Technical School of Charlottenburg before receiving a doctorate from the University of Leipzig. In 1899, Bosch became a chemist for Badische Anilin-und Soda-Fabrik (BASF), where he constructed plants for the indigo dye industry and, ultimately, for commercial nitrogen fixation. The latter work earned him a Nobel prize in chemistry in 1931.

Less than 5% of infected root hairs go on to form nodules. For nodules to form, the plant must also play a role. Even before rhizobia reach the cortex, the cortical cells divide. Rhizobia are released into the cortical cells and surrounded by a plant-produced membrane called the peribacteroid membrane. Within the peribacteroid membrane, rhizobia change shape to form cells called bacteroids—pleomorphic-shaped rhizobia. Up to 10,000 bacteroids are found per root cell and all are compartmentalized within peribacteroid membranes.

The nodule morphology is characteristic of the plant host. Clover nodules are club-shaped (Figure 27-8). Alfalfa and pigeon pea nodules are branched (Figure 27-9). Soybean nodules are spherical (Figure 27-10).

Figure 27-8   Nodules on a clover root. Note the club shape. (Photograph courtesy of M. S. Coyne)

Nodule number and size vary. Some nodules can approach the size of a baseball. Most are smaller—less than 0.5 cm in diameter or length. Grain legumes (pulses) have fewer nodules than do forage legumes, but the nodules are larger. As a general rule, the more nodules a plant has, the smaller they are and the less N_2 they fix. Effective nodules (N_2-fixing nodules) are larger than noneffective (non-N_2-fixing) nodules. The simple way to tell which is which is to cut the nodule open. A red color indicates the

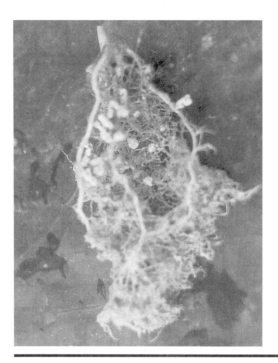

Figure 27-9   Nodules on a pigeon pea (*Cajanus cajan*) root. Note the branched shape. (Photograph courtesy of M. S. Coyne)

Figure 27-10   Nodules on a soybean root. Note the spherical shape. (Photograph courtesy of M. S. Coyne)

presence of leghemoglobin and also indicates an effective nodule (some nodules form dark pigments, so occasionally this simple method doesn't work). New nodules form throughout the growing season and old nodules slough off.

NODULE PHYSIOLOGY

The host plant supplies C and regulates O_2. Nitrogen fixation takes a lot of energy and it is often limited by the C supply. Theoretically, up to 22 moles of glucose are required for 1 mole of N_2 fixed, so a substantial amount of the plant's photosynthetic output— 7% to 12%—goes to sustaining its nodules (Paul and Kucey 1981). When grain filling starts, nodules die for want of C (Figure 27-11).

To refresh your memory, Figure 27-12 illustrates the components and the pathway of N_2 fixation by the enzyme complex nitrogenase in nodules. Waterlogging is detrimental to N_2 fixation in legumes. Although nitrogenase requires anaerobic or O_2-limited conditions, plants and rhizobia require O_2 for respiration. So, how does a plant juggle both of these requirements? Leghemoglobin accumulates between bacteroids and the peribacteroid membrane and gives cut nodules their red color. Leghemoglobin has a high affinity for O_2 and lowers the O_2 concentration in nodules sufficiently to allow respiration and N_2 fixation to occur simultaneously. Actually,

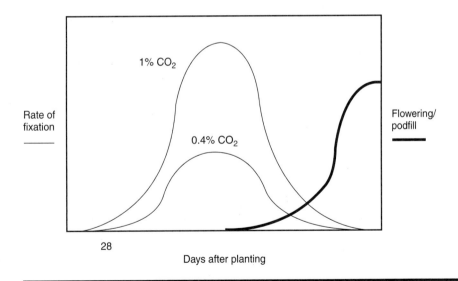

Figure 27-11   Change in the rates of N_2 fixation as a function of time and CO_2 concentration. When seed-filling begins, N_2-fixation declines because photosynthate goes to reproduction rather than to N_2 fixation. Elevating the level of CO_2 doesn't change this pattern, but the additional CO_2, which can be used to produce more photosynthate, permits greater fixation rates. (Adapted from Havelka et al. 1982)

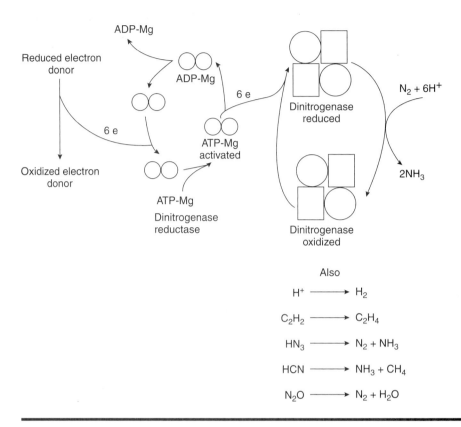

Figure 27-12   A schematic diagram of the interaction of dinitrogenase reductase and dinitrogenase in N_2 fixation.

leghemoglobin is too efficient, because N_2 fixation in environments containing more than 20% O_2 is greater than at atmospheric O_2 levels. This implies that the rhizobia are O_2 limited.

Rhizobia get C and shelter and supply N to plants in turn. The bacteroids excrete ammonia (NH_3) that is quickly fixed by glutamate dehydrogenase (GDH) or GS/GOGAT. The products leave the nodule via the xylem. The NH_3 that rhizobia produce is transformed to asparagine and glutamine in temperate legumes and allantoin and allantoic acid in tropical legumes (Figure 27-13).

What does it cost legumes to form nodules and fix N_2? There is the cost of forming nitrogenase and hydrogenase; the cost of assimilating NH_3; the cost of transporting the fixed N; and the cost to grow and maintain nodules. Overall, it costs the legume about 3 to 6 g of C for every g of N fixed. Consequently, when NO_3^- and organic N or NH_3 are present, nodule formation and nitrogenase activity are inhibited.

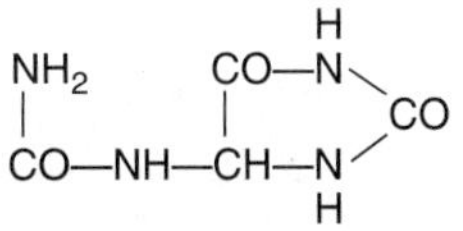

$$NH_2 \quad CO-N^{H} \quad CO$$
$$CO-NH-CH-N^{H}$$

Allantoin (C:N = 1)

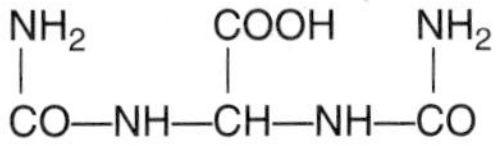

$$NH_2 \quad COOH \quad NH_2$$
$$CO-NH-CH-NH-CO$$

Allantoic acid (C:N = 1)

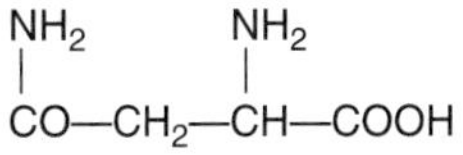

$$NH_2 \quad NH_2$$
$$CO-CH_2-CH-COOH$$

Asparagine (C:N = 2)

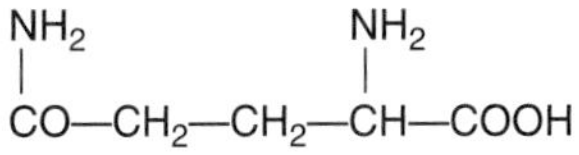

$$NH_2 \quad NH_2$$
$$CO-CH_2-CH_2-CH-COOH$$

Glutamine (C:N = 2.5)

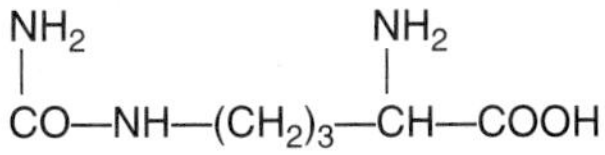

$$NH_2 \quad NH_2$$
$$CO-NH-(CH_2)_3-CH-COOH$$

Citrulline (C:N = 2)

$$NH_2 \quad CH_2 \quad NH_2$$
$$CO-C-CH_2-CH-COOH$$

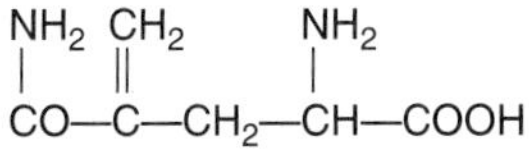

γ Methylene glutamine (C:N = 3)

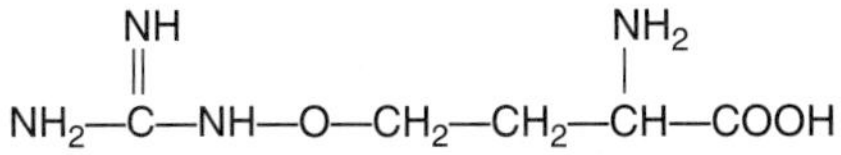

$$NH$$
$$NH_2-C-NH-O-CH_2-CH_2-CH-COOH$$
$$NH_2$$

Canavanine (C:N = 1.25)

Figure 27-13 N-rich compounds produced for export in N_2-fixing legumes.

Summary

No topic in soil microbiology has been studied more than the symbiotic association between higher plants and bacteria that leads to nodule formation. Nodules are specialized plant structures that specifically allow N_2 fixation. The most important nodule-forming associations are the actinorhizal nodules and the rhizobial nodules that form on legumes.

Actinomycetes of the genus *Frankia* form N_2-fixing nodules on trees and shrubs. Typical plants that *Frankia* nodulate are *Alnus, Myrica,* and *Ceanothus.* Three distinct host groups can be identified based on infection patterns: strains that form nodules on *Alnus,* strains that form nodules on *Elaeagnus,* and strains that form nodules on *Casaurina.* These nodulation groups are not mutually exclusive.

The best-studied and most significant source of biological N_2 fixation in agricultural ecosystems comes from the symbiosis of bacteria, collectively known as rhizobia, and higher plants. Taxonomists now recognize 3 genera and 11 species of rhizobia. The classical means of characterizing rhizobia–legume associations has been the cross-inoculation group (i.e., defining the species of rhizobia by the type of legume it infects).

Nodulation involves recognition, infection, and nodule formation. The nodule shape is characteristic of the plant species and the nodule number is characteristic of the fertility of the soil. Leghemoglobin is a compound jointly produced by the legume and the rhizobia that has a high affinity for O_2 and that lowers the O_2 concentration in nodules enough to allow respiration and N_2 fixation to occur simultaneously. Much of the plant's photosynthetic output (7% to 12%) goes to sustaining its nodules. When grain filling starts, nodules die for want of C and when NO_3^- and organic N or NH_3 are present, nodule formation and nitrogenase activity are inhibited.

Sample Questions

1. What is the distinction between *Rhizobium* and *Bradyrhizobium*?
2. How do *Frankia* differ from *Rhizobium?*
3. What is the basis for the cross-inoculation group?
4. What is the physiological role of leghemoglobin in the soybean root nodule?
5. Give an example of an actinorhizal association. Identify the participants. In what conditions or environment is your example important? Why is it important?
6. What are specific examples of free-living, actinorhizal, and rhizobial nitrogen-fixing organisms? What is the relative importance of each in agricultural systems?
7. Describe the infection process that leads to nodule formation in leguminous plants.
8. What is the pathway of electron flow during symbiotic N_2 fixation?
9. The relative efficiency of electron transfer to N_2 via nitrogenase may be defined in the following way by the Schubert-Evans equation. What is the basis for this equation?

$$\text{Relative efficiency} = 1 - \frac{\text{Rate of } H_2 \text{ evolution in air}}{\text{Rate of } C_2H_4 \text{ production}}$$

10. If the measured acetylene reduction rate is 40 nmoles C_2H_4 produced per hour and the relationship between moles N fixed and moles C_2H_2 reduced is 0.3, how many moles of N are fixed in 2 hours in this system?
11. Based on the data in the following table, which strain has the highest relative efficiency based on the Schubert-Evans equation?

Strain	Seed Yield (kg ha⁻¹)	N Content (%)	Bacteroid H₂ Uptake (mmol mg protein⁻¹ h⁻¹)	C₂H₄ Produced (mmol g nodule⁻¹ h⁻¹)
143	2,025	6.4	5.2	14.5
6	2,108	6.3	5.2	14.7
122	2,056	6.0	2.5	18.5
117	1,745	6.1	<0.05	13.3
135	1,976	5.84	<0.05	14.1
120	1,857	5.63	<0.05	10.2

12. The contribution of atmospheric N, soil N, and fertilizer N to soybean growth is illustrated below. What does this figure indicate about the soybean's N source as fertilizer N increases, and how can you account for this phenomenon?

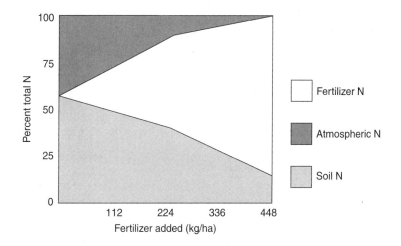

Thought Question

Would the world be a much different place today if Fritz Haber had remained a dye merchant and Carl Bosch had become an astrophysicist? Why?

Additional Reading

If you are interested in learning more about actinorhizal nodules, read *The Biology of Frankia and Actinorhizal Plants* by C. R. Schwintzer and J. D. Tjepkema (1990, Academic Press, San Diego, CA). Although it's becoming dated, a classic monograph on the rhizobia–legume symbiosis is *A Manual for the Practical Study of Root-nodule Bacteria* by J. M. Vincent (1970, Blackwell Scientific Publications, Oxford, England).

References

Ahmadjian, V., and S. Paracer. 1986. *Symbioses: An introduction to biological systems.* University Press of New England, Hannover.

Davey, M. R., E. C. Cocking, and Eileen Bush. 1973. Symbiosis between *Rhizobium* and non-legume, *Trema aspera. Nature* 244:459–61.

Dazzo, F. B., G. L. Truchet, J. E. Sherwood, E. M. Hrabak, M. Abe, and S. H. Pankratz. 1984. Specific phases of root hair attachment in the *Rhizobium trifolii*-clover symbiosis. *Applied and Environmental Microbiology* 48:1140–50.

Havelka, U. D., M. G. Boyle, and R. W. F. Hardy. 1982. Biological nitrogen fixation. In *Nitrogen in agricultural soils,* F. J. Stevenson (ed.), 365–422. Madison, WI: American Society of Agronomy.

Myrold, D. D. 1994. *Frankia* and the actinorhizal symbiosis. In *Methods of soil analysis, part 2: Microbiological and biochemical properties,* R. W. Weaver et al. (eds.), 291–328. Madison, WI: Soil Science Society of America.

Paul, E. A., and R. M. N. Kucey. 1981. Carbon flow in plant microbial associations. *Science* 213:473–74.

Silvester, W. B., S. L. Harris, and J. D. Tjepkema. 1990. Oxygen regulation and hemoglobin. In *The biology of* Frankia *and actinorhizal plants,* C. R. Schwintzer and J. D. Tjepkema (eds.), 157–76. San Diego, CA: Academic Press.

Weaver, R. W., and P. H. Graham. 1994. Legume nodule symbionts. In *Methods of soil analysis, part 2: Microbiological and biochemical properties,* R. W. Weaver et al. (eds.), 199–222. Madison, WI: Soil Science Society of America.

Chapter 28

Inoculation and Biocontrol

Overview

After you have studied this chapter, you should be able to:

- List the three key factors to successful inoculation.
- Discuss when rhizobia inoculation is used and explain how to maximize effective inoculation with rhizobia.
- Explain how to design successful tests to determine the need for inoculation and discuss the potential for success of inoculation.
- Define what "biocontrol" is and list the characteristics of a successful biocontrol program.
- Give several examples of biocontrol agents currently in use.

INTRODUCTION

Inoculation is the practice of applying microorganisms to agricultural or chemical systems to bring about desirable transformations such as enhanced N_2 fixation, increased P uptake, more rapid biodegradation, or improved disease resistance. Adding yeast to bread, for example, is a kitchen version of inoculation, because a microorganism is added to the dough to bring about a desirable transformation (getting the dough to rise).

Rhizobium, Azospirillum, Azotobacter, vesicular arbuscular mycorrhizal fungi, biological control agents, phosphobacteria, and genetically engineered organisms (GEMS) have all been used as inoculants. A genetically modified *Pseudomonas syringae,* for example, has been used to inoculate tomatoes and potatoes to improve frost resistance. In this chapter, we discuss some of the principles behind inoculant use. Toward the end of the chapter, we discuss how inoculation can be used in biocontrol.

SUCCESSFUL FACTORS IN INOCULATION

There are three key factors in successful inoculation: 1) the inoculum must be viable; 2) there must be sufficient inoculum to bring about the desired change; and 3) the ability of microorganisms in the inoculum to multiply and survive in soil must be ensured, either through selection of the inoculum itself, or through modification of the environment into which the inoculum is put. Alien populations—including inoculum—introduced into soil, usually decline due to antagonism from native populations.

RHIZOBIA INOCULANTS

The most important microorganisms used in inoculants today are the rhizobia that are added to legume seeds to ensure successful nodulation. Inoculation with rhizobia is needed when: 1) there are no native rhizobia; 2) the native rhizobia are ineffective; 3) the native rhizobia are harmful (some *Bradyrhizobium japonicum* strains produce toxins that causes chlorosis, yellowing in plants); or 4) the native rhizobia are uninfective because they don't recognize the host legume, they are easily outcompeted in the rhizosphere, they are killed by fungal and bacterial toxins, they are infected by bacteriophage, or they are simply unable to bind and colonize the root surface readily.

Correct inoculant use requires several considerations. The first is to use a species-specific inoculant. The second is to use an appropriate carrier to deliver and ensure survival of the inoculant; peat is one of the best. Some carriers contain antagonists to rhizobia. These antagonists can be reduced by sterilizing the carrier before adding the rhizobia, but sterilization, particularly steam sterilization, can create toxic substances in the carrier itself. Carriers are needed because inoculants survive only a short time on the seed coat. Adhesives that attach inoculum to the seed are useful. Gum arabic is one adhesive compound, but any sticky substance will do.

Another factor to consider is to ensure that the seed has not been treated with chemicals. Chemical treatment can cause osmotic shock or toxicity to rhizobia. Protecting the rhizobia from osmotic shock also includes separating the seeds from fertilizer because the salts in fertilizer increase the potential for osmotic shock. Some seed coats, unfortunately, contain their own inhibitory compounds—tannins—that are harmful to rhizobia.

Inoculants need to be kept as cool as possible until use. In storage and handling, heating and drying should be avoided. Optimum storage temperatures are near freezing, although storage at 21°C can still support reasonable rhizobia survival. However, it is clearly not desirable and it will reduce cell viability compared to storage at lower temperatures.

Once seeds are inoculated, they should be planted as soon as possible, usually within 2 to 3 weeks. Inoculants begin dying as soon as they are added to the carrier, so the longer the interval between preparation and planting, the less the viability of the intended inoculum. The inoculated seeds should also be planted into moist soil if at all possible. Although the seed can sprout readily, rhizobia are not particularly desiccation tolerant and planting in dry soil hastens their death, even if the seed has been in dry soil only for a short while.

Only tested and approved rhizobia cultures should be used for inoculation. The population of rhizobia per seed should be high; at least 2×10^5 rhizobia per seed are needed at planting for effective soybean nodulation. Greenhouse tests demonstrate that when rhizobia populations in inoculum decline below 1,000/seed, nodulation suffers or is absent.

TRADEMARKS OF SUCCESSFUL INOCULATION

To ensure successful rhizobial inoculation, the rhizobia must be infective, competitive, and effective (ICE). The rhizobia in inoculants must be able to infect the host legume. They also must be able to compete with native rhizobia for the infection sites on the host legume. Above all, the rhizobia in inoculants must be effective when they have infected the host legume, otherwise the inoculation practice is worthless.

Effective nodulation of legumes requires many rhizobia, even when field conditions are moderately good. Infective but ineffective native rhizobia already present in soil may require using even larger rhizobial numbers in the inoculum. Dunigan et al. (1984) discovered that 3 years of heavy soil inoculation were required to displace an indigenous *Bradyrhizobium japonicum* strain from soil. Depending on whether legumes were present at an earlier date, native rhizobia populations may range from 10 to 10^6 rhizobia per gram of soil.

The competitiveness of rhizobia strains differs markedly (Table 28-1). In some instances, 1% of the total rhizobial strains in soil forms 85% of the nodules. Relative competition is influenced by host compatibility, aeration, inoculum size, nutrient levels, the state of the inoculum, pH, and temperature.

In one study, multiple *Bradyrhizobium japonicum* strains were evaluated in the midwestern United States for their competitiveness with indigenous strains (Klubek et al. 1988). The average recovery of the introduced strains in year 1 of the study was 0.3% to 15.7%. In year 2 of the study, inoculum recovery in the soybean nodules was only 14.8% to 26.6%. The most successful strains were those that were adapted to, or had originally been isolated from, the soils into which they were introduced.

The control of competition between inoculum and indigenous rhizobia in soil is at both the biophysical and chemical levels. One fascinating experiment showed how rhizobial signals that stimulate nodule formation in their hosts affected competition between different strains (Cunningham et al. 1991). If *Bradyrhizobium japonicum*

Table 28-1   Competitiveness of rhizobial inoculants.

		Percent of Nodules Formed	
Strain #1:Strain #2	**Nodules/Plant**	**Strain #1**	**Strain #2**
5:3	92	100	0
1:6	81	98.3	1.7
1:60	88	80.0	12.5
1:600	144	26.1	72.8
1:6,000	109	1.3	98.7
1:60,000	113	0	100

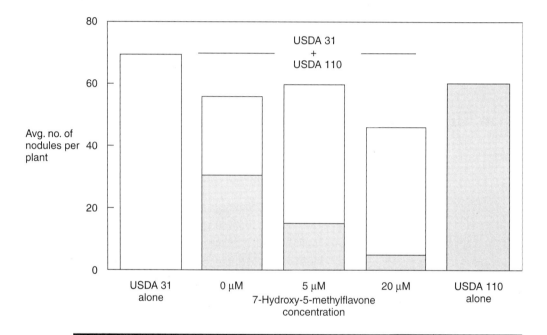

Figure 28-1   Effect of an exogenously applied nodulation inhibitor on interstrain competition between *Bradyrhizobium* strains USDA 110 and USDA 31. 7-hydroxy-5-methyl flavone inhibits induction of nodule formation by USDA 110. (Adapted from Cunningham et al. 1991)

strains USDA 31 and USDA 110 were added to soybeans individually, each formed about the same number of nodules (Figure 28-1). However, if they were added together, USDA 31 made up only about 46% of the nodules. USDA 110 was clearly the more competitive strain. When a chemical compound, a flavone, that specifically inhibited USDA 110's ability to induce nodule formation in the host was added, it enabled USDA 31 to form an increasing number of the nodules on each plant.

Table 28-2   Potential outcomes of inoculation trials.

Case	Uninoculated		Inoculated		Inoculated + Nitrogen	
	Nodulation	Plant Growth	Nodulation	Plant Growth	Nodulation	Plant Growth
1	None	Poor	None	Poor	None	Poor

Notes: No nodulation and poor plant growth suggest that no effective rhizobia are present in soil. Lack of growth in inoculated and fertilized plants suggests that some other factor is limiting plant growth.

| 2 | None | Poor | None | Poor | None | Good |

Notes: No nodulation and poor plant growth suggest that effective rhizobia are not present. Lack of growth and nodules in the inoculated sample suggests that the inoculum is wrong, or of poor quality. Growth of the fertilized control suggests that no other nutrient but N is limiting.

| 3a | None | Poor | Nodules | Good | Nodules | Good |

Notes: Same as Case 2, except that the inoculum strain appears to be effective.

| 3b | None | Poor | Nodules | Poor | None | Good |

Notes: Same as Case 3a except that the inoculum strain, while infective, is not effective and there is sufficient fertilizer N to inhibit nodulation.

| 4a | Nodules | Poor | Nodules | Good | Nodules | Good |

Notes: Native rhizobia in soil can infect plants but form ineffective nodules. The inoculant is effective and can outcompete native rhizobia.

| 4b | Nodules | Poor | Nodules | Poor | Nodules | Good |

Notes: Native rhizobia in soil can infect plants but form ineffective nodules. The inoculant strain is unable to outcompete the residents for nodule-forming sites.

| 5 | Nodules | Good | Nodules | Good | Nodules | Good |

Notes: Native rhizobia in soil are both infective and effective. The native rhizobia may be able to outcompete inoculum. Inoculation is not really necessary.

(Adapted from Vincent 1970)

INOCULATION TRIALS

Trials to determine the need for inoculation require at least three treatments: 1) an uninoculated control; 2) an inoculated treatment with effective rhizobia; and 3) an inoculated treatment with N (Table 28-2). The uninoculated control tests whether infective rhizobia are in the soil. It also tests whether effective native rhizobia are present. If the native rhizobia are both infective and effective, there may be no real benefit to inoculation unless a superior strain is used. The inoculated treatment with effective rhizobia tests whether the inoculum is competitive. It can also test whether the inoculum is infective and give an idea about its quality. The inoculated treatment with N tests whether other factors besides rhizobia or N limit plant growth. These factors can be nutritional (insufficient S, P, or a micronutrient) or environmental (acidic or alkaline conditions). Remember that ample organic and inorganic N in soil inhibit infection and also inhibit effectiveness if supplied to nodulated plants. If the soil contains a lot of N to begin with, nodulation and N_2 fixation will be impaired.

BIOCONTROL

Biocontrol represents the biological, rather than the chemical, control of pests. The basis for protection in plant populations may be biochemical, anatomical, or cultural. Phytoalexins are plant-produced compounds that have antimicrobial activity. Cuticle thickness also affects disease resistance. For pathogens that travel from plant to plant, sparse planting reduces disease levels by isolating diseased plants, and for pathogens that do not migrate, dense planting overcomes losses suffered from diseased plants. Rotation is a classical mechanism of disease protection that is achieved by removing the host plant long enough to reduce pathogen populations.

The principles of microbial amensalism and parasitism that we've already discussed can be used to control pathogen populations. This is often apparent in cases in which the environment has no natural resistance to a disease and no natural antagonists are present. For example, *Endothia parasitica* entered New York in 1904 and caused the nearly complete destruction of the American chestnut tree because there was no natural control present. Current research efforts are under way to find natural parasites of the fungus and its insect vector.

CHARACTERISTICS OF AN EFFECTIVE BIOCONTROL AGENT

There are four advantages to microbial biocontrol: 1) it appears to be ecologically safer than chemicals because control agents are not accumulated in the food chain; 2) some biocontrol agents can provide persistent control since more than a single mutation is required to adapt to them and because they can become an integral part of a pest's life cycle; 3) biocontrol agents should have slight effects on ecological balance; and 4) they are compatible with other control agents.

Bacteria, fungi, viruses, and protozoa are all used as biocontrol agents. However, there are some requirements for an effective biocontrol agent. The microbial pathogens should not themselves be pathogenized. They should be virulent. They should be insensitive to moderate environmental change, and should survive until infection of pests is achieved. The disease onset, or parasitism, they bring about should be rapid. They should be specific.

CONTROL OF MICROBIAL PATHOGENS

Saprophytic fungi can compete with pathogenic fungi, for example, *Trichoderma* competes with *Verticillium* and *Fusarium*. *Peniophora gigantea* can antagonize the pine pathogen *Heterobasidion annosum* by three mechanisms: 1) it prevents the pathogen from colonizing stumps and traveling down into the root zone; 2) it prevents the pathogen from traveling between infected and uninfected trees along interconnected

Table 28-3 Comparison of the properties of *Bacillus thuringiensis* and *Bacillus popilliae* as microbial biocontrol agents.

	Bacillus thuringiensis	**Bacillus popilliae**
Pest controlled	Lepidoptera (many)	Coleoptera (few)
Pathogenicity	Low	High
Response time	Immediate	Slow
Formulation	Spores and toxin crystals	Spores
Production	*In vitro*	*In vivo*
Persistence	Low	High
Resistance in pest	Developing	Reported but not confirmed

(Adapted from Deacon 1983)

roots; and 3) it prevents the pathogen from growing up to stump surfaces and sporulating (Deacon 1983).

Many saprophytic microorganisms that occur in the rhizosphere and phyllosphere protect plants against pathogens. Some may produce antibiotics, but the major mechanism for protection is probably competitive exclusion or preemptive colonization. *Agrobacterium radiobacter* antagonizes *Agrobacterium tumefaciens* (which causes crown gall). *Bacillus* and *Streptomyces* added to soils control damping-off disease of cucumber, peas, and lettuce caused by *Rhizoctonia solani*. *Bacillus subtilis* added to plant tissue also controls stem rot and wilt rot caused by the fungus *Fusarium*. *Mycobacteria* produce cellulolytic enzymes and their addition to young seedlings helps control fungal infection by *Pythium, Rhizoctonia,* and *Fusarium*. *Bacillus* and *Pseudomonas* produce chitinases that lyse fungal cell walls. This is an important basis for control of some fungal pathogens. Treatment of soil with chitin (waste crab and lobster shells) markedly reduces the severity of root-rot diseases caused by *Fusarium*. This is an example of selective enrichment for the chitinase-producing microorganisms.

BIOCONTROL OF INSECTS

The best example of a microbial insecticide is *Bacillus thuringiensis* endotoxin and exotoxin (Table 28-3). Bt toxin, as it is called, was first used in 1901. It has widespread commercial production and has been successfully tested on 140 insect species, including mosquitoes. *Bacillus thuringiensis* produces β and δ endotoxins during sporulation. These are heat-sensitive proteins that are unstable in the environment and degrade within 48 hours. *Bacillus thuringiensis* also produces exotoxins contained in crystalline parasporal bodies (Figure 28-2). These are heat-stable proteins. There are four types designated: α, β, γ, and δ. The parasporal crystals are insoluble in water, but readily dissolve in the alkaline conditions of an insect's gut. The proteolytic enzymes that are released paralyze the gut, and spores that have also been consumed germinate and overwhelmingly reproduce. Bt toxins have also been incorporated into corn and cotton DNA as part of intrinsic plant resistance.

Viruses have also been developed against insect pests such as *Lepidoptera, Hymenoptera,* and *Dipterans*. These viruses cause instances of hyperparasitism. Gypsy

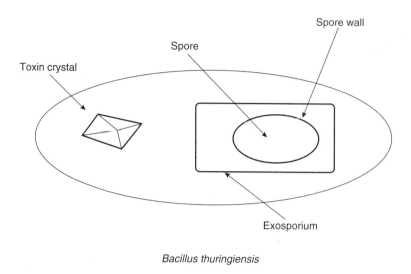

Bacillus thuringiensis

Figure 28-2   Toxic parasporal crystal formation in *Bacillus thuringiensis.*

moths and tent caterpillars, for example, are subject to epidemic infestations caused by viruses. Protozoa have occasionally been used as biocontrol agents, but their use has suffered from slow growth and complex culturing conditions.

BIOCONTROL OF WEEDS

There are several examples of biocontrol fungi such as *Gliocladium virens, Trichoderma hamatum, Trichoderma harzianum, Trichoderma viride,* and *Talaromyces flavus* (Churchill 1982). *Colletotrichum gloeosporoides* f. sp. *aeschynomene* is a biological herbicide. It controls northern jointvetch, a serious weed in rice and soybean fields in Arkansas. *Colletotrichum malvarum* is a pathogen of prickly sida. *Alternaria macrospora* is a pathogen of spurred anoda.

Summary

Inoculation is the practice of applying microorganisms to a system to bring about some desirable transformation. There are three key factors in successful inoculation: 1) the quantity of inoculum must be appropriate; 2) the number of viable cells applied must be sufficient to bring about the desired change; and 3) the microorganisms must multiply and survive in soil. The most important microorganisms used in inoculants today are the rhizobia that are added to legume seeds to ensure successful nodulation.

Successful inoculation requires an appropriate strain; an effective carrier; minimal environmental stress from salts, heat and desiccation; rapid planting; and high populations. Rhizobial inoculants must be able to infect the host legume, compete with native rhizobia for the infection sites on the host legume, and be effective when they have infected the host legume. The control of competition between inoculum and indigenous rhizobia in soil is both biophysical and chemical. Trials to determine the need for inoculation require at least three treatments: 1) an uninoculated control; 2) an inoculated treatment with effective rhizobia; and 3) an inoculated treatment with N.

There are four advantages to microbial biocontrol: 1) it appears to be ecologically safer than chemicals because control agents are not accumulated in the food chain; 2) some biocontrol agents can provide persistent control since more than a single mutation is required to adapt to them; 3) biocontrol agents should have slight effects on ecological balance; and 4) they are compatible with other control agents. The best example of a microbial insecticide is *Bacillus thuringiensis* endotoxin and exotoxin.

Sample Questions

1. Why is it sometimes necessary to use inoculum?
2. What conditions are necessary to successfully outcompete indigenous rhizobia?
3. What components are necessary for an effective rhizobia inoculation program?
4. Why is it important that *Bradyrhizobium* in soybean inoculum be competitive?
5. What does the term "cross-inoculation group" refer to, and why is it important?
6. In designing a soil inoculation program, what factors must you consider to ensure success?
7. What are the necessary treatments in an inoculation trial? Explain the rationale for each one.
8. Based on the following table, what is an appropriate inoculation rate to use if you want to maximize the number of nodules per plant, but minimize the amount of inoculum you use? Explain your reasoning.

Effect of *Bradyrhizobium japonicum* inoculation rates on soybean nodulation parameters at 53 days of growth.

B. japonicum (cells/cm row)	Taproot Nodules/Plant	Total Nodules/Plant	Nodule Weight/Plant
Uninoculated	0	0.1	4.1
Uninoculated + 100 kg N ha^{-1}	0	0.1	0.8
389	0.1	0.2	2.3
3,890	0.2	0.3	5.8
38,900	1.4	1.5	20.4
389,000	7.5	9.8	98.8
3,890,000	8.7	12.5	81.8
38,900,000	11.7	16.0	87.4
389,000,000	21.1	18.4	38.5
3,890,000,000	22.1	19.8	101.3
Least significant difference	2.9	2.6	29.0

9. Based on the following table for an inoculation trial, provide appropriate explanations for each case.

Possible Outcomes of Inoculation Trials

	Uninoculated		Inoculated		Inoculated + N	
Case	Nodules	Plant Growth	Nodules	Plant Growth	Nodules	Plant Growth
1.	No	Poor	None	Poor	None	Poor
2.	No	Poor	None	Poor	None	Good
3.	No	Poor	Yes	Poor	Yes	Good
4.	Some	Good	Yes	Good	Yes	Good

10. Given the data in the following table, is there an advantage in terms of a) total plant weight, and b) total N/g culture, to using inoculum that has hydrogenase (Hup$^+$) or that lacks it (Hup$^-$)?

Comparison of Hup$^-$ mutants of *Bradyrhizobium japonicum* with their Hup$^+$ revertants as inoculants for soybeans 40 days after planting.

Strain	Phenotype	H$_2$ evolution (μmol g^{-1} h^{-1})	Nodule Wt (g)	Shoot Wt (g)	Root Wt (g)	Total Plant Wt (g)	Total N/g Culture
PJ17-19	Hup$^-$	1.83	0.64	9.00	3.04	12.68	0.43
PJ17-1-20	Hup$^+$	0.03	0.85	11.50	3.49	15.84	0.54
PJ18	Hup$^-$	1.45	0.57	8.03	2.78	11.38	0.39
PJ18-1	Hup$^+$	0.03	0.67	9.19	3.18	13.04	0.44
Least significant difference	0.64	0.15	1.70	—	1.89	0.07	

11. Assuming the ratio of H$_2$ evolved to N$_2$ fixed is 1:3, how much N will strain PJ18-1 fix in one day?

12. Based on the data in the following table, what is the relationship between inoculum density and nodule dry weight? Do the days of growth influence this relationship?

Effect of *Bradyrhizobium* inoculation rates on nodule dry weight at 25 and 53 days of growth.

Bradyrhizobium (log cells/cm row)	Dry weight (g) /nodule	
	25 Days	53 Days
2	1.7	4.8
3	0.5	10.5
4	1.0	8.6
5	0.5	6.9
6	0.5	4.1
7	0.4	3.6
8	0.4	2.6
9	0.3	1.8

Thought Question

You have recently accepted a postdoctoral position in Agronomic Science at Miskatonic University. Shortly after you arrive, the regional extension agronomist comes to you for help with the following letter that she has received.

To Whom It May Concern:

I am removing 100 acres of land from the Conservation Reserve Program (CRP) when my 10-year contract expires in 1998 and I would like to grow organic soybeans for the local food co-op's tofu plant. I know that legumes don't need any fertilization because they can make their own nitrogen. Is there anything else that I need to do to this land other than clear away the black locust and fescue that have grown up in the past 10 years?

Thanking you in advance.

Yours,

Harold P. Love-Craft

The regional extension agronomist doesn't know any more about this farm than the letter implies, and is curiously reticent (more like averse) about visiting for herself. However, since it's an issue dealing with N fixation, she's come to you for advice.

Write a memo to the extension scientist indicating what should be done to assist Mr. Love-Craft in successfully growing organic soybeans. You need to look at this problem from the plant/soil perspective and consider all factors that may influence successful soybean growth. Provide the regional extension agronomist with potential factors to consider, and speculate about the potential result of employing, or not employing, those factors.

Additional Reading

An invaluable resource on inoculation technology is a monograph by J. M. Vincent, *A Manual for the Practical Study of Root-Nodule Bacteria* (1970, Blackwell Scientific Publications, Oxford, England). The American Society for Microbiology issued a monograph, *Microbial Control of Plant Pests and Diseases,* by J. W. Deacon (1983, American Society for Microbiology, Washington, DC) that is a very good introduction to the topic of biocontrol.

References

Churchill, B. W. 1982. Mass production of microorganisms for biological control. In *Biological control of weeds with plant pathogens,* R. Charudattan, and H. Walker (eds.), 139–56. New York: John Wiley and Sons.

Cunningham et al. 1991. Chemical control of interstrain competition for soybean nodulation by *Bradyrhizobium japonicum. Applied and Environmental Microbiology* 57:1886–92.

Dunigan et al. 1984. Introduction and survival of an inoculant strain of *Rhizobium japonicum* in soil. *Agronomy Journal* 76:463–66.

Klubek et al. 1988. Competitiveness of selected *Bradyrhizobium japonicum* strains in midwestern USA soils. *Soil Science Society of America Journal* 52:662–66.

Chapter 29

Mycorrhizae

Overview

After you have studied this chapter, you should be able to:

- Define what mycorrhizae are and discuss their importance.
- Explain the effect that soil fertility has on the prevalence of mycorrhizal symbioses.
- Identify the two major groups of mycorrhizae and tell how to distinguish them from one another.
- Describe the distribution and ecology of mycorrhizae in soil.
- Explain the role mycorrhizae play in plant mineral nutrition.
- Discuss several mechanisms that mycorrhizae use to improve plant P uptake.
- Summarize seven benefits that mycorrhizae provide to plants.
- Describe the cost of mycorrhizal colonization to plants.
- List six cultural practices that affect mycorrhizae populations in soil.

INTRODUCTION

Mycorrhizal symbioses play a critical role in the mineral nutrition of terrestrial plants. The term "mycorrhiza" (fungus root) was used by A. B. Frank in the late 1800s to characterize the association between higher plants and fungi. However, based on the fossil record, mycorrhizal infections are as old as terrestrial plants. That means mycorrhizae have been around for at least 370 million years. In this chapter we look at the ecology of mycorrhizal fungi in soil, some of the types of mycorrhizal symbioses that are found in nature, and how soil management affects mycorrhizal populations and subsequent plant nutrition.

THE PREVALENCE OF MYCORRHIZAL SYMBIOSES

Mycorrhizae occur on almost all terrestrial plants. The mycorrhizal symbioses are not as specific as the N_2-fixing symbioses. A plant may have several mycorrhizae that can infect and form symbioses with it. Some species, such as crucifers (i.e., broccoli) are unusual in that mycorrhizal symbioses are absent. Species with fine roots and many root hairs are not as dependent on mycorrhizae as are species with well-defined tap roots.

The extent of symbiosis depends on fertility. High fertility leads to low mycorrhizal infection and poor symbiosis, while low fertility leads to high mycorrhizal infection and effective symbioses if the appropriate mycorrhizae are present. One of the first clues to the importance of mycorrhizal symbioses in plants was the observation that establishing some tree species in new environments absolutely required mycorrhizae to be transferred from their old environments. Figure 29-1 shows dramatic examples of growth differences between a pine that has formed a mycorrhizal symbiosis (Figure 29-1a) and one that has not (Figure 29-1b).

TYPES OF MYCORRHIZAE

There are two types of mycorrhizae that we concern ourselves with in agricultural soils: ectomycorrhizae and endomycorrhizae. Some plants have both endo- and ectomycorrhizae, but most do not. Endomycorrhizae include eriaceous types (with characteristics of both ecto- and endomycorrhizae), orchid types (infected with basidiomycetes), and vesicular arbuscular mycorrhizae types (VA mycorrhizae, or VAM). We focus our attention on the VAM in this chapter because we know more about this group.

a   b

Figure 29-1 Differences in growth of pine in reclaimed mine spoil. (a) Mycorrhizal symbioses have formed; (b) Mycorrhizal symbioses are lacking. (Photographs courtesy of J. W. Hendrix)

Ectomycorrhizae

Ectomycorrhizal symbioses are mutually beneficial unions between fungi and the roots of vascular and nonvascular plants. The host of an ectomycorrhizal fungi is commonly a gymnosperm (evergreen or pine). The typical ectomycorrhizal fungus is a basidiomycete (*Agaricus* is one of the most common), an ascomycete, or a phycomycete. Ectomycorrhizal fungi are found primarily in temperate and subarctic regions and less commonly in the tropics. They can grow apart from the host on media containing simple sugars and vitamins, although spore germination is very poor on artificial media. They may also require NH_3, organic N, organic acids, and root exudates. Ectomycorrhizae have poor competitive saprophytic ability. So, apart from their hosts, ectomycorrhizal fungi have a tough time competing with other microorganisms in soil.

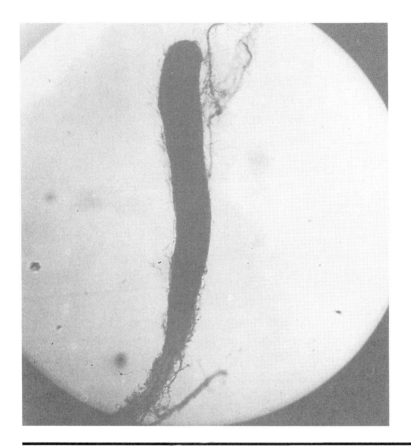

Figure 29-2 Plant root surrounded by a fungal mantle characteristic of the ectomycorrhizal symbioses. (Photograph courtesy of J. W. Hendrix)

During the infection process, ectomycorrhizal fungi in soil are stimulated to grow toward the root by root metabolites. The hyphae aggregate around the root, and ultimately penetrate between the root epidermis and the cortex. A structure called the Hartig net forms. The Hartig net is a fungal sheath surrounding the root in which fungal hyphae penetrate between the root cells. Eventually the root becomes surrounded by a fungal mantle (Figure 29-2). Hyphae can penetrate the host cells, but they usually do not. Another characteristic of ectomycorrhizal infections is that the infected roots, instead of elongating, become stunted and club shaped. Figure 29-3a,b shows the root system of an ectomycorrhizal symbiosis demonstrating this club-shaped root morphology.

a

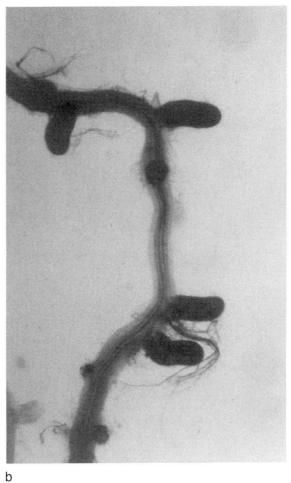

b

Figure 29-3 Club-shaped morphology of roots colonized by ectomycorrhizae (a). A closer look at stunted lateral roots (b). (Photographs courtesy of J. W. Hendrix)

a

b

Figure 29-4   Fruiting bodies of ectomycorrhizal fungi infecting pine (a) and strawberry (b).
(Photographs courtesy of J. W. Hendrix)

In undisturbed environments, the fruiting bodies of the ectomycorrhizal fungi can be
seen at the soil surface (Figure 29-4a, b).

VAM Mycorrhizae

Vesicular arbuscular mycorrhizae (VAM) have a wide host range. They infect most
agricultural crops and 90% of all vascular plants (Sylvia 1994). VAM are generally nonspecific. Unlike the ectomycorrhizae, they are nonculturable outside a plant host. Ger-

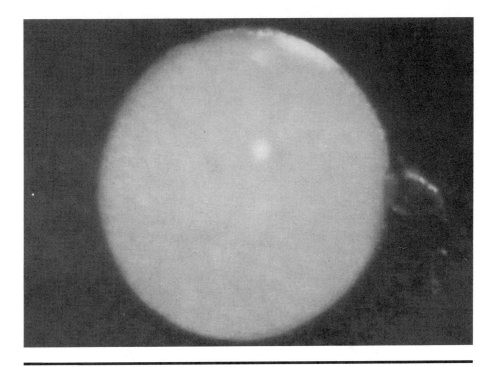

Figure 29-5   A spore from a VAM fungus. (Photograph courtesy of J. W. Hendrix)

minating VAM spores in soil infect root hairs to begin colonization. Since the direction of hyphal growth from the spore is random, it appears that plant root metabolites do not signal the direction in which the hyphae should grow. Consequently, infection is an accidental event.

VAM fungi are zygomycetes and phycomycetes. Important genera to remember are *Glomus* (commonly the most often VAM isolated from soil), *Gigaspora, Acaulospora, Entrophospora,* and *Scutellospora.* VAM species produce large, resting spores that can be 200 μm (0.2 mm) in diameter (Figure 29-5). Some VAM have aggregate spores in sporocaps. The morphology of the spores is the basis for identification since the fungus itself can't be cultured.

VAM fungi make two major structures that distinguish them from ectomycorrhizae—vesicles and arbuscles. Vesicles are 1- to 10- μm expansions of hyphae between plant cells (Figure 29-6). They are lipid filled and are apparently storage organs. Arbuscles are finely branched hyphae similar to haustoria (Figure 29-7). They do not penetrate the cell membrane, but form a boundary with it that has an enormous surface area. Arbuscles persist for 4 to 10 days and then are digested by the plant cells. The arbuscles are the site of metabolic exchange between mycorrhiza and plant.

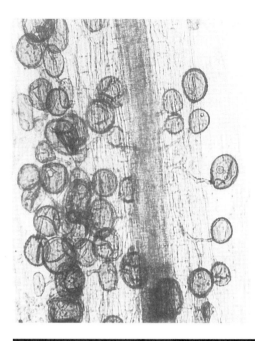

Figure 29-6   A heavily colonized plant root showing numerous vesicles. (Photograph courtesy of J. W. Hendrix)

Figure 29-7   The highly branched morphology of an arbuscle—site of metabolic exchange between VAM fungi and plant cells. (Photograph courtesy of J. W. Hendrix)

How Do You Study VAM?

You can't grow VAM on standard plate count media because they have an obligate association with plants (Sylvia 1994). So, how do you count them? The standard technique is a bioassay. A variation of this technique that works with VAM is to use soil as diluent. The diluent soil comes from the same location as the soil sample and is pasteurized rather than steam sterilized to remove any native mycorrhizae. Otherwise, the sequence for diluting the soil sample in diluent is the same. Maize is a good host plant to use because it has a high mycorrhizal dependency and a fibrous root system that is easy to stain for evidence of fungal colonization. Plants are usually grown for 6 to 8 weeks or until they are pot bound.

To visualize mycorrhizal colonization of the roots, heat them for 30 minutes in 1.8 M KOH at 80°C, rinse with tap water, rinse in 2.5% HCl twice, immerse in Trypan Blue stain at 80°C for 30 minutes, cool, rinse, and drain. The presence of distinctive vesicles and arbuscles that are characteristic of VA mycorrhizae is presumptive evidence that at least one infectious propagule (a spore, hyphae, or hyphal fragment) was present in the pot.

Alternatively, you can try to enumerate mycorrhizae spores directly. Fortunately, these spores are relatively large (up to 1 mm in diameter), which makes them readily observable with a dissecting microscope. First, mix 50 to 100 g of soil with 1 L of water and mix vigorously. Let the suspension settle for 30 seconds and decant the supernatant through sieves of different mesh size. Transfer the sieve contents to a centrifuge tube and pellet them. Resuspend the pellet in 1.17 M sucrose and centrifuge it again. The spores will stay in suspension and can be filtered, counted, and identified at a later date.

DISTRIBUTION AND ECOLOGY OF VAM IN SOIL

Most VAM are found within the first 20 cm of the soil profile, although they can be found at greater depths (70 to 100 cm). White et al. (1989) demonstrated that spores obtained from these greater depths had poor ability to colonize plant roots. Compared to total bacteria and fungi, VAM populations in soil are minuscule, ranging from perhaps 1 spore per gram of soil in uncultivated locations to 50 spores per gram of soil following growth of a VAM-colonized plant. Soil fumigation almost completely eliminates VAM from soil.

VAM populations are not constant during the growing season but vary on a cyclical basis. Populations are typically highest in spring, lowest in summer, and begin to rise again during fall. The greatest determinant of VAM population in soil is the growth of a VAM-colonized plant, as Figure 29-8 demonstrates. Poor plant growth of crops after long fallow periods (usually as a moisture-conserving practice in semiarid regions such as Australia) occurs because the long periods in the absence of a host serve to reduce the VAM populations in soil (Hayman et al. 1975). When crops that are colonized by

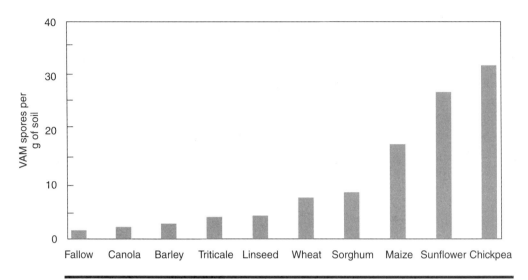

Figure 29-8   The number of VAM spores per gram of soil depends on the type of plant grown in it. (Adapted from Thompson et al. 1997)

VAM are grown, subsequent crops benefit from the higher number of VAM spores that result. The soil supports a community of VAM fungi rather than a single species, although continuous cultivation of a single crop causes some species to predominate. Over a dozen species of VAM are typically found in most unmanaged or agricultural soils. These generally belong to the Order Glomales, of which the genus *Glomus* is representative. Not all of these species are equally numerous. Some are continually isolated and some are only periodically found.

Crop rotation disrupts the stable mycorrhizal community that develops in a monoculture and allows some VAM species to increase in number at the expense of others. Annually rotated corn and soybeans typically yield up to 10% more after rotation than corn and soybeans in continuous cultivation (Crookston et al. 1991). One theory used to explain this phenomenon is that stable mycorrhizal communities in continuous crops eventually accumulate parasitic, rather than beneficial, species. Rotation, by disrupting the mycorrhizal populations, allows beneficial species to predominate again.

NUTRITIONAL SIGNIFICANCE

What is the role of mycorrhizae in plant nutrition? One way to look at mycorrhizae is as nutritionally specialized host-dependent fungi. Mycorrhizae increase the nutrient-absorbing capacity of roots for P, K, Fe, Cu, N, S, and Zn (Table 29-1).

The data show that the two *Glomus* species have considerably higher nutrient contents per plant than the nonmycorrhizal control. With the exception of iron, *Glomus*

Table 29-1 Effect of mycorrhizae on nutrient uptake in onion.

| Treatment | Total Nutrient Uptake Per Plant | | | | | | | |
| | P | Ca | Mg | Na | K | Zn | Mn | Fe |
	mg					μg		
Control	0.39	8.7	0.46	0.25	10.9	38	69	171
Glomus fasciculatus	4.42	25.2	2.49	2.76	35.9	112	106	412
Glomus monosporus	3.26	14.4	1.46	1.36	22.5	79	71	432

(Adapted from Ojala et al. 1983)

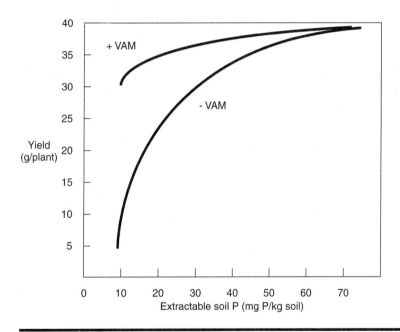

Figure 29-9 Growth response of mycorrhizal and nonmycorrhizal cotton to increasing levels of extractable soil P. (Adapted from Thompson et al. 1997)

fasciculatus has greater nutrient uptake than *Glomus monosporus,* which illustrates that different mycorrhizal–plant symbioses have different capacities for nutrient uptake.

The greatest benefit to the plant from mycorrhizal colonization occurs at low soil fertility levels. Figure 29-9 illustrates that cotton colonized by VAM sustains higher yield than nonmycorrhizal cotton when extractable soil P levels are low. However, at higher P levels, there is essentially no yield difference. The symbiosis is not as sensitive to high

Table 29-2 Top growth and acetylene reduction activity of two alfalfa cultivars in the presence and absence of mycorrhizae and 1344 kg/ha P.

Treatment	Top Growth (g/plant)	C_2H_2 Reduction (μmoles plant^{-1}hr^{-1})
Cultivar 1		
Control	5.72	3.79
+ Mycorrhizae	6.99	4.38
+ P	5.76	5.10
+ Mycorrhizae and P	7.18	5.10
Cultivar 2		
Control	6.15	3.90
+ Mycorrhizae	4.50	4.01
+ P	7.11	3.71
+ Mycorrhizae and P	7.54	4.62

(Adapted from Satterlee et al. 1983)

fertility levels as is the legume–rhizobia symbiosis. Even when nutrient availability is high, there may still be added benefit to a plant's nutrition by mycorrhizal colonization. Table 29-2 shows an experiment in which top growth and N_2 fixation in alfalfa (as measured by the acetylene reduction assay) still benefited from mycorrhizal inoculation despite being fertilized with P.

The alfalfa response in Table 29-2 is cultivar dependent. Top growth in cultivar 1 is stimulated more by mycorrhizae and P together than it is by either amendment alone. Acetylene reduction does not demonstrate a similar synergistic response. Top growth in cultivar 2 responds to P fertilization but not mycorrhizal addition when the two amendments are added separately. However, both top growth and acetylene reduction in cultivar 2 increase when the two amendments are added together.

At the highest soil fertility levels, mycorrhizal colonization may be detrimental to plant growth. Figure 29-10 is typical of numerous experiments that show the diminishing returns of additional P fertility (for example) on plant growth of mycorrhizal plants. The data in Figure 29-10 suggest that if P fertility increases further, plant yield starts to decrease.

The role of mycorrhizal fungi in absorbing soil P may be summarized as follows: Mycorrhizal plants absorb and accumulate more P than do nonmycorrhizal plants especially when plants are grown in soils low in this nutrient. The increased P absorption is due to more efficient mining of soil P. There is more efficient mining of soil P because the surface area and volume of the roots increases; first, because the roots are healthier (better nourished) and second, because the fungal hyphae act as extensions of the plant root. The longevity of absorbing roots increases because the hyphae may extend from regions of the root that are no longer efficient at nutrient uptake. The length of root over which rapid absorption takes place increases and, consequently, the exploitation of soils increases.

Mycorrhizae use the same P sources as do plants. It is also possible that mycorrhizae are better able to exploit soluble P by virtue of having a lower K_m for its uptake. Solu-

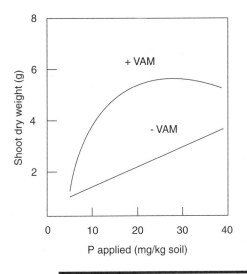
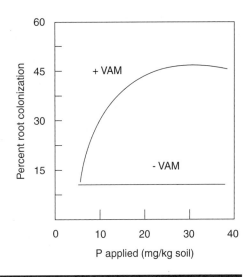

Figure 29-10   Diminishing returns of P fertilization on shoot and root weight of mycorrhizal plants. (Adapted from Medina et al. 1988)

bilizing insoluble forms of P increases overall P availability, which must aid plant nutrition if most soil P is insoluble and unavailable. Mycorrhizae increase P solubility by mechanisms such as excreting phosphatases, carbonic acid, and organic acid, and extending the surface area of root exposed to P.

BENEFITS TO THE PLANT

How do mycorrhizae benefit plants? 1) They improve overall plant growth by improving P and Zn nutrition; 2) they allow more efficient use of P and Zn in fertilizer; 3) they also stimulate N_2 fixation in nodulated plants by increasing P flow through plant roots; 4) they increase disease tolerance in plants by improving the plant nutrition and by competing with pathogenic microorganisms for space on the plant root; 5) they immobilize some heavy metals such as Zn, Cd, and Mn; 6) they improve water use and drought tolerance; and 7) they improve soil structure by helping to bind soil aggregates together.

The last two benefits of mycorrhizal symbioses deserve more comment. It makes sense that mycorrhizae aid plant water uptake. Fungi, as a group, maintain metabolic activity at lower water potentials than do most other soil organisms. The same physical reasons that apply to mycorrhizal exploitation of soil P should also apply to water uptake: They increase the area of soil explored for water, they increase the effective surface area of the root for absorbing water, and they increase the longevity of roots for water absorption.

Greater exploitation of soil water in mycorrhizal plants shows up as greater turgor pressure in plants (plants wilt because of low turgor pressure). A graphical

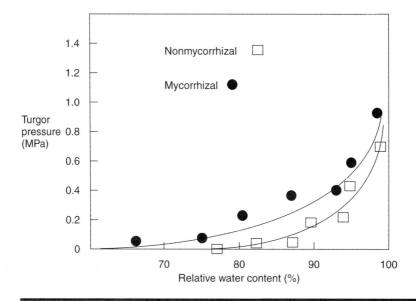

Figure 29-11 The relationship between turgor pressure and water content in mycorrhizal and nonmycorrhizal *Rosa hybrida* plants experiencing a 21-day drought. (Adapted from Augé 1989)

representation of this idea is shown in Figure 29-11. At high water content, mycorrhizal and nonmycorrhizal plants have about the same turgor pressure because water is readily available. At low water content, mycorrhizae are of little benefit to the plants because even the mycorrhizae can't extract water from soil. However, at intermediate water contents in drying soil, mycorrhizal plants consistently maintain higher turgor pressure than do nonmycorrhizal plants.

As mycorrhizal hyphae grow through soil, they intertwine around soil particles and help form them into aggregates. The mycorrhizae surface also acts as a charged surface on which clay particles can bind. Consequently, mycorrhizal plants increase soil aggregation and improve soil structure around the plant roots (Figure 29-12). With improved soil aggregation and structure come the benefits of greater aeration, more rapid water infiltration, and easier paths through which plant roots can move, all of which contribute to better plant growth.

BENEFITS TO THE FUNGI

Aside from being an obligate association in some instances, what do mycorrhizal fungi get from the symbiosis? Mycorrhizal fungi get plant carbon. VAM and ectomycorrhizal fungi make up 3% to 16% of total root weight and get 4% to 14% of the photosynthate. In VAM fungi, three-fourths of this plant carbon is respired. Of the retained carbon, VAM glycolipids accumulate and ectomycorrhizae produce trehalose and mannitol as storage sugars.

Figures 29-13 and 29-14 show the approximate partitioning of plant-derived (fixed) carbon in the *Glomus*/soybean symbiosis and the *Glomus*/soybean/*Bradyrhizobium* symbiosis.

Figure 29-12   Mycorrhizal plants improve soil aggregation around plant roots. (Photograph courtesy of J. W. Hendrix)

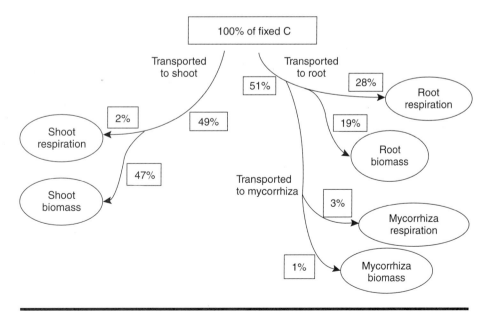

Figure 29-13   Carbon allocation in the *Glomus*–soybean symbiosis. (Adapted from Paul and Kucey 1981)

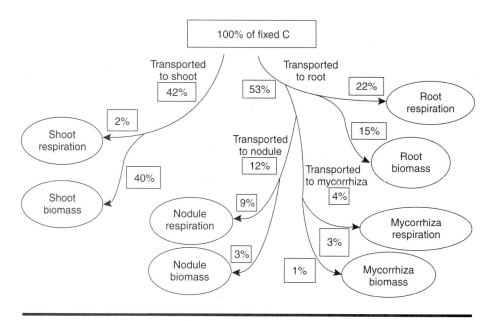

Figure 29-14   Carbon allocation in the *Glomus–Soybean–Bradyrhizobium* tripartite symbiosis.
(Adapted from Paul and Kucey 1981)

CULTURAL EFFECTS

Different crop plants have different dependencies on mycorrhizae (Table 29-3). For example, maize and sorghum have high mycorrhizal dependencies while wheat, barley, and oats have low mycorrhizal dependencies. Consequently, the order of crop rotation has a significant effect on P and other nutrient fertility because the mycorrhizal spore population decreases in soils while plants with little or no mycorrhizal dependency grow on them.

We've already seen fallow length as a cultural effect on mycorrhizae. Poor growth of some plants after long fallow periods may simply be because the mycorrhizal populations have diminished during the interim between crops. One management strategy is to follow long fallow periods with crops that have low mycorrhizal dependency.

Tillage can reduce mycorrhizal colonization and function. The tillage operations severely disturb the soil and break up mycorrhizal networks that have formed. No-tillage is less disruptive. In addition, since most mycorrhizae and their spores form close to the soil surface, cultivation, which buries surface residue at greater soil depths, reduces mycorrhizal infection by burying spores out of reach of the plant roots early in a growing season. Tillage also stimulates surface soil erosion, which can carry off mycorrhizal spores.

Table 29-3   Mycorrhizal dependency of plants.

Crop	Mycorrhizal Dependency				
	Very High	**High**	**Low**	**Very Low**	**Independent**
Winter crop	Linseed Faba bean	Chickpea	Field pea Oats Wheat Triticale	Barley	Canola Lupines
Summer crop	Cotton Maize Pigeon pea Lablab	Sunflower Soybean Navy bean Mung bean Sorghum			

(Adapted from Thompson et al. 1997)

Other management practices affect mycorrhizae. Burning stubble as a management practice for residue removal and disease control has a detrimental effect on mycorrhizae since the heat generated in the first few cm of soil kills the existing spores. Mycorrhizae are fungi and most fungi are obligate aerobes. Consequently, poor drainage and water-logging of soil decrease mycorrhizal populations. Fungicides are extremely toxic to mycorrhizae, particularly those broad spectrum fungicides that are used to reduce pathogenic fungi in soil. However, most other herbicides and pesticides, applied at recommended rates, have little effect on mycorrhizae.

Summary

Mycorrhizae occur on almost all terrestrial plants. They benefit plants by improving P and Zn nutrition, allowing more efficient use of P and Zn in fertilizer, stimulating N_2 fixation in nodulated plants, increasing disease tolerance in plants, immobilizing some heavy metals, improving water use and drought tolerance, and improving soil structure. High fertility means low mycorrhizal infection while low fertility means high mycorrhizal infection.

Two basic types of mycorrhizae are ectomycorrhizae and endomycorrhizae. Ectomycorrhizae aggregate around the root and form a fungal sheath surrounding the root called the Hartig net. Another characteristic of ectomycorrhizal infections is that the infected roots become stunted and club shaped. Vesicular arbuscular mycorrhizae (VAM) have a wide host range. They infect most agricultural crops. Important genera to remember are *Glomus, Gigaspora, Acaulospora, Entrophospora,* and *Scutellospora*. The VAM make two structures that distinguish them from ectomycorrhizae—vesicles and arbuscles. Vesicles are lipid-filled storage organs. Arbuscles are the site of metabolic exchange between mycorrhiza and plant. Most VAM are found within the first 20 cm of the soil profile, although they can be found at greater depths. The soil supports a community of mycorrhizae rather than a single species. Mycorrhizae increase P solubility by excreting acids and extending the surface area roots exposed to soil P.

Sample Questions

1. What are endomycorrhizae and ectomycorrhizae? In what ways are they different?
2. As a generality, most fungi involved in ectomycorrhizal associations are probably incapable of leading an independent saprophytic existence in soil. What observations would lead you to conclude this?
3. Under what conditions do you expect mycorrhizae to be of greatest importance in plant nutrition?
4. Briefly discuss two mechanisms by which VAM affect P availability to plants.
5. In terms of mycorrhizae, what is the difference between vesicles and arbuscles?
6. What theories have been advanced to explain why mycorrhizae are beneficial to plants? What are the relevant mechanisms?
7. Can mycorrhizae ever be harmful to plant growth? Explain the circumstances.
8. Draw a graph that shows the ideal relationship between the P concentration in soil and the extent of mycorrhizal infection of plant tissue.
9. Based on the data in the following table, what is the percent change in root P concentration in *Araucaria cunninghami* when it is infected with mycorrhiza?

Effect of vesicular-arbuscular mycorrhiza on yield and nutrient status of 2-year-old *Araucaria cunninghami* seedlings growing in steam-sterilized soil.

Parameter	Mycorrhizal Status		Significance of Difference
	Infected	Uninfected	
Dry weight (g)	78.0	72.6	***
Root P concentration (%)	0.122	0.050	***
Root N concentration (%)	0.90	0.89	NS

*** = Highly significant
NS = Not significant
(From Richards 1987)

10. Identify A, B, and C in the following diagram. What important part of the mycorrhizae is not shown? What is the function of each part?

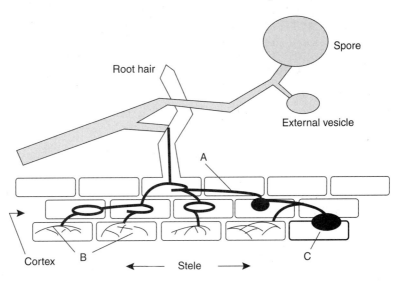

11. In the following table, what can you say about the effects of mycorrhizae on P uptake by *Ceanothus?*

Effects of mycorrhizae and N-fixing nodules on growth and N$_2$ fixation in *Ceanothus velutinus* seedlings.

	Control	+ Mycorrhizae	+ Nodule	+ Nodules + Mycorrhizae
Mean shoot weight (mg)	72.8	84.4	392.9	1028.8
Mean root weight (mg)	166.4	183.4	285.0	904.4
Root/shoot ratio	2.29	2.17	0.73	0.88
Nodules/plant	0	0	3	5
Mean nodule weight (mg)	0	0	10.5	44.6
Acetylene reduction (mg/nodule/hour)	0	0	27.85	40.46
Percent mycorrhizal colonization	0	45	0	80
Nutrient content (% of shoot by weight)				
N	0.32	0.30	1.24	1.31
P	0.08	0.07	0.25	0.25
Ca	–	–	1.07	1.15

(Adapted from Rose and Youngblood 1981)

12. In the data from the following table, what is the relationship between P content and mycorrhizal infection of cowpea and soybean?

Relationship between P level in soil and VA mycorrhizal infection of cowpea and soybean.

mg P per Liter	Cowpea	Soybean
	Percent of Cortical Tissue Infected	
0.003	75	50
0.006	75	75
0.012	50	50
0.025	10–50	10–50
0.05	50	50
0.1	10–50	10–50
0.2	50	<10
0.4	<10	<10
0.8	<10	<10
1.6	50	0

(Adapted from Yost and Fox 1982)

Thought Question

How would you design an experiment to show that mycorrhizae were actually transporting nutrients to plant roots?

Additional Reading

An excellent chapter on the various types of mycorrhizae appears in B. N. Richard's *Microbiology of Terrestrial Ecosystems* (1987, Longman Scientific and Technical, Essex, England).

References

Augé, R. M. Osmoregulation in plant roots. *Tennessee Farm and Home Science,* No. 152. University of Tennessee Experiment Station, Knoxville.

Crookston, J. E., J. E. Kurle, P. J. Copeland, J. H. Ford, and W. E. Lueschen. 1991. Rotational cropping sequence affects yield of corn and soybean. *Agronomy Journal* 83:401–13.

Hayman, D. S., A. M. Johnson, and I. Ruddledin. 1975. The influence of phosphate and crop species on endogone spores and vesicular-arbuscular mycorrhizae under field conditions. *Plant and Soil* 43:489–95.

Medina, O. A., D. M. Sylvia, and A. E. Kretschmer. 1988. Response of siratro to vesicular-arbuscular mycorrhizal fungi: II. Efficacy of selected vesicular-arbuscular fungi at different phosphorus levels. *Soil Science Society of America Journal* 52:420–23.

Ojala, J. C., W. M. Jarrell, J. A. Menge, and E. L. V. Johnson. 1983. Influence of mycorrhizal fungi on the mineral nutrition and yield of onion in saline soil. *Agronomy Journal* 75:255–59.

Paul, E. A., and R. M. N. Kucey. 1981. Carbon flow in plant microbial associations. *Science* 213:473–74.

Rose, S. L., and C. T. Youngblood. 1981. Tripartite associations in snowbrush (*Ceanothus velutinus*): Effect of vesicular-arbuscular mycorrhizae on growth, nodulation, and nitrogen fixation. *Canadian Journal of Botany* 59:34–39.

Satterlee, L., B. Melton, B. McCaslin, and D. Miller. 1983. Mycorrhizae effects on plant growth, phosphorus uptake, and N_2 (C_2H_4) fixation in two alfalfa populations. *Agronomy Journal* 75:715–16.

Sylvia, D. M. 1994. Vesicular-arbuscular mycorrhizal fungi. In *Methods of soil analysis, part 2: Microbiological and biochemical properties,* R. W. Weaver et al. (eds.), 351–78. Madison, WI: Soil Science Society of America.

Thompson, J., N. Seymour, D. Peck, and T. Clewett. 1997. VAM—a natural way to healthy crops. Croplink, Information Series ISSN 0727-6273 Q197053. The State of Queensland, Australia, Department of Primary Industries.

White, J. A., L. C. Munn, and S. E. Williams. 1989. Edaphic and reclamation aspects of vesicular-arbuscular mycorrhizae in Wyoming red desert soils. *Soil Science Society of America Journal* 53:86–90.

Yost, R. S., and R. L. Fox. 1982. Influence of mycorrhizae on the mineral content of cowpea and soybean grown in an oxisol. *Agronomy Journal* 74:457–81.

6

SOIL MICROORGANISMS AND ENVIRONMENTAL QUALITY

*I*n this last section of the text, we look at some ways in which soil microorganisms and their activity affect environmental quality. We begin by examining composting as a model for microbial ecology, decomposition, and pretreatment of biodegradable wastes. Composting is an ancient technology receiving renewed interest as a benign method of recycling. One of the hottest topics in environmental microbiology is the use of microorganisms to biodegrade and bioremediate toxic wastes that are present in the environment. We begin our discussion of this topic by examining the mechanisms by which soil microorganisms decompose toxic chemicals. Later, we examine the practical application of these mechanisms and the methods by which soil microbiologists manipulate both microorganisms and the soil environment to bioremediate toxic wastes. Finally, we briefly discuss the interaction of soil microorganisms and heavy metals.

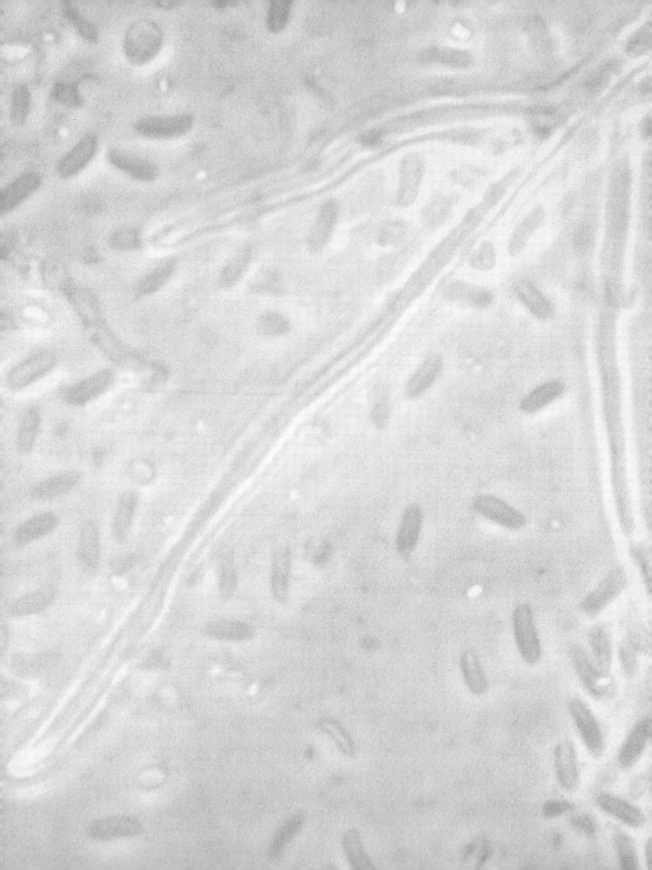

Chapter 30

Composting

Overview

After you have studied this chapter, you should be able to:

■ Describe what "composting" is and list five benefits of composting.
■ Explain why composting has become an increasingly popular means of waste treatment.
■ Discuss three composting processes.
■ Summarize seven conditions or considerations for optimum composting.
■ Explain what "vermicomposting" is.

INTRODUCTION

In traditional small-scale farm operations, most solid organic waste is composted and recycled as fertilizer. Composting is simply an acceleration of natural processes of organic matter mineralization. Composting is gaining in popularity as an alternative to incinerating or landfilling organic wastes from cities and industry. Composting accomplishes several beneficial goals for its practitioners: 1) It reduces the bulk of waste; 2) it lowers the biological oxygen demand (BOD) of waste; 3) it improves the waste's physical characteristics and makes it easier to handle; 4) it reduces human, animal, and plant pathogens and eliminates weed seeds; and 5) it reduces land use for landfilling and for surface application of waste. In this chapter we take a brief look at some microbiological characteristics of composting.

COMPOSTING

From a practical point of view, composting is a pretreatment process that converts organic wastes into forms suitable for land application. Composting is a microbial process in which noxious organic waste is converted into stable humus-like substances that have reduced bulk. One disadvantage of composting is that lots of land is required, particularly if static windrow systems are used. This affects neighboring residents who will support most environmentally conscious waste management systems as long as they're in someone else's backyard.

In composting, initial sorting of waste separates organic and inorganic fractions (compostable and noncompostable waste). Ferrous material can be removed by magnets; mechanical separators can partition glass, aluminum, and plastic into light and heavy fractions; new technologies allow plastics to be separated into recyclable and nonrecyclable fractions. Recycling these fractions helps subsidize the composting operation (a minor advantage) and reduces the amount of solid waste disposal in landfills (a major advantage). One of the principal motivations that cities have for composting is to extend the working lives of their landfills, and not necessarily to be ecologically sound.

The remaining organic waste is ground up to increase the surface area, amended with sludge, soil, or old compost (an inoculum), mixed bulking agents such as shredded newspaper, wood chips, or pecan shells (anything that decomposes slowly) to provide porosity, and is then composted.

Composting is accomplished in windrows, aerated piles, and continuous feed reactors. The windrow process is simple but slow. It is a large-scale version of the backyard compost pile. Windrow composting requires mixing at intervals for even composting and for several months for stabilization. Odor is controlled by soil or compost covering.

The aerated pile method composts wastes faster than does the windrow system (Figure 30-1). Perforated pipes are buried in a pile or windrow and air is either pumped inside or drawn through the piles by vacuum. The air stream oxygenates and cools the compost. This is important, since composting is an aerobic rather than an anaerobic process. The heat generated by a compost pile can be used to dry the final product.

Figure 30-1 An aerated static pile in the process of composting. (Photograph courtesy of W. O. Thom)

Continuous feed reactors are the quickest and costliest method of composting, and the most technologically sophisticated. A continuous stream of waste enters the reactor where it is agitated and composted. The retention time of the waste in the reactor is adjusted to meet the goals that the operator has for the compost's final chemical and physical condition. Typically, the final product requires some additional maturation and stabilization because its degradation is not complete.

THE MICROBIOLOGY OF COMPOSTING

Composting is initiated by mesophilic chemoheterotrophs. As they respire, the temperature in the compost pile increases and they are replaced by thermophilic organisms. The heat is produced by aerobic oxidation of the waste. The greater the biological availability of the waste, the more rapid the temperature rise of the compost. If you remember the sequence of decomposition in nature—large polymers are broken down into smaller subunits and then reutilized by microorganisms and plants and incorporated into soil—then you can diagram the microbiological and biochemical changes that occur in compost as it decomposes. Eventually, the temperature of the compost declines as the readily available substrates are consumed and mesophiles reestablish themselves.

The temperature in self-heating compost piles may rise to 76° to 80°C. This is why compost piles frequently steam on a cool day when poked open. A temperature of 80°C is too high for optimum composting because the heat begins to kill the organisms responsible for composting. However, high temperatures are more effective at killing pathogens and weed seeds in the compost than are lower temperatures. Maximum thermophilic activity in a compost is between 60° and 65°C. Aerating, watering, or turning a compost pile prevents excessive self-heating. Compost should be kept in thermophilic conditions for as long as possible. Not only does this speed up the composting process

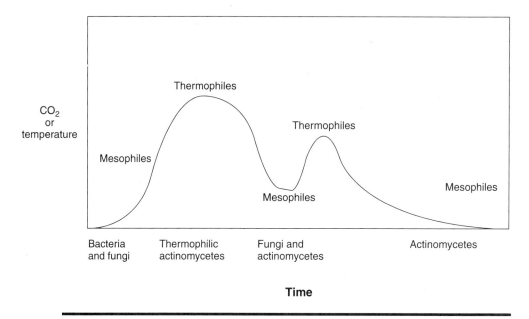

Figure 30-2   Succession of populations in a compost with time.

(remember that for every 10°C rise in temperature, microbial activity increases two- to threefold), but it also kills pathogens in the composted material. This is particularly important for composted sewage sludge and manure.

Turning the compost also makes the final product more uniform, since the thermophilic decomposition would otherwise be restricted to the core of the compost. By turning the compost, a second burst of decomposition often occurs as readily degradable material from the outside of the original compost is turned into the inside. The second thermophilic phase is followed by several months of maturation at a mesophilic temperature. During this period, resistant organic compounds are decomposed and the moisture content of the compost declines. Both processes make the final product easier to handle and reduce the compost's reactivity once it is used (Figure 30-2). Important bacteria in compost are *Bacillus stearothermophilus, Clostridium thermocellum, Thermomonospora,* and *Thermoactinomyces.* Important fungi are *Geotrichum, Aspergillus,* and *Mucor.*

OPTIMAL COMPOSTING

The principal components or considerations associated with optimal composting are: 1) the type and composition of the organic waste; 2) the availability of microorganisms; 3) aeration; 4) the C, N, and P ratios; 5) moisture content; 6) temperature; 7) pH; and 8) time. We discuss the microbial basis for each of these considerations in turn.

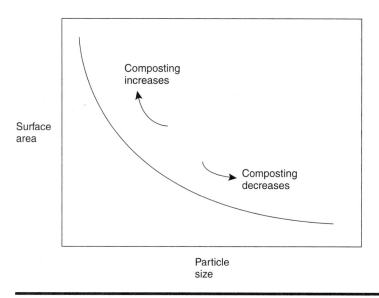

Figure 30-3 The influence of particle size on composting.

The Type and Composition of the Organic Waste

Organic wastes vary considerably in terms of their chemical and physical characteristics. The surface area, hydrophobicity, and type and complexity of chemical bonds in the waste are key issues. The decomposition rate is proportional to the surface area and inversely proportional to size. So, as particle size decreases, surface area increases, and decomposition rate increases (Figure 30-3). A particle size that varies from 0.65 to 2.54 cm is recommended as the optimal range for composting. Smaller particles interfere with aeration; larger particles are too unreactive.

Insoluble compounds are more resistant to decomposition than are soluble compounds, and hydrophobic compounds are more resistant to decomposition than are hydrophilic compounds. Soluble compounds, such as monosaccharides and amino acids, decompose more readily than do linear macromolecules and these, in turn, decompose more readily than do branched macromolecules. Branched macromolecules decompose more readily than compounds such as lignin that have multiple bond types and irregular structure.

Another key issue is whether contaminants, such as heavy metals or toxins, are carried in the waste. The contaminants can make the compost unfit for distribution after the composting process is done, which is a serious problem with composted municipal sludge, or may cause the composting itself to be impeded by poisoning the microbial population.

The Availability of Microorganisms

It's assumed that microorganisms are in the compost. Whether they came from the composting material or were added as part of the composting process is another question. Rapid composting presupposes that a large, active microbial population is present. The microbial population consists of a mixed group of mesophilic and thermophilic bacteria, actinomycetes, and fungi. Inoculating compost with material from a previous compost pile can be beneficial, because it adds a representative mix of microorganisms in high numbers.

Aeration

Microbial decomposition of organic compounds proceeds fastest under aerobic conditions. More energy is made available from aerobic respiration than from fermentation. This means that for the same amount of C, higher microbial populations can develop and if higher microbial populations develop, waste decomposition can be accelerated. Fungi and actinomycetes are obligate aerobes for the most part, so the less O_2 available, the slower they metabolize waste.

The complete oxidation of volatile compounds reduces the potential for noxious odors arising from composting. Aerobic conditions lead to oxidized compounds such as NO_3^- and CO_2. Anaerobic conditions, in contrast, lead to odorous volatile compounds being released. Consequently, the O_2 concentration of the compost air space should be kept at 5% or above. The porosity of the composting mix should be about 30%. Bulking agents that resist decomposition are used to maintain porosity.

The C, N, and P Ratios

A C:N ratio of 30:1 to 40:1 is appropriate for the initial compost material, but ratios as low as 20:1 have been recommended. The higher ratios may cause some immobilization of N, but because compost is frequently used as a potting or horticultural mix, immobilizing N in the compost is a good way of keeping it available for a later date. The C:N ratio should not be lower than 20:1 because that could cause N loss through leaching and volatilization as the organic N mineralizes. C:P ratios of 100:1 to 150:1 are recommended.

At all costs avoid extreme C:N ratios. This should be intuitive, based on whether the composting material is obviously nutrient rich or nutrient poor. High C:N ratios do not allow a large microbial population to form without additional nutrients. The result is a slower composting rate. Low C:N ratios lead to NH_3 volatilization, which causes N loss and odor problems. For that reason, home composters are urged to avoid putting meat and bones in their compost unless suitably C-rich materials are also added, or unless they have understanding neighbors. You can lower the C:N ratio by mixing high C:N material with low C:N material. Computer programs exist to help in designing these mixing ratios (Brodie 1994). In their absence, the skills of a compost operator include blending the appropriate amounts of N-rich, readily decomposable "green" material with slowly decomposing, C-rich "brown" material to maintain an appropriate C:N ratio.

Moisture Content

Adequate moisture is critical for composting; 50% to 60% water content (water-filled porosity) is about right. The compost should be moist but not soggy. Water helps cool the compost. Microbial growth and activity also require enough moisture to keep water films on solid surfaces for movement and metabolism and for the diffusion of soluble compounds. Too much water impedes O_2 diffusion, creates anaerobic conditions, and slows decomposition rates.

Temperature

Composting occurs in two temperature ranges: a mesophilic range varying from 10° to 43°C (50° to 109°F) and a thermophilic range varying from 55° to 60°C (109° to 151°F). Self-heating static piles experience a temperature rise of 55° to 60°C in 2 days. After a few days at peak temperature, the temperature gradually declines, and turning the compost at this point causes a secondary temperature increase.

Temperature control is one of the critical aspects of composting that is distinctly different from the way in which decomposition occurs in soil. It accounts for one of the reasons why decomposition in compost is much faster than decomposition in soil. Composting occurs at temperatures much higher than can be generated in soil. Since microbial activity doubles or triples for every 10°C rise in temperature (up to a point), the warmer the compost, the faster things will decompose. Ultimately, a temperature is reached that becomes inhibitory, but this varies among different groups of microorganisms.

During the early stages of composting, mesophilic microorganisms are responsible for decomposition. Microbial metabolism generates heat, which causes the temperature of the compost to rise. The rising temperature allows thermophilic bacteria, actinomycetes, and fungi to grow and metabolize. A self-heating compost that has a temperature rise of 70° to 80°C becomes inhibitory. This is one reason why fauna find the compost inhospitable. An optimal temperature is around 60° to 65°C. Composting at this temperature supports a large thermophilic population.

When composting manures and sludge, it is often important to maintain the temperature above 70°C for 72 hours. Pathogens are not thermophilic and this temperature effectively kills most of them (but not all). Temperatures as high as 63°C are needed to kill weed seeds.

Aerating and watering compost are ways to maintain optimum temperatures. Frequent turning also helps to cool the compost. When readily decomposable material is used up, metabolism slows and the temperature falls, allowing mesophilic populations to redevelop. Turning the compost at this stage causes a second temperature increase as undecomposed material is metabolized.

When turning no longer causes increases in temperature, the compost is ready for stabilization. This can be accomplished in as little as 20 days, but usually takes longer. Stabilization, or maturation, decreases the water content and allows metabolism of more resistant organic fractions.

Table 30-1 Composting times for combinations of methods and materials.

Method	Materials	Composting Time		Curing Time (Maturation)
		Range	Typical	
Static pile	Leaves	2–3 years	2 years	—
	Well-bedded manure	6 months–2 years	1 year	—
Aerated static pile	Sludge/wood chips	3–5 weeks	4 weeks	1–2 months
Windrow, infrequent turning	Leaves	6 months–1 year	9 months	4 months
	Manure	4–8 months	6 months	1–2 months
Passive aerated windrow	Manure bedding	10–12 weeks	—	1–2 months
	Fish waste/peat moss	8–10 weeks	—	1–2 months
Agitated bed	Sludge/yard waste or manure/sawdust	2–4 weeks	3 weeks	1–2 months
Rotating drum	Sludge and/or solid waste	3–8 days	—	2 months

(From Carr et al. 1995).

pH

Optimum pH is near neutral for most microorganisms and slightly alkaline for actinomycetes important in composting. This translates into a range of about 5.5 to 8.5, but the pH extremes should be avoided. A pH of 6.5 to 7.2 is ideal.

Time

One final consideration is time, which will be a function of all of these other factors and the mechanical manipulation of the compost. Curing is the slow maturation of compost after active composting is done. It allows volatile compounds to escape, some additional decomposition to occur, and the moisture of the composted material to decline. In Table 30-1, the effects of composting method and composting material are shown. Table 30-1 reflects the different amounts of time it takes to compost different materials.

Compost is a good soil conditioner because it is humus-like. Thus, it promotes aggregation and improved aeration in the soil to which it is applied. It is not, however, a substitute for fertilizer because it only reflects the fertility status of the parent material that was composted. If the parent material had a balanced fertility level, it is reflected in the compost. If not, as will likely be the case, the nutrient status of the finished compost may be lacking in one or more key plant nutrients.

VERMICOMPOSTING

Vermicomposting is a recycling method that uses earthworms to consume and process organic wastes. Vermicomposting is an increasingly popular method of recycling household food scraps and has even been tried in the decomposition of animal wastes and municipal sewage sludges. Worms are bought or collected from a garden and placed in a ventilated wooden box to which bedding material such as newspaper and household scraps are added. The worms digest the scraps and the bedding material helps to maintain the worm casts in a moist condition. Periodically, the worms are sieved from the box to prevent the population from exceeding the food supply.

Vermicomposting Municipal Biosolids

Vermicomposting uses earthworms to decompose wastes such as biosolids (the residual material produced by municipal water treatment). The advantages of vermicomposting, compared to other forms of biosolid treatment, are that it is all natural, it requires minimum technology, it reduces pathogen levels in the compost, and it produces a nutrient-rich fertile material in the worm casts.

How do you vermicompost biosolids? In pilot projects, composters prepare small, 6.1×9.1 m plots (20×30 ft) of level land and install an impermeable layer (plastic or clay) at the base of a windrow. A layer of sand is placed on top to help drainage since biosolids are still fairly wet and contain only about 20% solids. Fifty pounds of red worms or manure worms (*Eisenia foetida*) are placed at the bottom of a 1.5×9.1 m (5×30 ft) windrow that is 15 cm (6 inches) high. When the worms eat their way to the top of the pile, another 15-cm-high layer is added. This process continues until the windrow is 45 cm high. Then, a fresh windrow is placed next to the first and migrating worms begin the process of vermicomposting anew.

Vermicomposting, like traditional composting, requires aerated environments, protection from the environment, and sufficient moisture to maintain the biology of the system. Vermicomposts also require periodic turning to uniformly mix the decomposing wastes with the earthworms. Unlike traditional composting, however, windrows in vermicomposts are kept shallow to prevent heat buildup. The raised temperatures that accelerate decomposition and kill pathogens in traditional composts kill any earthworms that were present.

Summary

Composting is simply an acceleration of organic matter mineralization. Composting is gaining in popularity as an alternative to incinerating or landfilling organic wastes from cities and industry. Composting reduces the bulk of waste; it lowers the biological oxygen demand (BOD) of waste; it improves the waste's physical characteristics and makes it easier to handle; it reduces human, animal, and plant pathogens and eliminates weed seeds; and it reduces land use for landfilling and for surface application of waste.

From a practical point of view, composting is a pretreatment process that converts organic wastes into forms suitable for land application. Composting is a microbial process in which noxious organic waste is converted into stable humus-like substances that have reduced bulk. Composting is performed in windrows, aerated piles, and continuous feed reactors.

The principal components or considerations associated with optimal composting are the type and composition of the organic waste; the availability of microorganisms; aeration; the C, N, and P ratios; moisture content; temperature; pH; and time.

Vermicomposting is a recycling method that uses earthworms to consume and process organic wastes. Vermicomposting is an increasingly popular method of recycling household food scraps and has even been tried in the decomposition of animal wastes and municipal sewage sludges.

Sample Questions

1. Why is it necessary to "turn over" compost?
2. What are the optimal environmental conditions for composting?
3. Why does CO_2 evolution and temperature increase in a compost pile just after it has been turned?
4. Why is aeration important to a well-functioning compost pile?
5. Will attempting to compost wood chips from old lumber be any different than composting wood chips from freshly cut trees?
6. Why do compost piles occasionally catch fire?
7. Why do professional composters sometimes talk about getting the correct mix of "green" and "brown"?
8. Can everything be composted?
9. Draw a graph that shows the change in C:N ratio in a compost pile with time.
10. Draw a figure that illustrates the changing composition of composted material with time.
11. What effect will adding inorganic N have on the rate of CO_2 evolution from a new compost pile that has a C:N ratio of approximately 50:1?
12. Will adding additional inorganic N stimulate CO_2 evolution rates very much if it is added to a maturing compost pile?
13. How can you tell if a compost pile is "mature"?

Thought Question

Beginning with yard clippings added to a compost pile, describe or illustrate the pathway that those clippings will take until they are ultimately incorporated into soil organic matter. At each point, indicate how they will change chemically, and for each step, give some sense of the microbial communities involved.

Additional Reading

A delightful little book on the procedures for home vermicomposting is called *Worms Eat My Garbage* by Mary Appelhof (1982, Flower Press, Kalamazoo, MI). For serious composters, the Solid Waste Composting Council (601 Pennsylvania Avenue NW, Suite 900, Washington DC 20004) published *Compost Facility Planning Guide,* 1st ed. (1991, Solid Waste Composting Council, Washington DC).

References

Brodie, H. L. 1994. Multiple component compost recipe maker. ASAE paper 94-3037. International ASAE Meeting. Kansas City, MO.

Carr, L., R. Grover, B. Smith, T. Richard, and T. Halbach. 1995. Commercial and on-farm production and marketing of animal waste compost products. In *Animal waste and the land-water interface,* K. Steele (ed.), 485–92. Boca Raton, FL: CRC Lewis Publishers.

Chapter 31

Bioremediation

Overview

After you have studied this chapter, you should be able to:

- Explain some of the basics of biodegradation and xenobiotic compounds.
- Identify two approaches to bioremediation.
- Compare the advantages and disadvantages of bioremediation with conventional waste treatment.
- Discuss how to use substrate analog enrichment to increase the cometabolism of xenobiotics.
- Explain some of the methods used to bioremediate petroleum waste in soil.
- Identify criteria and techniques used to bioremediate wastes in aquifers.
- Describe how enzymes are used to bioremediate toxic compounds.
- Identify criteria for seeding microorganisms in bioremediation.

INTRODUCTION

The principle of microbial infallibility is that all natural compounds are biodegraded given favorable environmental conditions. The term "xenobiotics" refers to artificial compounds that are foreign to biological systems and contain structures and bonds that don't occur in biological systems. Xenobiotics may be polymers, gases, polychlorinated or polybrominated compounds, or pesticides. Bioremediation is the process of cleaning up contaminated sites. Biodegradation refers to the processes in bioremediation by which xenobiotics are transformed to less toxic states. Just because a compound is biodegraded doesn't mean that the breakdown products are any less toxic or undesirable. Mineralization usually means decomposition of a xenobiotic to inorganic ions and CO_2. This is the most desirable situation because the end products are usually nontoxic.

Recalcitrant molecules are fossil organic matter (humus), polyaromatic compounds (tanins and lignins), persistent microorganisms (endospores and melanin-rich fungi), synthetic molecules (fungicides, nematicides, herbicides, insecticides), polyhalogenated biphenyls (flame retardants and solvents), plastics, and detergents. The persistence of xenobiotics ranges from days to years and minor alterations in biodegradable compounds can render them recalcitrant.

In this chapter we discuss biodegradation by looking at some specific examples of technologies and practices that are used for bioremediation.

APPROACHES TO BIOREMEDIATION

There are two general approaches to bioremediation:

1. Environmental modification—for example, improving the potential activity of existing microorganisms through fertilization and aeration
2. Addition of appropriate and sometimes selective microorganisms

Bioremediation can be done on-site (*in situ*) or at special facilities (*ex situ*).

There are several advantages to bioremediation compared to conventional waste disposal. The end products are generally nontoxic if complete mineralization occurs. Other biological activity in the contaminated site is left relatively undisturbed. Bioremediation is inexpensive compared to physical methods for cleanup, and bioremediation requires simple equipment.

There are also several disadvantages to bioremediation. A high waste concentration is sometimes needed to stimulate growth of the bioremediating microorganisms. There are limits to the materials that can be treated because they may be inherently recalcitrant or toxic. Bioremediation is limited by the environmental conditions existing at the contaminated site. The xenobiotic's fate depends on the intrinsic properties of the xenobiotic and the microbial population and extrinsic factors inherent in the environment. If it is too hot, too cold, too wet, too dry, too acidic, or too alkaline, bioremediation is slowed or stopped completely. Bioremediation is limited by the time available for treatment; it takes longer to bioremediate a contaminated site than to treat it by chemical or physical means.

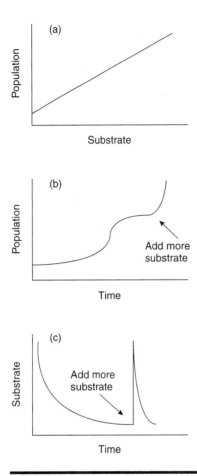

Figure 31-1   Enrichment culture. If the xenobiotic concentration is not toxic, there is usually a direct relationship between the amount of xenobiotic you add and the amount of microbial growth you observe (a). Every time you add more xenobiotic, you can stimulate additional growth (b). One consequence of stimulated growth is that each addition of xenobiotic should disappear more quickly because there are more cells that degrade the xenobiotic and because the enzymes involved in biodegradation are already induced (c).

METABOLISM, COMETABOLISM, AND BIOREMEDIATION

Bioremediation is based on microbial metabolism. Xenobiotics can serve as substrates for microbial growth and energy or they can be cometabolized. The xenobiotic supports microbial growth if it is metabolized (Figure 31-1). The xenobiotic becomes a source of C, N, S, and energy. If the xenobiotic is added to soil, the microbial population increases. Microorganisms that specifically grow on the xenobiotic can be isolated by enrichment culture.

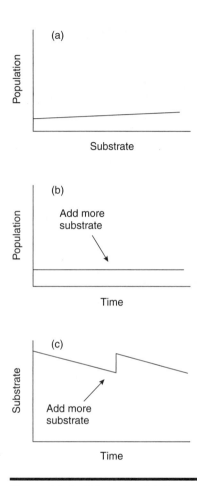

Figure 31-2   Failure to enrich for microorganisms that cometabolize. Since additional xenobiotic does not support additional microbial growth (a), there is no effect of the xenobiotic addition on a cometabolizing population (b). The xenobiotic will be cometabolized at a constant rate regardless of whether more is added and you will never be able to obtain more of the cometabolizing microbes unless you add something else to stimulate their growth (c).

The xenobiotic may be modified but does not serve as a nutrient source. This is called cometabolism. The microorganisms do not obtain energy from the transformation reaction and require another substrate for growth. This usually results in simple modifications that can increase or decrease the toxicity of the xenobiotic. No growth occurs and no population develops because the substrate has been added. The degradation rate is constant and depends on whatever factors control the degrading population. You cannot isolate cometabolizers by enrichment (Figure 31-2). Xenobiotics that are cometabolized are usually relatively persistent.

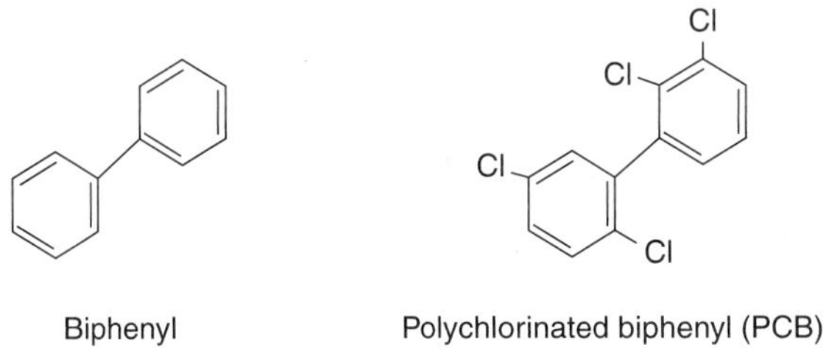

Biphenyl   Polychlorinated biphenyl (PCB)

Figure 31-3   Biphenyl is an analog of PCB that can be used to increase its cometabolism by analog enrichment.

If a material is cometabolized you can enrich for organisms that speed its degradation by using a substrate analog (Figure 31-3). For example, biphenyl addition increases the population of microorganisms that cometabolize PCBs (Brunner et al. 1985) and aniline can be added to increase the cometabolism of dichloroaniline.

BIOREMEDIATION OF PETROLEUM

Bioremediation has seen its greatest application in treating petroleum contamination. The *Exxon Valdez* oil spill is a case in point. On March 24, 1989 there was a massive oil spill in Prince William Sound, Alaska that contaminated thousands of miles of pristine beaches. Some beaches were physically cleaned but the U.S. Environmental Protection Agency also tested several hundred sites for bioremediation of petroleum.

Fertilization approaches appeared to be much more successful at removing oil from beaches in Alaska than was mechanical cleaning. It stimulated a small but active hydrocarbon-degrading community in the soil. The experiment with bioremediation was so successful that the U.S. EPA recommended that Exxon (the company assuming responsibility for the waste spill) expand its bioremediation efforts to more of the affected beaches. Venture capital firms now specialize in developing specific microbial strains for oil degradation in soil and water. These companies supply mixtures of hydrocarbon-degrading pseudomonads to bioremediate contaminated sites. In winter, they provide a mixture of psychrophilic hydrocarbon degraders.

Undegraded hydrocarbons do not readily leach, so bioremediation has the potential for bioremediating fuel-contaminated soil. Although remediation of gasoline in surface soil probably occurs through volatilization as much as anything, less volatile fuel compounds can be reduced by 90% in 300 days through land farming (Kincannon 1972). In one demonstration, oily sludge was applied at a rate of 5% hydrocarbon concentration (% wt/wt) in the upper 15 to 20 cm of soil. This rate was used because >10% oily sludge inhibited degradation. This application rate was equivalent to 100,000 L of hydrocarbon per ha. The hydrocarbon:N ratio was about 200:1 and the hydrocarbon:P ratio was about 800:1.

This example illustrates three important considerations of land farming petroleum to bioremediate it: 1) The material is incorporated shallowly because this is where the

PCBs—Useful Product Turned Environmental Contaminant

A PCB, or polychlorinated biphenyl, is a biphenyl molecule (a pair of benzene rings joined by a single carbon-to-carbon bond; benzene itself has the chemical formula C_6H_6) on which chlorine atoms substitute for some of the hydrogen atoms on the biphenyl structure. Polybrominated biphenyl (PBB) uses bromine instead of chlorine. Each ring on the biphenyl structure can have up to 5 chlorine atoms, so it is theoretically possible to make 209 different kinds (congeners) of PCBs. In practice, most PCBs are mixtures of 70 or more of these different congeners (Gustafson 1970).

PCBs were first made in 1881 and became commercially available in the 1920s. Monsanto Company sold PCBs with the trade name "Aroclor" in the United States. Aroclors were usually liquids or resins and had a four-digit identification code. The last two digits in the code indicated the percentage of chlorine (by weight) in the total mix. For example, Aroclor 1254 contained 54% chlorine by weight; higher numbers meant more chlorine.

PCBs had attractive chemical properties such as high boiling points, low water solubility, and low conductivity. PCBs were hard to burn, resisted acids and bases, and were mostly inert. Consequently, PCBs were soon used in adhesives, ballasts for fluorescent light fixtures, carbonless copy paper, insulation in transformers, high-pressure hydraulic fluids, machine tool-cutting oils, specialized lubricants and gasket sealers, plasticizers, and protective coatings for wood, metal, and concrete. Some varnishes and epoxy paints contained PCBs. Braided cotton and asbestos in electric wire insulation were impregnated with PCBs to improve their fire resistance.

The properties that made PCBs so useful to industry also made them potentially long-lasting and widespread pollutants. In 1966, Sören Jensen, a Swedish chemist from the University of Stockholm, reported evidence that PCBs had entered the food chain as early as 1944 (Anonymous 1966). Scientists soon found PCBs everywhere, from the Arctic to the Antarctic and, most disturbingly, in human hair and fat.

Scientists and environmentalists worried about the environmental consequences of PCBs. PCBs cause thin-shelled eggs in birds because (like DDT) they stimulate enzymes that make estrogen more soluble and readily excreted. The hormone estrogen controls the calcium level in breeding females. Low estrogen levels keep calcium reserves low and make little calcium available for eggshell formation. Although the acute toxicity of PCBs is low, chronic exposure has adverse health effects, and as PCBs become more chlorinated, they also become more carcinogenic.

Although PCB concentrations are highest around industrial or urban areas, the global spread of PCBs seemed odd based on their low solubility. It turns out that burning PCB-contaminated waste releases PCBs into the atmosphere. It is virtually impossible to find environmental samples that do not contain some trace amounts of PCBs and other chlorinated chemicals now that they are widely distributed in the environment. Even foods grown in "chemical free" fields can receive trace amounts of PCBs in atmospheric deposits from incinerator exhausts.

PCB and PBB production in the United States was ultimately halted by several notorious events (Carter 1976). In 1974, 30,000 or more cattle and other farm animals in Michigan were quarantined and destroyed because they were contaminated with PBBs. The contamination occurred when Michigan Chemical Corporation of

Continued

St. Louis, Michigan accidentally included 10 to 20 bags of "Firemaster," a PBB-containing flame retardant, with a truckload of "Nutrimaster," a magnesium oxide feed additive, that they were delivering to a feed mill. The PBBs were mixed with feed that was distributed around the state. As a result, contaminated milk and eggs exposed most of the population in Michigan to PBBs. The contamination, animal disposal, and aftermath became a multimillion dollar environmental catastrophe.

Since that time, researchers have learned that microorganisms slowly dechlorinate PCBs under anaerobic conditions and make the PCBs less carcinogenic. Consequently, PCB deposits in many aquatic environments will become naturally attenuated with time if left undisturbed.

maximum microbial activity is and this placement also makes it easier to mix in additional fertilizer N and P as well as aerate the soil by tillage; 2) the oily sludge has an extremely high hydrocarbon-to-nutrient ratio, so, unless more fertilizer N is added, the hydrocarbon degraders will immobilize all of the available N and P in soil and the biodegradation of the oily waste itself will be slowed; and 3) the application rate is kept below 10%. Application rates that start to affect the physical structure of soil, and aeration in particular, will decrease biodegradation. At higher application rates, toxic components of the waste inhibit the microbial population.

BIOREMEDIATION OF WASTE GASES

Bioremediation of air is not something everyone thinks about, but it does occur and it is a common practice (Bohn 1977). To remove volatile compounds from air, biofilters, trickle filters, and bioscrubbers are used. These scrubbers remove H_2S, dimethyl sulfide, terpene, organo-sulfur gases, ethyl benzene, tetrachloroethylene, and chlorobenzene from air streams. Adsorption of these gases into biofilms or beds is also used. Classic biofilter materials in use include peat, compost, bark, and soils. The soil layer placed on top of most backyard compost heaps, for example, is there to absorb foul odors that may be produced within the compost. It's nothing more than a biofilter. The waste gases are filtered by adsorption followed by biodegradation since the biofilter material becomes a rich microbial community.

BIOREMEDIATION OF SUBSURFACE WASTE

What can be done to bioremediate subsurface waste—waste buried so deep that excavation of contaminated soil is prohibitively expensive, or waste so diluted in groundwater that pumping and remediation are impractical? *In-situ* bioremediation is possible for some sites. Microbial bioremediation of chemicals in groundwater and subsurface soil is limited by oxygen availability. One solution is to pump in O_2 or a diluted H_2O_2 solution. Decomposition of H_2O_2 provides O_2, which supports aerobic metabolism. A concentration of 100 ppm is often used. This method seems to work for BTEX (benzene, toluene, ethyl-benzene, xylene) bioremediation. The BTEX contains contaminants with aromatic rings that require O_2 for ring cleavage by ortho- and meta-cleavage pathways. At the same time O_2 is being pumped in, either by compressed air or by H_2O_2 treatment, nutrients can be added to speed the growth of native bioremediating microorganisms. For specific wastes, mixtures of specialty microorganisms can be simultaneously added to enhance bioremediation by seeding. For compounds that are cometabolized, such as TCE (trichloroethylene), methane can be simultaneously injected to stimulate the growth of methanotrophs that cometabolize TCE (Chapelle 1993).

BIOTREATMENT OF METALS

Is it possible to bioremediate metals? Phytoremediation refers to using plants to bioremediate an environment. Plants are typically used when the environment is contaminated by heavy metals such as lead (Pb), mercury (Hg), or selenium (Se). *Astragalus* (loco weed), for example, accumulates Se in its tissue. Indian mustard is a Pb accumulator. The plants can be harvested, the tissue burned, and the metal-contaminated ash, now that it is a much smaller volume, can be stored in a hazardous waste facility.

Metals in an undesirable chemical state can be transformed to more easily handled forms by adjusting the redox potential of the environment. Chromium (VI) (Cr^{6+}) is water soluble and highly toxic. If it is reduced to (Cr^{3+}), it becomes less soluble and less toxic. Arsenic (III) (As^{3+}), on the other hand, is oxidized to As^{5+}, making it more easily precipitated by lime and phosphate. Manganese (II) (Mn^{2+}), likewise, can be oxidized to Mn^{4+}, which precipitates as an insoluble manganese oxide. In a nonacidic environment, oxidation of Fe^{2+} to Fe^{3+} causes iron hydroxide to precipitate and pulls other metals out of solution with it. In anaerobic, SO_4^{2-}-rich environments, formation of reduced S^{2-} promotes formation of insoluble metal sulfides.

ENZYME TECHNOLOGY

Bioremediation and detoxification can occur through the use of exoenzymes (Bollag and Liu 1990). Enzymes are not affected by inhibitors of microbial metabolism and a wider variety of environmental conditions are supported. Enzymes are effective at lower substrate concentrations and are active in the presence of toxins and microbial predators. Enzymes are very substrate specific and there are no diffusion limitations through cell membranes when enzymes are used. The disadvantages of using enzymes are that extracting and purifying enzymes is time-consuming and expensive. Many enzymes may also require cofactors and are susceptible to microbial proteases. Some of these disadvantages can be overcome by using immobilized enzymes. Immobilized enzymes are more resistant to degradation than are free enzymes.

Some examples of exoenzymes used in bioremediation are esterases, acylamidases, phosphatases, lyases, lipases, proteases, and phenol oxidases. Peroxidases, for example, catalyze phenolic aromatic or amine radical formation, and then react to form polymers. The polymers are insoluble and can be removed by filtration. Since cell growth is not required, the pH can range from 3 to 12. Laccases catalyze similar reactions to peroxidases. Laccases, peroxidases, and tyrosinases catalyze oxidative-coupling reactions of organic compounds to themselves and to other organic material. The toxicity of the bound compounds decreases, as well as their availability.

SEEDING

Seeding is the introduction of foreign microorganisms into the natural environment for the purpose of increasing the rate or extent of bioremediation. To be successful, the seeded microorganisms must be able to biodegrade most compounds. They must be genetically stable and remain viable during storage, yet grow rapidly after storage. Seeded organisms must not be pathogens nor should they produce toxic metabolites. Finally, they must be able to compete with native microorganisms.

There have been mixed results with seeding. Indigenous soil microbial populations that are highly adapted to a particular environment prevent seed organisms from successfully competing and surviving. Contaminants also typically have a low concentration—high enough to cause a health concern but too low to permit growth of the seeded organism. Inhibitory substances that prevent growth of the seeded microorganism may be present in the contaminant. The seeded microorganisms may prefer to grow on alternative organic material. If the seeded organisms are immobile, they may be prevented from reaching the contaminant if the contaminant is poorly soluble and the seeded organism is not well dispersed. The best results from seeding have occurred where the environment has been controlled. It is much more difficult to maintain suitable environmental conditions in the field.

Summary

Bioremediation is the practice of using microorganisms, plants, and animals to biodegrade and detoxify harmful chemicals in the environment. Two general approaches to bioremediation are *in situ* bioremediation by enhancing the activity of native organisms, and seeding contaminated sites with organisms known to bioremediate particular wastes.

There are several advantages to bioremediation compared to conventional waste disposal. The end products are generally nontoxic, other biological activity in the contaminated site is left relatively undisturbed, it is inexpensive compared to physical methods for cleanup, and it requires simple equipment. There are also several disadvantages. A high waste concentration is sometimes needed to stimulate growth, the wastes may be inherently recalcitrant or toxic, and bioremediation may be inhibited by lack of oxygen, an unfavorable pH or temperature, or by lack of nutrients and water. It takes longer to bioremediate a contaminated site than to treat it by chemical means.

Bioremediation has seen its greatest application in the cleanup of petroleum wastes. Fertilization approaches appeared to be much more successful at removing oil from beaches in Alaska after the *Exxon Valdez* oil spill in 1989 than was mechanical cleaning. Fertilization stimulated a small but active hydrocarbon-degrading community in soil. Several companies now specialize in developing specific microbial strains for oil degradation in soil and water. Biofilters, trickle filters, and bioscrubbers are used to bioremediate volatile compounds in air. Bioremediation of petroleum and by-products in groundwater and subsurface soil is most often limited by oxygen availability.

Phytoremediation refers to using plants to bioremediate an environment. Some plants can accumulate heavy metals such as Pb, Hg, or Se. For compounds that are cometabolized, such as TCE (trichloroethylene), methane can be added simultaneously to stimulate the growth of methanotrophs that cometabolize TCE. Metals in an undesirable chemical state can be transformed to more easily handled forms by adjusting the redox potential of the environment.

Bioremediation and detoxification can occur through the use of exoenzymes. Seeding introduces foreign microorganisms into the environment to increase the rate or extent of bioremediation. There have been mixed results with seeding because native soil microorganisms compete well against seeded organisms. The best results from seeding have occurred in controlled environments.

Sample Questions

1. What does the term "xenobiotic" imply?
2. What intrinsic and extrinsic factors can result in recalcitrance of xenobiotic compounds?
3. What is "seeding" with respect to biodegradation?
4. Why is seeding often unsuccessful?
5. What steps would you take to bioremediate petroleum on a deserted island?
6. What are the advantages of using plants and/or enzymes for bioremediation?
7. Rank the following compounds in order of their ease of biodegradation. Explain your system of ranking.

   a. Catechol
   b. Protocatechuic acid
   c. Gentisic acid
   d. Inositol phosphate
   e. Glucose
   f. N-acetyl glucosamine
   g. Phenyl propane
   h. 2,4-D
   i. Fatty acid
   j. Urea

8. Of the chemicals shown below, which will be most likely to biodegrade at an accelerated rate if it is applied to a soil continuously treated with 2,4-D? Explain your reasoning. Is there another chemical that will also biodegrade more rapidly than normal in a 2,4-D-treated soil? Why might this happen?

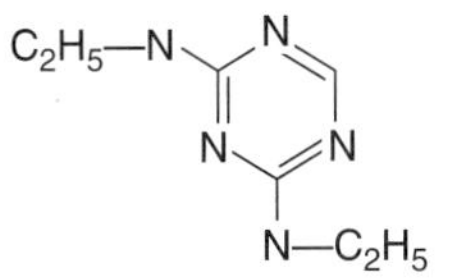

Simazine

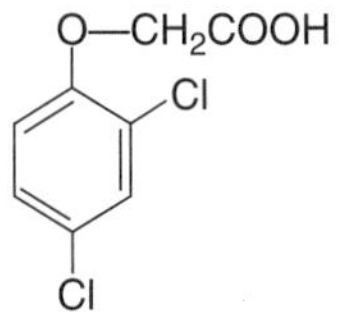

2, 4-D

p-Nitrophenol

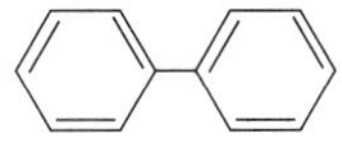

Biphenyl

9. What biodegradation process is being illustrated in the figure below? What are the characteristics of this process?

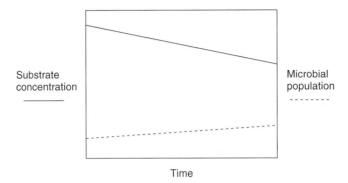

10. On March 24, 1989, there was a massive oil spill in Prince William Sound, Alaska. Many beaches were coated with hydrocarbons. Based on soil microbiological principles, the remediation effort included spray application of oleic acid, urea, and an organic P compound. The oleic acid makes the additive stick to oil. Laboratory incubations show that addition of this mix increases hydrocarbon-degrading populations and accelerates ^{14}C-labeled hydrocarbon degradation, yet in the field, results are mixed. On rock surfaces it is effective, but with depth in sediment it seems less effective. You are called as an expert witness and are asked the following questions. Answer these questions based on specific microbial processes and reactions.
   a. Why are you using the mix and how does it work?
   b. Why do you think it is of variable effectiveness on the beaches?

Thought Question

Acme Bioengineering Corporation (ABC) wants to win a $100,000 contract from your city government to release genetically engineered microorganisms onto a waste-contaminated land site in your community. The microorganism they plan to release biodegrades the waste in question within weeks in the laboratory; biodegradation of the waste normally takes years in the field. However, ABC has never field-tested its product. The waste itself is relatively insoluble and causes chronic health problems only with persistent exposure. The contaminated site is on city land that is also ideally situated for a housing development that could ultimately generate millions of revenue dollars for the city through its sale and future taxes.

As a concerned citizen, write a letter to your city council that takes and defends a position on the following question: Should Acme Bioengineering Corporation (ABC) be hired to release genetically engineered microorganisms onto a waste-contaminated land site in your community?

Additional Reading

A basic primer on bioremediation is *Biodegradation and Bioremediation* by Martin Alexander, 1994 (Academic Press, San Diego, CA). Alexander also concisely described chemical reactions in bioremediation in "Biodegradation of Chemicals of Environmental Concern" (1980, *Science,* Vol. 211, No. 9). Horace Skipper and Ron Turco edited a collection of reports on bioremediation treatment technologies in *Bioremediation: Science and Applications* (1995). *Understanding Bioremediation: A Guidebook for Citizens* (1991, EPA/540/2-91/002, U.S. Government Printing Office, Washington, DC) is a short pamphlet that explains bioremediation at a very basic level.

References

Anonymous. 1966. Report of a new chemical hazard. *New Scientist* 32:612.

Bohn, H. L. 1977. Soil treatment of organic waste gases. In *Soils for management of organic wastes and waste waters,* L. F. Elliot et al. (eds.), 607–18. Madison, WI: Soil Science Society of America.

Bollag, J. M., and S.-Y. Liu. 1990. Biological transformation processes of pesticides. In *Pesticides in the soil environment. Processes, impacts, and modeling,* H. H. Cheng (ed.), 169–211. Madison, WI: Soil Science Society of America.

Brunner, W., F. H. Sutherland, and D. D. Focht. 1985. Enhanced biodegradation of polychlorinated biphenyls in soil by analog enrichment. *Journal of Environmental Quality* 14:324–28.

Carter, L. 1976. Michigan's PBB incident: Chemical mix-up leads to disaster. *Science* 192:240–43.

Chapelle, F. H. 1993. *Ground-water microbiology and geochemistry.* New York: John Wiley & Sons.

Gustafson, C. 1970. PCBs—prevalent and persistent. *Environmental Science and Technology* 4:214–20.

Kincannon, C. B. 1972. *Oily waste disposal by soil cultivation process.* USEPA, EPA-R2-72-110. U.S. Government Printing Office, Washington, DC.

chapter

32

Heavy Metals

Overview

After you have studied this chapter, you should be able to:

■ List elements that are regarded as heavy metals.

■ Explain what happens to microbial populations in metal-contaminated soil.

■ Describe how microorganisms respond to heavy metals.

■ Identify key mechanisms by which microorganisms resist heavy metals.

■ Discuss important considerations in studying heavy metals.

INTRODUCTION

Heavy metals (the elements, not the music) are the metals in the periodic table that have a molecular weight greater than 55. While this includes Mn and Fe, heavy metals that are an environmental concern are usually regarded as chromium (Cr), nickel (Ni), copper (Cu), zinc (Zn), arsenic (As), selenium (Se), strontium (Sr), molybdenum (Mo), technetium (Tc), cadmium (Cd), mercury (Hg), and lead (Pb). It should be apparent that some of these metals (Zn, Cu, and Mo being examples) are micronutrients absolutely required for plant and animal growth. All cells need trace quantities of heavy metals. Heavy metals are often cofactors in enzyme-catalyzed reactions. The distinction between heavy metal deficiency and heavy metal excess is a fine one. In this, the last chapter, we discuss how microorganisms deal with environments in which heavy metal excess has occurred and is an environmental problem.

MICROBIAL RESPONSE TO HEAVY METALS IN THE ENVIRONMENT

What happens to microorganisms when the environment is contaminated with heavy metals? You typically find that the surviving microorganisms can withstand higher heavy metal concentrations than can similar organisms from uncontaminated soil. Table 32-1 shows data from soil at a mining site illustrating this point.

The metal-contaminated soils in Table 32-1 have significantly more Pb and Zn than do the control soils. Nevertheless, in the absence of any additional metals, there are about as many bacteria in the control soil as in the metal-contaminated soil from both sites. As additional Cd is added to soils from Site 1, the bacterial population declines

Table 32-1   Response of bacteria in metal-contaminated soil to Cd addition.

Soil Type	Soil Metal Concentration (ppm)			Cadmium Addition (ppm)			
	Pb	Zn	Cd	0	10	100	500
				Population (cells g^{-1})			
Site 1							
Control	96	220	2	4.8×10^7	4.5×10^7	3.0×10^4	2.0×10^3
Contaminated	8,000	55,200	468	4.8×10^7	2.5×10^7	1.7×10^6	3.2×10^5
Site 2							
Control	120	117	2	2.4×10^7	2.1×10^7	1.4×10^4	7.0×10^2
Contaminated	1,304	384	5	3.1×10^7	9.0×10^6	1.2×10^5	4.7×10^3

(Olsen and Thornton 1982)

Table 32-2   Background metal concentration of biosolids-amended soils and an off-site control.

Metal	Biosolids amended	Control
	Metal concentration (mg kg^{-1} soil)	
Cd	5.8	0.5
Cr	1,016.2	25.7
Cu	601.5	7.0
Hg	ND†	ND
Ni	495.0	11.1
Pb	205.0	40.4
Zn	1,016.7	44.9

† ND = Not determined
(L. W. Jacobs, unpublished)

more slowly in the metal-contaminated soil than in the control soil, which had no prior Cd exposure. In Site 2, the metal-contaminated soil has more Pb and Zn than does the control soil, but the Cd concentrations in the control and metal-contaminated soil are about the same. Consequently, as the microorganisms from Site 2 are exposed to increasing Cd, the decline in their numbers is about the same as the control's. Prior exposure to one heavy metal does not mean better survival when a different heavy metal is present.

Heavy metal contamination need not occur because soils are adjacent to mining regions. Municipal biosolids disposal can contaminate soils with heavy metals, as the example from a Michigan soil shows in Table 32-2. There's an important distinction between the extractable and biologically available heavy metal content in soil. Heavy metals can be immobilized in soil organic matter or as metal carbonates. Both fates render the metals biologically unavailable, so the soil microbial population experiences a lower heavy metal concentration. Heavy metal extraction from soil recovers both available and nonavailable metal forms. Although the extraction process may accurately reflect the heavy metals in soil, it may not reflect their biological effect.

MICROBIAL RESISTANCE TO HEAVY METALS

When pollutants are applied at levels that kill most cells, the pollutants select for the few cells that have evolved resistance mechanisms and are able to persist. Heavy metals follow the same pattern—exposure to heavy metals selects for resistance to heavy metals in the surviving microorganisms. One resistance mechanism is extracellular detoxification. An organic compound may chelate and bind the heavy metal so that it is not taken up. For example, oxalic acid excreted by *Penicillium* binds Cu and prevents its uptake. Microorganisms may have altered permeability, which means that the heavy metal is simply not taken up. This is usually a plasmid-associated trait that alters the cell's membrane permeability to a metal such as Cd. *Micrococcus* and *Azotobacter* can rapidly immobilize Pb in the cell wall so that it is not transported across the cell membrane.

The heavy metal may be removed from the cell and cell environment. There are several plasmid-associated traits that confer resistance to Hg in bacteria such as *Escherichia coli* and *Pseudomonas aeruginosa*. For example, methyl transferases methylate Hg to methylmercury, which is volatile.

$$Hg^0 + CH_3\text{-B12 (Cobolamine)} \xrightarrow{\text{Methyl transferase}} CH_3\text{-Hg (Methylmercury)}$$

$$CH_3\text{-Hg} + CH_3\text{-B12 (Cobolamine)} \longrightarrow CH_3\text{-Hg-CH}_3 \text{ (Dimethylmercury)}$$

Methyl- and dimethylmercury are lipophilic and concentrate in fish fat, which is one reason why fish from some metal-contaminated environments are declared to be inedible. A tragic case of heavy metal poisoning occurred in Minamata Bay, Japan in the 1950s when families consuming fish and shellfish from the bay were poisoned by Hg compounds, such as methylmercury, that a nearby chemical company dumped into the water.

Some cells reduce Hg^{2+} to Hg^0, which is volatile:

$$Hg^{2+} + NADPH + H^+ \longrightarrow Hg^0 \text{ (g)} + 2H^+ + 2NADP$$

STUDYING HEAVY METALS AND MICROORGANISMS

To study how heavy metals affect microorganisms, the lethal heavy metal concentrations and the concentrations that affect growth rate are typically measured. Also likely to be determined are the concentrations that cause morphological abnormalities, for example, inhibition of bacterial cell division, or concentrations that inhibit specific physiological or metabolic processes such as spore germination. In typical studies, a metal salt is added to culture medium and inoculated with the test organism. It's important to realize that the physiological state of the test organism affects its response to the metal pollutant. Yeast susceptibility to Cd injury is greatest for P-starved cells, intermediate for N-starved cells, and least for C-starved cells. Most cells in soil are C-starved in the environment. It is also important to consider whether the heavy metal concentration is realistic and what form of heavy metal you are using. The form of heavy metal affects its chemical and biological behavior.

MERCURY (Hg)

The Hg concentration in soil is usually < 1 ppm. Localized Hg deposits of higher concentrations occur that permit mining. Mercury has no beneficial effects, unlike Cu and Ni. High Hg concentrations are biocidal. Low Hg concentrations are mutagenic. Mercury is a serious environmental contaminant because of our blatant disregard for its environmental effects. Around 3,000 metric tons of Hg (3×10^6 kg) are released into the

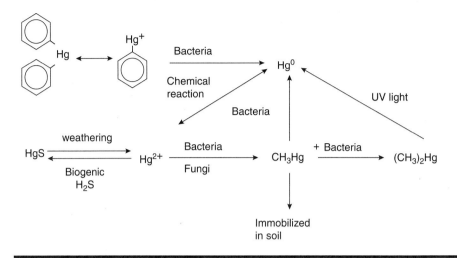

Figure 32-1   Pathways of Hg transformation in soil.

atmosphere from the combustion of fossil fuels and Hg-containing compounds (National Research Council 1978). Some of the biologically mediated fates of Hg are illustrated in Figure 32-1.

Mercury is toxic to microorganisms, and to survive in its presence, they must have resistance mechanisms. Resistance to Hg is fairly common, though not as much to organic Hg compounds as to inorganic Hg compounds. Microorganisms transform Hg^0 into methylated compounds such as:

$$CH_3Hg \text{ (Methylmercury)}$$
$$CH_3HgCH_3 \text{ (Dimethylmercury)}$$

These methylated forms of Hg are volatile, which benefits the microorganisms since the Hg concentration decreases. However, methylated Hg compounds are readily adsorbed and retained by human tissue, where they cause neurological disorders. The character of the Mad Hatter in Lewis Carroll's book *Alice in Wonderland* is based on real hat-makers in England who treated felt hats with Hg-containing compounds and then acted "mad" because of Hg toxicity.

Both bacteria and fungi transform Hg. Some representative microbial species include *Bacillus, Clostridium, Mycobacterium, Pseudomonas, Aspergillus,* and *Neurospora.* Along with methylation, these microorganisms release organically-bound Hg during their heterotrophic activity.

$$RH\text{-}Hg\text{-}CL + 2H^+ \Rightarrow RH + Hg^{2+} + HCL$$

Demethylation and Hg precipitation as mercuric sulfides in anaerobic environments also occurs:

$$CH_3Hg + 2H^+ \Rightarrow Hg^{2+} + CH_4 + H^+$$
$$Hg^0 \text{ or } Hg^{2+} + H_2S \Rightarrow HgS \text{ (ppt)}$$

Mercury resistance is at least one trait carried on plasmids that are exchanged between gram-negative bacteria. The hypothesized mechanism of this plasmid-borne trait is quite elegant. The plasmid codes for a protein that initially binds Hg^{2+} in the periplasm. A second protein transfers the Hg^{2+} to the membrane. A third protein transfers the Hg^{2+} to the cytoplasm where it is reduced by mercuric reductase to elemental mercury (Hg^0), which is lost from the cell by enhanced diffusion.

ARSENIC (As)

Arsenic is used in pesticides and herbicides. Volatile arsenic compounds are made by microorganisms. Both *Pseudomonas* and *Alcaligenes* reduce arsenate (AsO_4^{3-}), which is soluble to arsine (AsH_3), which is volatile. As with Hg, volatile As compounds are often methylated forms of the metal:

$$
\begin{array}{cccc}
O & O & O & \\
\| & \| & \| & \\
OH-As-OH \Rightarrow CH_3-As-OH \Rightarrow CH_3-As-CH_3 \Rightarrow CH_3-As-CH_3 \\
| & | & | & | \\
OH & OH & OH & CH_3 \\
\text{Arsenate} & \text{Methylarsenate} & \text{Dimethylarsenate} & \text{Trimethylarsine} \\
(As^{5+}) & (As^{3+}) & (As^+) & (As^{3-})
\end{array}
$$

Arsenic compounds are also oxidized in soil, primarily because of heterotrophic metabolism in which arsenic is not an energy source (Figure 32-2).

SELENIUM (Se)

Selenium illustrates the point that "the dose is the poison." Blood concentrations of 0.1 mg Se per liter are nutritionally sound. However, the minimum lethal concentration of Se in tissue is only 1.5 to 3.0 mg Se per kg of body weight. Selenium is a widely distributed metal. Nonseleniferous soils have concentrations of 0.1 to 2.0 ppm Se. Seleniferous soils contain 2 to 200 ppm Se. Some areas of the United States have seleniferous soils. For example, the average soil concentration of Se in South Dakota is 17 ppm. Plants such as milk vetch (*Astragalus,* also called "Loco Weed"), goldenweed (*Haplopappus*), prince's plume (*Stanleya*), and woody aster (*Xylorhiza*) accumulate Se. The milk vetch (*Astragalus*) accumulates up to 20 g of Se per kg of plant tissue. These plants are often indicators that the soil may contain high Se concentrations.

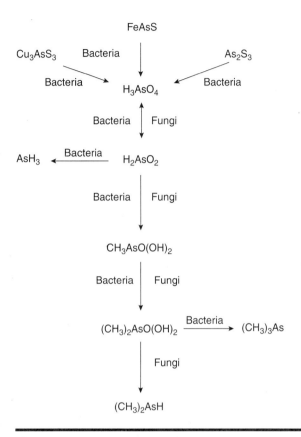

Figure 32-2   Biological transformations of arsenic in soil.

Selenium is a concern in areas that use irrigation in Se-rich soils. For example, from 1981 to 1986, 9,000 kg of Se were deposited in the Kesterson Reservoir from agricultural drainage water in northern California. The Se came from soil leachate of the western San Joaquin Valley. By 1983, bird death, deformity, and poor reproduction were apparent and were attributed to Se toxicity.

Microbial transformations of Se are essentially oxidation and reduction reactions and Se mobility and biological availability depend on its oxidization state. Selenium is similar to S in terms of its organic and inorganic forms:

Selenate SeO_4^2 (+6)—Mobile, moves to groundwater
Selenite SeO_3^2 (+4)—Mobile, but adsorbed by iron oxides
Elemental Se^0 (0)—Immobile or reduced amorphic precipitate
Selenide HSe^- (–2)—Volatile

Selenium and the Kesterson Reservoir: Environmental Contamination by Heavy Metals

Se is a nutritional supplement in domesticated animals such as poultry, cattle, and swine. It is also used as a dietary supplement in humans. It has a crucial role as an antioxidant in the enzyme glutathione peroxidase and contributes to vitamin E activity. But Se can also be an environmental toxin, as was discovered at the Kesterson Reservoir in California (Ohlendorf 1989).

The Kesterson Reservoir was a series of 12 shallow ponds serving as an evaporation basin for the drainage waters of the western San Joaquin Valley. Almost all of the water entering the Kesterson Reservoir was subsurface agricultural drainage water from irrigated agricultural fields. Starting in 1981, some of the water that entered the reservoir was diverted and used to preserve wetlands in the marsh management program at the adjacent Kesterson National Wildlife Refuge in Merced County, California.

By 1983, embryo deformity and mortality among the aquatic birds nesting in the Kesterson Reservoir were alarmingly high. Wildlife managers little suspected that the drainage water contained almost 4.2 mg of Se per liter, a Se concentration 1,000 times more concentrated than found in the naturally occurring drainage of this region. Phytoplankton in the reservoir were accumulating Se to levels 100- to 2,600-fold greater than normal. These plankton formed the base of the food chain in the reservoir, so the Se concentrations in fish, frogs, snakes, birds, and mammals also increased to levels up to 120 times greater than normal (20 to 170 mg Se per kg). Migratory birds that fed on plants, invertebrates, and fish in the reservoir bioaccumulated up to 24 times the normal level of Se in their tissue and died as a result. To protect the migratory birds from future Se exposure, the reservoir was drained and filled in 1988 to bury and isolate the excess Se.

The organic forms of Se are selenomethionine and selenocystine. Like S, microorganisms are dominantly responsible for converting Se between different forms. When flooded soils containing Se^0 and Se^{2-} are drained, oxidization is rapid. Much of the Se^0 and Se^{2-} oxidation is microbially mediated.

Selenium can be reduced to methylated organic Se:

$$Se(VI)(IV) \Rightarrow CH_3SeCH_3 \text{ (Dimethylselenide) or } CH_3SeSeCH_3 \text{ (Dimethyldiselenide)}$$

This reduction can be catalyzed by bacteria, but is primarily due to the fungi *Acremonium, Alternaria, Fusarium, Penicillium,* and *Ulocladium* (Karlson and Frankenberger 1988). Under normal soil conditions, probably only a small amount of Se is volatilized this way. Sulfate-reducing bacteria are capable of reducing SeO_4^{2-}. However, this probably doesn't have much practical significance because SeO_4^{2-} reduction is inhibited by SO_4^{2-}, and low levels of Se are toxic to the microorganisms studied.

Summary

Heavy metals that are an environmental concern are usually regarded as chromium (Cr), nickel (Ni), copper (Cu), zinc (Zn), arsenic (As), selenium (Se), strontium (Sr), molybdenum (Mo), technetium (Tc), cadmium (Cd), mercury (Hg), and lead (Pb). All cells need some heavy metals, but the distinction between heavy metal deficiency and heavy metal excess is a fine one.

Exposure to heavy metals selects for resistance to heavy metals in the surviving microorganisms. Prior exposure to one heavy metal does not mean better survival when a different heavy metal is present. There are several mechanisms of resistance to heavy metals: extracellular detoxification, altered permeability, pollutant removal from the cell environment, intracellular detoxification, and abiotic binding or precipitation. Microorganisms may persist in soils contaminated with extremely high heavy metal concentrations because those heavy metals may be extractable but not biologically available. The physiological state of the microorganism and the form of heavy metal are the two most important considerations when studying heavy metal effects on microbial populations.

Heavy metals may be remediated in metal-contaminated soil by volatilization, precipitation, and binding to organic matter. Methylmercury produced by microbial activity may lead to contamination of higher organisms because the methylmercury is readily adsorbed. Selenium behaves much like S in the environment; it undergoes oxidation and reduction reactions that can cause it to be leached. Selenium leaching led to serious environmental problems in the Kesterson Reservoir, which used agricultural drainage water from soils naturally high in Se.

Sample Questions

1. What are some of the mechanisms that microorganisms use to tolerate environments high in heavy metals?
2. What are microbial transformations of selenium important in the containment/cleanup of agricultural drainage waters in the Central Valley of California?
3. What are some of the elements regarded as heavy metals?
4. How is phytoremediation used to control heavy metal contamination in soil?
5. What properties of selenium make it possible to remediate it by oxidation and reduction mechanisms?
6. Why is methylation of heavy metals significant?
7. Where do heavy metals come from?
8. Why are heavy metals indispensable for life?
9. What is the difference between biological and chemical extractability of heavy metals?
10. How do heavy metals or other environmental pollutants change the composition of microbial communities?

Thought Question

Wallpaper can be a rather significant investment in home remodeling, so its preservation is important. Wallpapers were once treated with heavy metals such as arsenic to help preserve them. Can you explain why this might not be such a good idea, especially in a damp climate?

Additional Reading

An up-to-date treatment of selenium is contained in *Selenium in the Environment* (1994, W. Frankenberger and S. Benson, eds., Marcel Dekker, Inc., New York). Environmental effects of selenium are covered in *Selenium in Agriculture and the Environment*, edited by Lee Jacobs (1989, Soil Science Society of America, Madison, WI).

References

Karlson, U., and W. T. Frankenberger. 1988. Effects of carbon and trace element addition on alkylselenide production by soil. *Soil Science Society of America Journal* 52:1640–44.

National Research Council. 1978. *An assessment of mercury in the environment.* Washington, DC: National Academy of Science.

Ohlendorf, H. M. 1989. Bioaccumulation and effects of selenium in wildlife. In *Selenium in agriculture and the environment,* L. W. Jacobs (ed.), 133–77. Madison, WI: Soil Science Society of America.

Olsen, B. H., and I. Thornton. 1982. The resistance patterns to metals of bacterial populations in contaminated land. *Journal of Soil Science* 33:271–77.

Glossary of Soil Microbiology and Biochemistry Terms[a]

Abiontic enzymes—Enzymes (exclusive of live cells) that are (1) excreted by live cells during growth and division; (2) attached to cell debris and dead cells; (3) leaked into soil solution from extant or lysed cells but whose original functional location was on or within the cell. Synonymous with **exoenzymes.**

Acetylene-block assay—A technique for demonstrating or estimating denitrification by measuring nitrous oxide (N_2O) released from acetylene-treated soil. Acetylene inhibits N_2O reduction to dinitrogen (N_2) by denitrifying bacteria.

Acetylene-reduction assay—A technique for demonstrating or estimating nitrogenase activity by measuring the rate of acetylene (C_2H_2) reduction to ethylene (C_2H_4).

Adenylate energy charge ratio (EC)—A measure of the metabolic state of microorganisms and microbial communities. The energy charge ratio is calculated using the formula: EC = (ATP + 1/2ADP)/(ATP + ADP + AMP). The denominator represents the total adenylate pool; the numerator, the portion charged with high-energy phosphate bonds. The greater the EC, the more active the cells.

Aerobic—(1) Having molecular oxygen as a part of the environment. (2) Growing only in the presence of molecular oxygen (such as aerobic organ-

isms). (3) Occurring only in the presence of molecular oxygen (said of chemical or biochemical processes such as aerobic decomposition).

Aerobic digestion—The partial biological decomposition of suspended organic matter in wastewater or sewage in aerated conditions.

Allochthonous flora—Organisms that are not indigenous to the soil but that enter in precipitation, diseased tissues, manure, sewage, etc. They may persist but do not notably contribute to ecologically significant transformations orinteractions.

Amensalism—An interaction between two organisms in which one organism is suppressed by the other (such as suppression of one organism by toxins produced by the second).

Ammonification—The biological process leading to ammonium formation from nitrogen-containing organic compounds.

Anaerobic respiration—The metabolic process whereby electrons are transferred from a reduced compound (usually organic) to an inorganic acceptor molecule other than oxygen. The most common acceptors are carbonate, sulfate, and nitrate. See **denitrification.**

Antagonism—Production of a substance by one organism that inhibits one or more other organisms. The terms antibiosis and allelopathy have also been used to describe such cases of chemical inhibition.

Antibiotic—An organic substance produced by one organism that in low concentrations kills or inhibits the growth of other organisms.

[a]Glossary of Terms courtesy of the Soil Science Society of America, 677 South Segoe Road, Madison, WI 53711-1086.

Antibody—A protein produced by the body in response to the presence of an antigen to which it can specifically combine.

Antigen—A substance that incites specific antibody production.

Arbuscule—Specialized dendritic (highly branched) structure formed within root cortical cells by endomycorrhizal fungi. See **vesicular-arbuscular.**

Archaea—(1) Prokaryotes with cell walls that lack murein, have ether bonds in their membrane phospholipids, and are characterized by growth in extreme environments. (2) A primary biological kingdom distinct from both Bacteria and Eucarya. Formerly called "Archae bacteria".

Aseptic—Free from pathogenic or contaminating organisms.

Associative dinitrogen fixation—A close interaction between a free-living nitrogen-fixing organism and a plant that results in enhanced dinitrogen fixation rates.

Associative symbiosis—A close but relatively casual interaction between two dissimilar organisms or biological systems. The association may be mutually beneficial but is not required to accomplish specific functions. See also **commensalism** and **mutualism.**

Autochthonous flora—(1) That portion of the microflora presumed to subsist on the more resistant soil organic matter and is little affected by the addition of fresh organic materials. (2) Microorganisms indigenous to a given ecosystem; the true inhabitants of an ecosystem; referring to the common microbiota of the body of soil microorganisms that tend to remain constant despite fluctuations in the quantity of fermentable organic matter. Contrast with **zymogenous flora.** Also termed **oligotrophs.**

Autotroph—An organism that has the capacity to gain its carbon for biosynthesis from CO_2 or bicarbonate dissolved in solution.

Autotrophic nitrification—Oxidation of ammonium to nitrate through the combined action of two chemoautotrophic bacteria, one forming nitrite from ammonium and the other oxidizing nitrite to nitrate.

Bacteroid—An altered form of bacterial cells. Refers particularly to the swollen, irregular vacuolated cells of *Rhizobium* and *Bradyrhizobium* in legume nodules.

Batch culture—A method for culturing organisms in which the organism and supporting nutritive medium are added to a closed system. Contrast with **chemostat.**

Bioassay—A method for quantitatively measuring a substance by its effect on the growth of a suitable microorganism, plant, or animal under controlled conditions.

Biodegradable—A substance able to be decomposed by biological processes.

Biological denitrification—See **denitrification.**

Biological immobilization—See **immobilization** and **biological interchange.**

Biological interchange—The interchange of elements between organic and inorganic states in a soil or other substrate through the action of living organisms. It results from the biological decomposition of organic compounds with the liberation of inorganic materials (mineralization) and the utilization of inorganic materials with synthesis of microbial tissue (immobilization).

Biomass—(1) The total mass of living organisms in a given volume or mass of soil. (2) The total weight of all organisms in a particular environment. See **microbial biomass.**

Bioremediation—The use of biological agents to reclaim soil and water polluted by substances hazardous to the environment or human health. The biological agents may be enzymes, microbial cells, or plants.

Biotic enzymes—Enzymes associated with viable proliferating cells located (1) intracellularly in cell protoplasm; (2) in the periplasmic space; (3) at the outer cell surfaces.

BOD (biochemical oxygen demand)—The quantity of oxygen used in the biochemical oxidation of organic and inorganic matter in a specified time, at a specified temperature, and in specified conditions. It is an indirect measure of the concentration of biologically degradable material present in organic wastes.

Bradyrhizobia—Collective common name for the genus *Bradyhizobium*. See **rhizobia**.

Carbon cycle—The sequence of transformations whereby carbon dioxide is converted to organic forms by photosynthesis or chemosynthesis, recycled through the biosphere (with partial incorporation into sediments), and ultimately returned to its original state through respiration or combustion.

Carbon-nitrogen ratio—See **carbon-organic nitrogen ratio**.

Carbon-organic nitrogen ratio—The ratio of the mass of organic carbon to the mass of organic nitrogen in soil, organic material, plants, or microbial cells.

Chemoautotroph—See **chemolithotroph**.

Chemodenitrification—Nonbiological processes leading to the production of gaseous forms of nitrogen (molecular nitrogen or an oxide of nitrogen).

Chemolithotroph—An organism able to use CO_2 or carbonates as the sole source of carbon for cell biosynthesis, and deriving energy from the oxidation of reduced inorganic compounds. Used synonymously with **chemoautotroph**.

Chemoorganotroph—An organism for which organic compounds serve as both energy and carbon sources for cell synthesis. Used synonymously with **heterotroph**.

Chemostat—A device for the continuous culture of microorganisms in which growth rate and population size are regulated by the concentration of a limiting nutrient in incoming medium.

Chemotaxis—The oriented movement of a motile organism with reference to a chemical agent. May be positive (toward) or negative (away) with respect to the chemical gradient.

COD (chemical oxygen demand)—A measure of the oxygen-consuming capacity of inorganic and organic matter present in water or wastewater. It is expressed as the amount of oxygen consumed from a chemical oxidant in a specific test.

Coliform—A general term for a group of bacteria that inhabits the intestinal tract of humans and other animals. Their presence in water constitutes presumptive evidence for fecal contamination. Includes all aerobic and facultatively anaerobic, gram-negative rods that are nonspore-forming and ferment lactose with gas formation. *Escherichia coli* and *Enterobacter* are important examples of coliforms.

Colonization—Establishment of a community of microorganisms at a specific site or eco-system.

Cometabolism—Transformation of a substrate by a microorganism without deriving energy, carbon, or nutrients. The microorganism can transform the substrate into intermediate degradation products but fails to multiply at its expense.

Commensalism—Interaction between two species in which one species derives benefit while the other is unaffected.

Community—All of the organisms that occupy a common habitat and that interact with one another.

Competition—A rivalry between two or more species for a limiting factor in the environment.

Compost—Organic residues, or a mixture of organic residues and soil, that have been mixed, piled, and moistened, with or without addition of fertilizer and lime, and generally allowed to undergo thermophilic decomposition until the original organic materials have been substantially altered or decomposed. Sometimes called "artificial manure" or "synthetic manure." In Europe, the term may refer to a potting mix for container-grown plants.

Conjugated metabolites—Metabolically produced compounds that are linked together by covalent binding (complex formation).

Copiotrophs—See **zymogenous flora**.

Cryophile—See **psychrophilic organism**.

Culture—A population of microorganisms cultivated in an artificial growth medium. A pure culture is grown from a single cell; a mixed culture consists of two or more microorganisms growing together.

Denitrification—Reduction of nitrogen oxides (usually nitrate and nitrite) to molecular nitrogen or nitrogen oxides by bacterial activity (denitrification) or by chemical reactions involving nitrite (chemodenitrification). Nitrogen oxides are used by bacteria as terminal electron acceptors in place of oxygen in anaerobic or microaerophilic respiratory metabolism.

Diatom—Algae having siliceous cell walls that persist as a skeleton after death. Any of the microscopic unicellular or colonial algae constituting the class Bacillariaceae. They are abundant in fresh- and salt waters and their remains are widely distributed in soils.

Diatomaceous earth—A geologic deposit of fine, grayish siliceous material composed chiefly or wholly of the remains of diatoms. It may occur as a powder or as a porous, rigid material.

Diazotroph—A microorganism or association of microorganisms that can reduce molecular nitrogen (N_2) to ammonia (i.e. fixes nitrogen).

Digestibility (as applied to organic wastes)—The potential degree to which organic matter in wastewater or sewage can be broken down into simpler and/or more biologically stable products.

Dinitrogen fixation—Conversion of molecular nitrogen (N_2) to ammonia and subsequently to organic nitrogen utilizable in biological processes.

Direct counts—In soil microbiology, a method of estimating the total number of microorganisms in a given mass of soil by direct microscopic examination.

Dissimilation—The release from cells of inorganic or organic substances formed by metabolism.

Ectomycorrhiza(e)—A mycorrhizal association in which the fungal mycelia extend inward, between root cortical cells, to form a network ("Hartig net") and outward into the surrounding soil. Usually the fungal hyphae also form a mantle on the surface of the roots.

Electron acceptor—A compound that accepts electrons during metabolism or chemical reactions and is thereby reduced.

Electron donor—A compound that donates or supplies electrons during metabolism or chemical reactions and is thereby oxidized.

Endomycorrhiza(e)—A mycorrhizal association with intracellular penetration of the host root cortical cells by the fungus as well as outward extension into the surrounding soil. See **arbuscule, vesicle,** and **vesicular-arbuscular.**

Endophyte—An organism (e.g., fungus, bacteria) growing within a plant. The association may be symbiotic or parasitic.

Enrichment culture—A technique in which environmental (including nutritional) conditions are controlled to favor the development of a specific organism or group of organisms through prolonged or repeated culture.

Enzyme—Any of numerous proteins that are produced in the cells of living organisms and function as catalysts in the chemical processes of those organisms.

Eubacteria—Old terminology for prokaryotic organisms other than Archaea.

Eukaryote—Cellular organisms having a membrane-bound nucleus within which the genome of the cell is stored as chromosomes composed of DNA; includes algae, fungi, protozoa, plants, and animals.

Eutrophic—Having concentrations of nutrients optimal, or nearly so, for plant, animal, or microbial growth. (Said of nutrient or soil solutions and bodies of water.) The term literally means "self-feeding."

Exoenzyme—Enzymes that are excreted by organisms into the surrounding environment and carry out their metabolic or catabolic activity in that location.

Exudate—Low molecular weight metabolites that enter the soil from plant roots.

Facultative organism—An organism that can carry out both options of a mutually exclusive process (e.g., aerobic and anaerobic metabolism). It may also be used in reference to other processes, such as photosynthesis (e.g., a facultative photosynthetic organism is one that can use either light or the oxidation of organic or inorganic compounds as a source of energy).

Faecal (fecal) pellets—The excreta of fauna.

Fermentation—The metabolic process in which organic compounds serve as both electron donor and final electron acceptor.

Fluorescent antibody—An antiserum conjugated with a fluorescent dye (e.g., fluorescein or rhodamine). Fluorescent-labeled antiserum can be used to stain buried slides or other preparations and visualize the specific microorganism (antigen) of interest by fluorescence microscopy. See also **immunofluorescence.**

Fungistat—A compound that inhibits or prevents fungal growth.

Habitat—The place where a given organis lives.

Heterotroph—An organism able to derive carbon and energy for growth and cell synthesis by utilizing organic compounds. Used synonymously with **chemoorganotroph.**

Heterotrophic nitrification—Biochemical oxidation of ammonium and/or organic nitrogen to nitrate and nitrite by heterotrophic microorganisms. See **nitrification.**

Hybridization—The binding or annealing of two complementary, single strands of nucleic acid.

Hypha (pl., hyphae)—Filament of fungal cells. Many hyphal filaments (hyphae) constitute a mycelium. Bacteria of the order Actinomycetales also produce branched mycelium.

Immobilization—The conversion of an element from inorganic to organic form in microbial or plant tissues.

Immunofluorescence—Fluorescence resulting from a reaction between a substance and a specific antibody that is bound to a fluorescent dye.

Indigenous—Native to an area.

Inoculate—To treat, usually seeds, with microorganisms to create a favorable response. Most often refers to the treatment of legume seeds with *Rhizobium* or *Bradyrhizobium* to stimulate dinitrogen fixation, but also refers to the introduction of microbial cultures into sterile growth medium.

K-selected—In ecological theory, that group of microorganisms in soil living at or near the carrying capacity of the soil environment. Analogous to **autochthonous** microorganisms.

Labile—Readily transformed by microorganisms or readily available to plants.

Labile pool—The sum of an element in the soil solution and the amount of that element readily solubilized or exchanged when the soil is equilibrated with a salt solution.

Land farming—A process of bioremediation or biodegradation in which wastes are incorporated into soil and allowed to decompose via naturally occurring microbial activity.

Lectins—Plant proteins that have a high affinity for specific sugar residues.

Leghemoglobin—An iron-containing, red pigment(s) produced in root nodules during the symbiotic association between *Bradyrhizobium* or *Rhizobium* and legumes. Leghemoglobin is an oxygen-binding compound functionally equivalent to mammalian hemoglobin.

Mesobiota—See **mesofauna.**

Mesofauna—Nematodes, oligochaete worms, smaller insect larvae, and microarthropods.

Mesophile—See **mesophilic organism.**

Mesophilic organism—An organism whose optimum temperature for growth falls in an intermediate range of approximately 15° to 35°C. Synonymous with **mesophile.**

Microaerophile—An organism that requires a low concentration of oxygen for growth. Sometimes used to indicate an organism that will carry out metabolic activities under aerobic conditions but will grow much better under anaerobic conditions.

Microbial biomass—(1) The total mass of living microorganisms in a given volume or mass of soil. (2) The total weight of all microorganisms in a particular environment.

Microbial population—The sum of living microorganisms in a given volume or mass of soil.

Microbiota—Microflora and protozoa.

Microfauna—Protozoa, nematodes, and arthropods of microscopic size.

Microflora—Bacteria (including actinomycetes), fungi, algae, and viruses.

Mineralization—The conversion of an element from an organic form to an inorganic state as a result of microbial activity.

Most probable number—A method for estimating microbial numbers in soil based on dilution to extinction.

Mucigel—The gelatinous material at the surface of roots grown in nonsterile soil. It includes natural and modified plant exudates (more specifically mucilages), bacterial cells, and their metabolic products (e.g., capsules and slimes) as well as colloidal mineral and organic matter from the soil.

Mutualism—See **symbiosis.**

Mycelium—A mass of interwoven filamentous hyphae, such as that of the vegetative portion of the thallus of a fungus.

Myco—Prefix designating an association or relationship with a fungus (e.g., mycotoxins are toxins produced by a fungus).

Mycorrhiza (pl., mycorrhizae)—Literally "fungus root." The association, usually symbiotic, of specific fungi with the roots of higher plants. See **endomycorrhiza** and **ectomycorrhiza.**

Neutralism—A lack of interaction between two organisms in the same ecosystem.

Niche—(1) The particular role that a given species plays in the ecosystem. (2) The physical space occupied by an organism.

Nitrate reduction (biological)—The process whereby nitrate is reduced by plants and microorganisms to ammonium for cell synthesis (nitrate assimilation, assimilatory nitrate reduction) or to nitrite by bacteria using nitrate as the terminal electron acceptor in anaerobic respiration (respiratory nitrate reduction, dissimilatory nitrate reduction). Sometimes used synonymously with **denitrification.**

Nitrification—Biological oxidation of ammonium to nitrite and nitrate, or a biologically induced increase in the oxidation state of nitrogen.

Nitrogenase—The specific enzyme system required for biological dinitrogen fixation.

Nitrogen cycle—The sequence of biochemical changes undergone by nitrogen wherein it is used by a living organism, transformed upon the death and decomposition of the organism, and converted ultimately to its original oxidation state.

Nitrogen fixation—See **dinitrogen fixation.**

Nodule bacteria—The bacteria that fix dinitrogen (N_2) within organized structures (nodules) on the roots, stems, or leaves of plants. Sometimes used as a synonym for **rhizobia** or **bradyrhizobia.**

Nodules—Specialized tissue enlargements, or swellings, on the roots, stems, or leaves of plants, such as those caused by nitrogen-fixing microorganisms.

Oligotrophic—Environments in which the concentration of nutrients available for growth is limited. Nutrient poor habitats.

Oligotrophs—Organisms able to grow in environments with low nutrient concentrations.

Organotroph—See **heterotroph.**

Oxidative phosphorylation—Conversion of inorganic phosphate into the energy-rich phosphate of adenosine 5'-triphosphate.

Parasitism—(1) Feeding by one organism on the cells of a second organism, which is usually larger than the first. The parasite is, to some extent, dependent on the host at whose expense it is maintained. (2) An association whereby one organism (parasite) lives in or on another organism (host) and benefits at the expense of the host.

Partial sterilization—The destruction by heat, chemical, or physical treatment of part of the biological populations of soil, peats, composts, and other natural substrates.

Pasteurization—Partial sterilization of soil, liquid, or other natural substances by temporary heat treatment.

PCR (polymerase chain reaction)—An *in vitro* method for amplifying defined segments of DNA. PCR involves a repeated cycle of oligonucleotide hybridization and extension on single-stranded DNA templates.

Phosphobacteria—Bacteria able to convert organic phosphorus into orthophosphate.

Photolithotroph—An organism that uses light as a source of energy and CO_2 or carbonates as the source of carbon for cell biosynthesis. See **autotroph.**

Plasmids (episomes)—Extrachromosomal DNA.

Plate count—A count of the number of colonies formed on a solid culture medium when inoculated with a small amount of soil. The technique has been used to estimate the number of certain organisms present in the soil sample.

Predation—A relationship between two organisms whereby one organism (predator) engulfs and digests the second organism (prey).

Priming effect—Stimulation of microbial activity in soil, usually organic matter decomposition, by the addition of labile organic matter.

Prokaryotes—Single-celled organisms lacking a nuclear membrane.

Propagule—Any cell unit capable of developing into a complete organism. For fungi, the unit may be a single spore, a cluster of spores, hyphae, or a hyphal fragment.

Psychrophile—See **psychrophilic organism.**

Psychrophilic organism—An organism whose optimum temperature for growth falls in the approximate range of 5° to 15°C. Synonymous with **cryophile.**

Protocooperation—An association of mutual benefit to two or more species but without the cooperation or without being obligatory for their existence or the performance of some function.

Pure culture—A population of microorganisms composed of a single strain. Such cultures are obtained through selective laboratory procedures and are rarely found in a natural environment.

Respiratory quotient (RQ)—The number of molecules of CO_2 liberated for each molecule of O_2 consumed.

Restriction enzyme—A class of highly specific enzymes that make double-stranded breaks in DNA at specific sites near where they combine.

Rhizobia—Bacteria able to live symbiotically in roots of leguminous plants, from which they receive energy and often utilize molecular nitrogen. Collective common name for the genus *Rhizobium.*

Rhizobia free—Any material that does not contain rhizobia able to nodulate leguminous plants of interest. The material need not be void of all rhizobia. See **rhizobia populated.**

Rhizobia populated—Any material that contains rhizobia able to nodulate leguminous plants of interest. Contrast with **rhizobia free.**

Rhizocylinder—The plant root plus the adjacent soil that is influenced by the root. See **rhizosphere.**

Rhizoplane—Plant root surfaces usually including the adhering soil particles.

Rhizosphere—The zone of soil immediately adjacent to plant roots in which the kinds, numbers, or activities of microorganisms differ from that of the bulk soil.

r-selected—In ecological theory, that group of organisms in soil that rapidly proliferate in response to an abundance of resources. Analogous to **zymogenous** microorganisms.

Saprophyte—An organism that lives on dead organic material.

Saprophytic competence—The ability of a nodule symbiont or pathogenic microorganism to establish itself and live in soil as a saprophyte.

Secondary metabolite—A product of intermediary metabolism released from a cell.

Selective enrichment—A technique for specifically encouraging the growth of a particular organism or group of organisms. See **enrichment culture.**

Siderophore—A nonporphyrin metabolite secreted by certain microorganisms that forms a highly stable coordination compound with iron. There are two major types: catecholate and hydroxamate.

Soil biochemistry—The branch of soil science concerned with enzymes and the reactions, activities, and products of soil microorganisms.

Soil microbiology—The branch of soil science concerned with soil-inhabiting microorganisms, their functions, and activities.

Soil organic residue—Animal and vegetative materials of recognizable origin in soil.

Soil population—(1) All the organisms living in the soil, including plants and animals. (2) Members of the same taxa.

Soil quality—The capacity of a soil to function within ecosystem boundaries to sustain biological productivity, maintain environmental quality, and promote plant and animal health.

Specific activity—The number of enzyme activity units per mass of protein. Often expressed as micromoles of product formed per unit time per milligram of protein. Also used in radiochemistry to express the radioactivity per mass of material (radioactive + nonradioactive).

Spores—Specialized reproductive cells. Asexual spores germinate without uniting with other cells, whereas sexual spores of opposite mating types unite to form a zygote before germination occurs.

Sterilization—Rendering an object or substance free of viable microbes.

Substrate—(1) That which is laid or spread under an underlying layer, such as the subsoil. (2) The substance, base, or nutrient on which an organism grows. (3) Compounds or substances that are acted upon by enzymes or catalysts and changed to other compounds in the chemical reaction.

Sulfur cycle—The sequence of transformations undergone by sulfur wherein it is used by living organisms, transformed upon death and decomposition of the organism, and ultimately converted to its original oxidation state.

Symbiosis—The obligatory cohabitation of two dissimilar organisms in intimate association. Often, but not always, mutually beneficial.

Synergism—The nonobligatory association of organisms that is mutually beneficial. Both populations can survive in their natural environment on their own although, when formed, the association offers mutual advantages.

Thermophile—See **thermophilic organism.**

Thermophilic organism—An organism whose optimum temperature for growth is above 45°C.

Threshold moisture content—The minimum moisture condition, measured either in terms of moisture content or moisture stress, at which biological activity just becomes measurable.

Vegetative cell—The growing or feeding form of a microbial cell, as opposed to a resistant resting form.

Vesicle—A storage organ produced by endomycorrhizal fungi. See **Vesicular-arbuscular.**

Vesicular-arbuscular—A common endomycorrhizal association produced by phycomycetous fungi of the family Endogonaceae. Host range includes most agricultural and horticultural crops.

Xenobiotic—A compound foreign to biological systems. Often refers to human-made compounds that are resistant or recalcitrant to biodegradation and/or decomposition.

Zymogenous—So-called opportunistic organisms found in soils in large numbers immediately following addition of a readily decomposable organic substrate. Synonymous with **copiotrophs.**

Index

NOTE: Page numbers of tables, figures, and photographs are in italics.

SUBJECTS

NAMES

ORGANISMS